ILLUSTRATED PROGRAMMED GROSS ANATOMY

by Verlee E. Gross

Former Instructor of Medical Terminology
Adult Education Division
Los Angeles City Schools

**with pronunciation, English translation
and removable study section.**

Anatomical Illustrations
by Verlee E. Gross

Standard Book Number 0-912256-05-2

Published by Halls of Ivy Press, 13050 Raymer Street, North Hollywood, California 91605, Phone (213) 875-3050

PRINTED IN U.S.A.

CONTENTS

ACKNOWLEDGEMENTS

From the moment that an idea for a text is born to the moment of publication an author depends on numerous experts in various fields.

Because of the unusual concept of this book the knowledge furnished by my brother, Edmund J. Gross, was indispensible. As a graphic arts specialist he helped to develop the method of presentation for correct reproduction.

All material was reviewed by medical specialists in the various areas of anatomy. Each of the doctors has an extremely busy practice and I am sincerely indebted to them for the conscientious manner in which they reviewed each chapter and for their valuable suggestions.

To avoid any show of preference they are listed in the order of the text material each reviewed in his specialty: George L. Kraft, M. D., orthopedics; William H. Faeth, M.D., neurosurgery; Richard E. Anderson, M.D., internal medicine; J. P. Myles Black, M.D., thoracic surgery; Warren S. Line, M.D., E.N.T.; G. Ellis Doty, M.D., general practice; Michael Marchese, D.D.S., dentistry; Ralph H. Walker, M.D., gynecology; John H. Burbidge, M.D., urology; John J. McGroarty, M.D., ophthalmology.

My thanks to each and every one.

Verlee E. Gross

North Hollywood, California
March, 1973

What we have to learn to do,
we learn by doing.

—*Aristotle*

CHAPTER I

INTRODUCTION

PURPOSE OF TEXT: This text is an introduction and review of basic anatomy. It might almost be called a "primer" in the study of the human body. It was developed because education in the paramedical field is bringing about many new demands and needs in teaching material. This fairly new area of training in the health sciences has also brought about the necessity for an easier method for students to acquire a basic knowledge of medical terminology and anatomy.

Today's requirements for paramedical personnel demand they have some formal education in this rapidly growing area of medicine. In addition, the high degree of specialization in medicine makes education in anatomy and medical terminology a definite "must." On-the-job training is almost a thing of the past—there isn't enough available time.

During the 20 years in which I taught medical terminology I was constantly frustrated because I could never find an uncomplicated, easy to understand text on basic anatomy. Out of this frustration and the pleas of students for a book which they could understand and in which they could participate in a study method, grew the idea for this text.

As with my previous texts this book also contains pronunciation and English translation of the terms. The translation is important because it often gives a clue to the area in which the body component occurs or describes it in such a manner as to make it easy to identify.

Certain liberties had to be taken with the drawings for production purposes. The method offered here required simple, easy to identify illustrations. The basic drawings were developed to make the human anatomy easier to understand. Written material has been kept to a minimum.

The study method is easy and requires concentration more than anything else. Many of the drawings show and identify areas found in other illustrations thus offering repetition for easier learning. The body components can be seen as "individuals" rather than as "one of a crowd." Yet, they are still a part of the whole area.

It is suggested that students using this text for initial learning of basic anatomy read the descriptive material in my previous text "Mastering Medical Terminology." They can then "graduate" to any of the more involved anatomical texts available in bookstores. For students already studying anatomy this book will serve as a study aid. Not every school, hospital, insurance company, etc., can offer anatomical dissection. Therefore, the programmed method used here is probably the closest thing to dissection for students.

The most difficult part of preparing this text was the elimination of material. In organizing the contents a definite line had to be drawn so the material would be comprehensive but still basic to facilitate learning. The book was planned to cover many fields in medical work for study purposes and also as a constant reference for brush-up work. The work book, when neatly filled in, will serve as a complete review of each anatomical system and its components. In addition, the programmed material in the text furnishes the opportunity to check on specific locations of various parts of the anatomy.

It is not the aim of this book to develop skilled anatomists. The goal to be achieved from study of this material is a good basic knowledge of the components of the body and the systems to which they belong. With this knowledge paramedical personnel will not only find their work more interesting but will derive greater gratification from knowing more about what they are doing.

STUDY METHOD: The method for studying this text is very simple. There are no trick questions to figure out. Everything is exactly as it is shown.

At the back of the book is a removable study section. Contained in this study section are illustrations of anatomical areas which appear in the text. These same anatomical drawings shown in the text in duplicate illustrations highlight only one particular area or organ in each drawing. When the study page is removed from the back of the book it is easy for the student to correctly identify each area in the study section illustration by referring to the highlighted areas in the duplicate individual drawings. The most important factor to remember when filling in the study section is "TAKE TIME!" Little will be gained by speed. Knowledge is only acquired through concentration. And, concentration is the best asset in completing the drawings in the study section.

It is suggested that identifications (lines indicate where they should be written) be made lightly in pencil and then checked out. They can then be typed (using a red typewriter ribbon is very effective) or copied neatly in ink or ball point pen (red again is suggested). Felt pens should not be used as they often "bleed" through the paper and will show on the other side spoiling the drawings on that page.

Writing in the phonetics given in the text is not required. It would consume too much space and make the study section drawings confusing when used for reference.

For those who have a copy of "Mastering Medical Terminology" it is suggested they read the chapter of the anatomical system being studied. This will provide continued repetition and also increase better understanding of that portion of the anatomy.

The study section, which is perforated and drilled for a three-ring binder, should be kept neat as it will reflect the type of work a student can do. This is often important on a job interview when the interviewer asks to see (or the student may wish to show) the type of material used in instruction. Another important factor for neatness is that the study section, when completed, will be a constance source of reference.

On the following pages are shown a series of drawings from the Gastro-Intestinal System. These illustrations with their highlighted areas show the method to be used for finishing the drawing in the study section.

Study each highlighted area carefully to clearly identify where it belongs. In these drawings it is imperative to correctly locate the various biliary ducts—cystic duct, right and left hepatic ducts, common hepatic duct, common bile duct and pancreatic duct. Understanding where the sphincter of Oddi and the ampulla of Vater are located can prove very useful, particularly for transcribers who constantly hear these terms in surgical reports.

After studying the areas and being fairly certain of their locations, the student should then remove the page bearing that particular drawing from the study section in the back of the book. By referring to the highlighted area and comparing its location to the identification lines drawn on the study lesson illustration the student will then write in the name of that particular area or organ. Concentration and careful deliberation rather than working rapidly will do more to orient the student of the correct location of areas and organs than other methods of study. And, as the student is actually working with the anatomy through illustration there is a feeling of participation it is difficult to obtain by reading about the subject. However, after working with the illustrations it is wise to refer to anatomy books for description of specific organs and areas and their function.

GALLBLADDER AND DUCTS

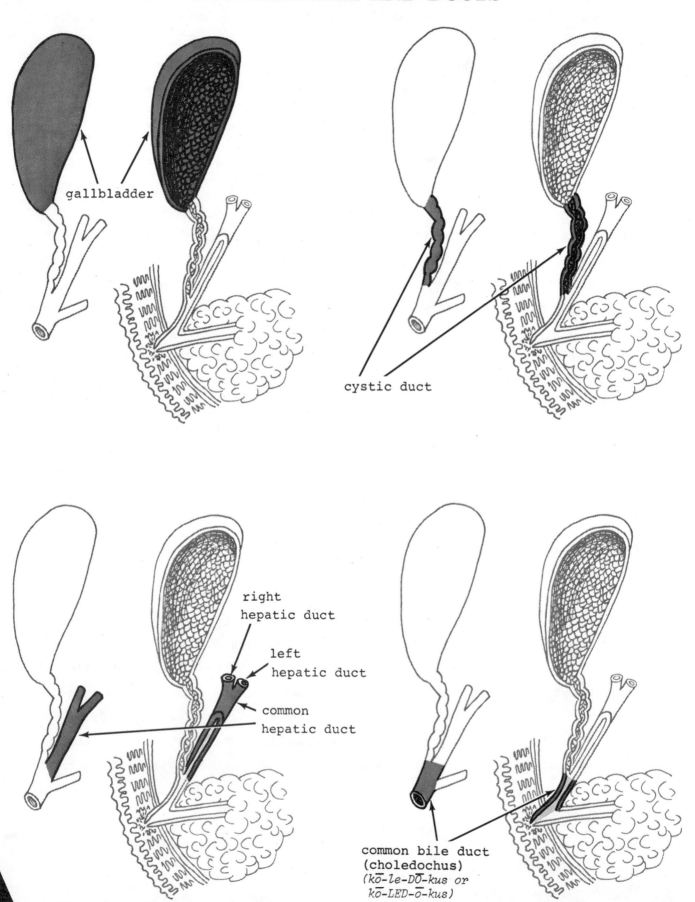

gallbladder

cystic duct

right
hepatic duct

left
hepatic duct

common
hepatic duct

common bile duct
(choledochus)
(kō-le-DŌ-kus or
kō-LED-ō-kus)

GALLBLADDER AND DUCTS

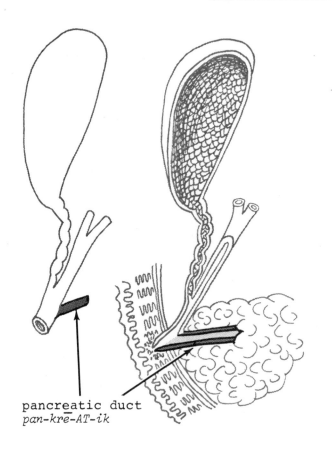

pancreatic duct
pan-krē-AT-ik

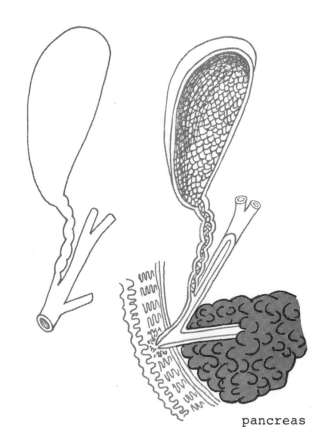

pancreas

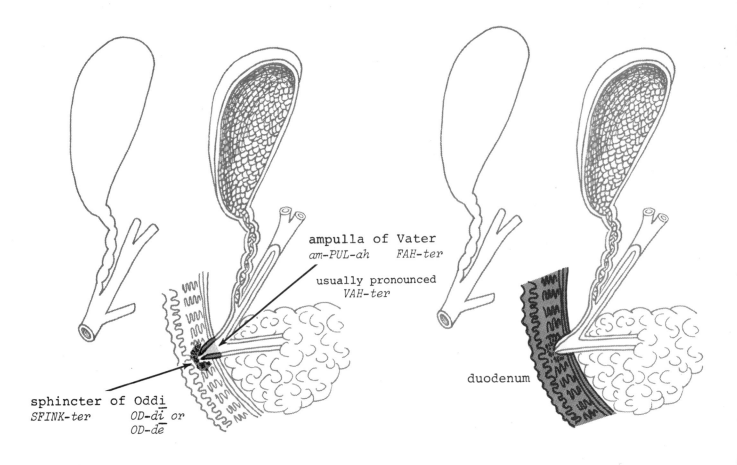

ampulla of Vater
am-PUL-ah   *FAH-ter*

usually pronounced
VAH-ter

sphincter of Oddi
SFINK-ter   *OD-dī* or
                 OD-dē

duodenum

GALLBLADDER AND DUCTS

After each area is checked carefully the terms are then written in on the lines provided. Note how effective the red appears and how much easier it is to locate specific areas.

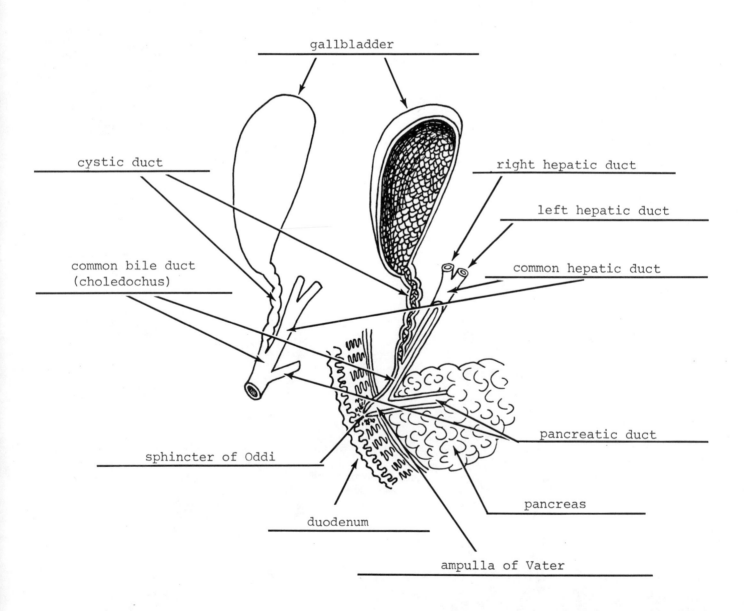

PRONUNCIATION: The principle followed for pronunciation is to present the actual, current use rather than what is supposed to be correct. These pronunciations are not necessarily the only ones used. This is impossible as there is much conflict between doctors and dictionaries and also between doctors and doctors. A uniform pronunciation of medical terms is almost impossible today. Therefore, optional pronunciations are given wherever the change in pronunciation is quite noticeable. Pronunciation is perhaps most important to medical transcribers who constantly encounter a variety of versions in the pronunciation of a large number of medical terms.

Phonetics have been simplified for easy recognition. There should be little trouble for students in reading the phonetics and developing correct pronunciation. The following instructions will explain the method used for the phonetics.

1. The strongest accented syllable is given in capital letters, for example: RĒ-nal, bī-LAT-er-al

2. If there is a secondary accent it will be noted by a double quote mark ("), for example: glom-er"ū-lō-ne-FRĪ-tis

3. If there are several secondary accents, as may occur in very long terms, all of these secondary accents will be noted by the double quote mark ("), for example: sal-ping"gō-ō"oo-fōr-EK-tō-mē

4. The secondary accent (") has some degree of emphasis but is not as strongly accented as the syllables noted in capital letters which are often called the heaviest accent.

5. Pronunciation of vowels is as follows:

   Where vowels are pronounced with a long sound this is signified by a line above the vowel. Therefore, these vowels would be pronounced as follows:

   ā as in māke, ē as in bē, ī as in īvy, ō as in pōle, ū as in pūre

   Unless otherwise shown in the phonetics any vowels not so marked will be pronounced with the short sound as follows:

   i as in bit and sip, e as in met and bet, o as in not, u as in bud

   Additional phonetics are as follows:

   a as in father is written ah and the diphthong of oi is written as oy

SUFFIX ENDINGS: Medical terms have the habit of using the same suffix endings over and over again. The following suffix endings are consistently used in medical terms occurring in this text. To avoid repetitive translation of them it is suggested this list be referred to on all endings not translated in the Index. After a short while it will not be necessary to refer to this list as the endings will have been learned.

SUFFIX	ENGLISH TRANSLATION
-al	pertaining to (adjective)
-alis	pertaining to (adjective)
-ar	pertaining to (adjective)
-aris	pertaining to (adjective)
-ary	one who, or that which (adjective)
-ate	possessing or characterized by (adjective)
-ation, -ion, -tion	act or state of
-form	shape, form
-ic, -icus	pertaining to or connected with (adjective)
-ode, -oid	resemble, like, form
-ory	place or thing where (adjective)
-ose	to be full of (adjective)
-osis(sing.), -oses (pl.)	condition or state of
-ous	to be full of (adjective)

CHAPTER INDEX: At the end of each chapter is an alphabetical index of the terms occurring in that chapter. Each word is separated into its component parts, for example: clav/icle. Under each word component the English translation is given, i.e. clav/icle.
key little

It is important to learn the English translation of medical terms. As an extra memory aid for identifying body components, and as an aid to spelling, this portion of medical terminology is invaluable. Translation shows how many areas of the body were given terms which relate to shape and/or resemblance to objects or animals. This translation, therefore, often provides a clue and helps in the identification of many difficult to remember body components. The index is an integral part of each chapter and should be referred to frequently.

BOOK INDEX: A complete index appears at the end of the book. It tabulates every page on which each term occurs, including each Chapter Index. This greatly facilitates locating specific body areas. Pages on which pronunciation of the term are given appear in *italic* type. Pages on which words are translated are shown in larger (Gothic) type. Therefore, there are three faces of type. To make it easier to understand their use they are shown here. This is explained again at the beginning of the Index.

medical term *12*, 20, **25**

CHAPTER II

SKELETAL SYSTEM

There are 206 bones in the human skeleton. In the following pages each of these bones is shown in its proper location as well as the skeletal division to which it belongs.

The bones provide a firm framework to give shape to the body and support the parts. They protect the vital organs (heart, brain and lungs); facilitate body movements by their action with muscles attached to them by tendons; store reserves of calcium (nearly all of the body's calcium is stored in the skeleton); manufacture red blood cells in the bone marrow.

All bones consist of an outer, dense material known as compact bone and an inner, spongy, more porous material known as cancellous bone.

A new born child's skeleton is practically all cartilage. As growth takes place most of the cartilage cells are replaced by osseous (bone) cells. With progressing age more calcium is deposited so that elderly persons very often have fragile, easily fractured bones. Thus, we begin life as "jellyfish" and end up as "coral."

DIVISIONS OF SKELETAL SYSTEM
CRANIAL AND FACIAL BONES

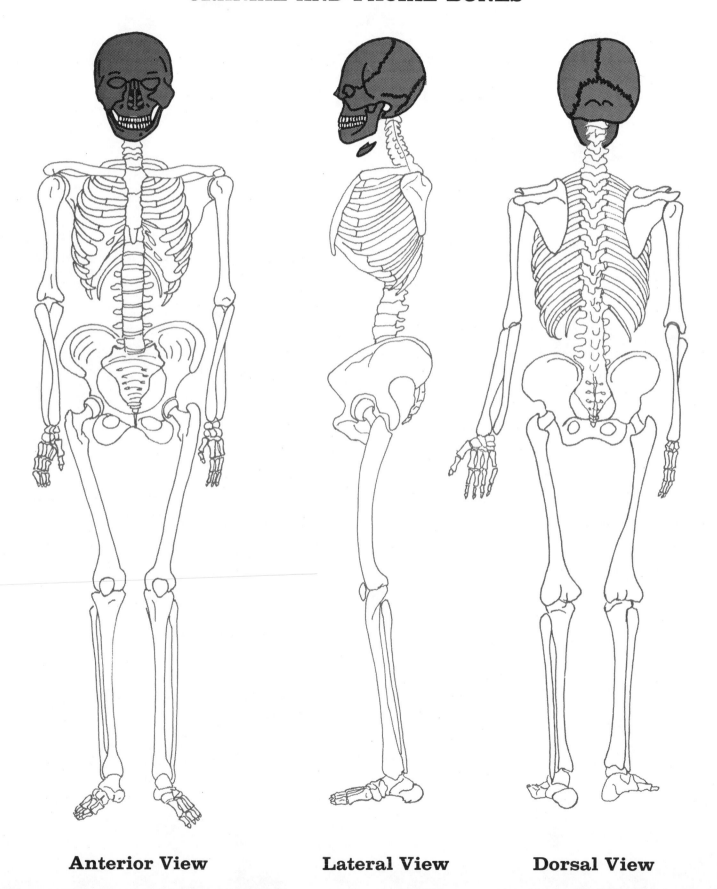

Anterior View   **Lateral View**   **Dorsal View**

DIVISIONS OF SKELETAL SYSTEM
UPPER EXTREMITIES

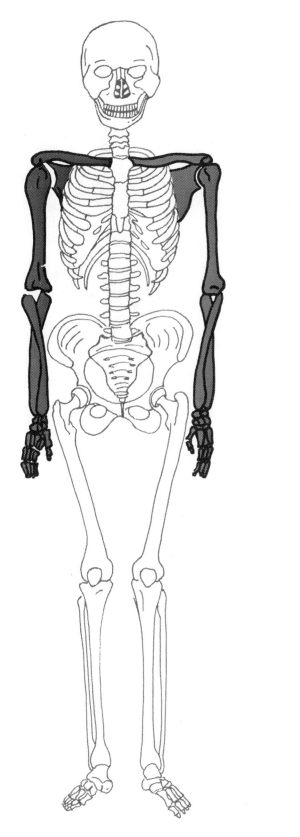
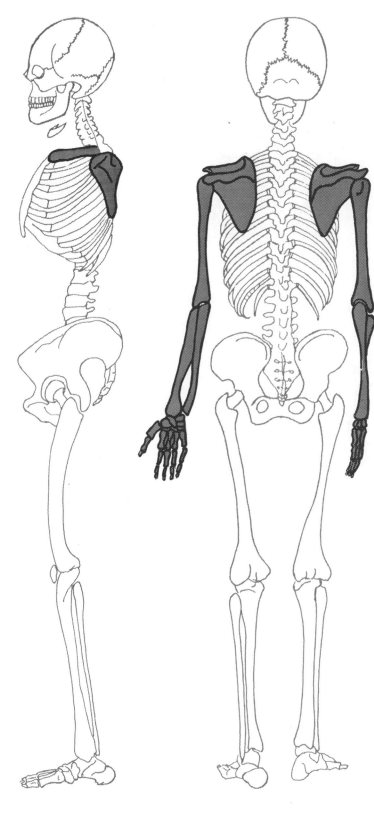

Anterior View   **Lateral View**   **Dorsal View**

DIVISIONS OF SKELETAL SYSTEM
TRUNK

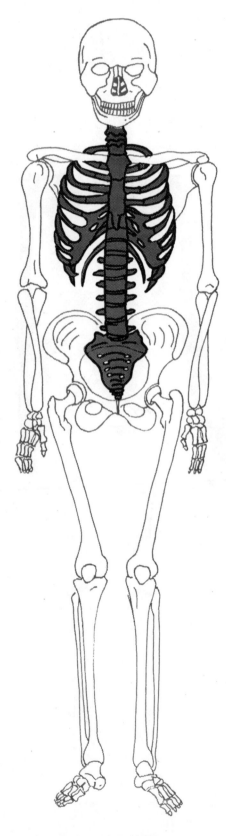
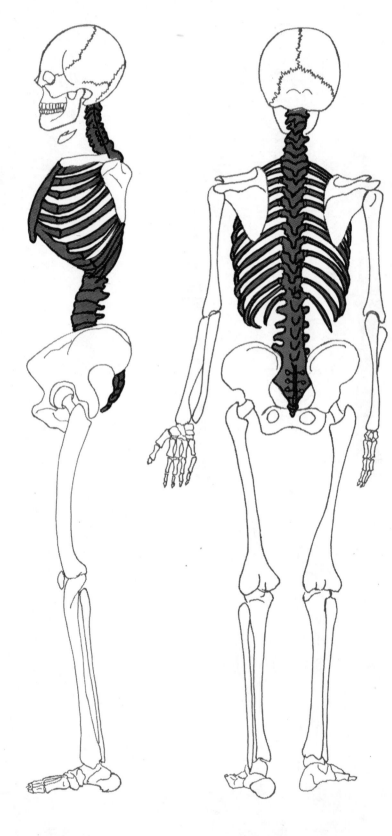

Anterior View   **Lateral View**   **Dorsal View**

DIVISIONS OF SKELETAL SYSTEM
LOWER EXTREMITIES

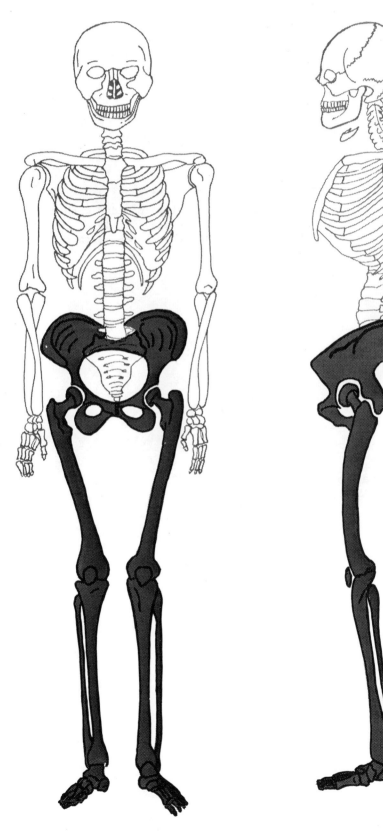
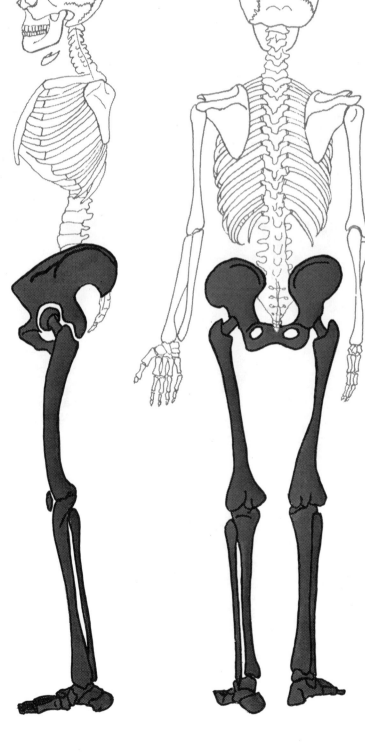

Anterior View   **Lateral View**   **Dorsal View**

CRANIAL AND FACIAL BONES Anterior View

BONES OF THE CRANIUM

occipital.................................. 1
parietal.................................. 2
frontal................................... 1
temporal.................................. 2
sphenoid.................................. 1
ethmoid................................... 1

Total   8

BONES OF THE FACE

nasal..................................... 2
vomer..................................... 1
nasal concha.............................. 2
lacrimal.................................. 2
zygomatic (malar)......................... 2
palatine.................................. 2
maxilla................................... 2
mandible.................................. 1

Total   14

TOTAL BONES OF SKULL - 22

BONES OF THE EAR

malleus................................... 2
incus..................................... 2
stapes.................................... 2

TOTAL BONES OF EAR   6

hyoid bone in the neck.................... 1

(The hyoid bone is usually included with the bones of the skull. Actually, it is a "loner" as it connects to no other bones, either in the skull or trunk.)

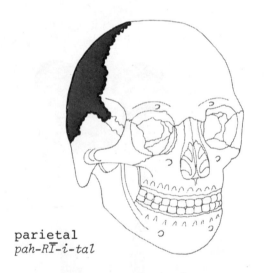

parietal
pah-RĪ-i-tal

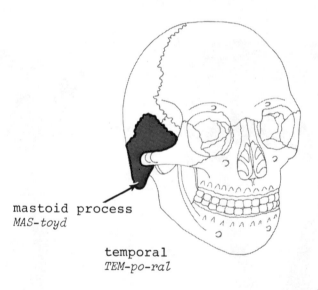

mastoid process
MAS-toyd

temporal
TEM-po-ral

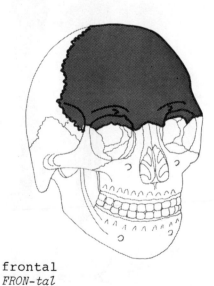

frontal
FRON-tal

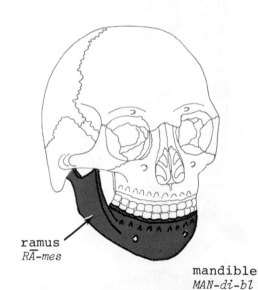

ramus
RĀ-mes

mandible
MAN-di-bl

CRANIAL AND FACIAL BONES (Anterior View)

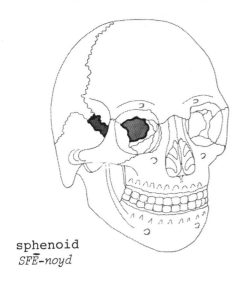

sphenoid
SFĒ-noyd

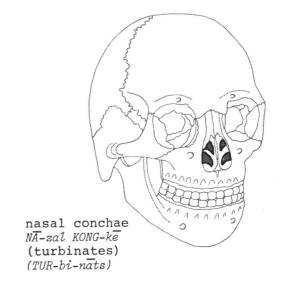

nasal conchae
NĀ-zal KONG-kē
(turbinates)
(TUR-bi-nāts)

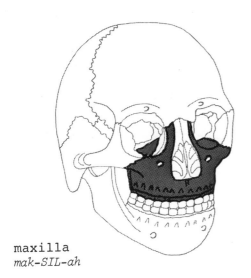

maxilla
mak-SIL-ah

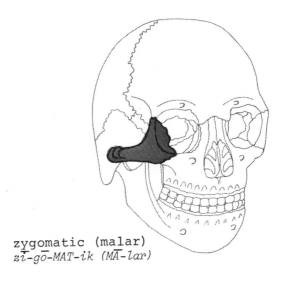

zygomatic (malar)
zī-gō-MAT-ik (MĀ-lar)

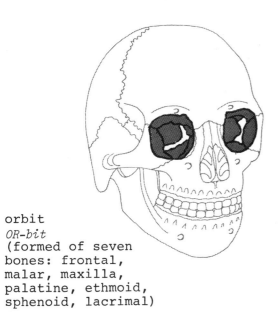

orbit
OR-bit
(formed of seven
bones: frontal,
malar, maxilla,
palatine, ethmoid,
sphenoid, lacrimal)

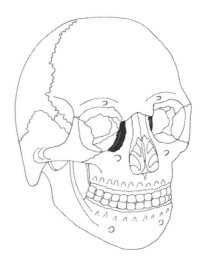

lacrimal
LAK-re-mal

CRANIAL AND FACIAL BONES (Anterior View)

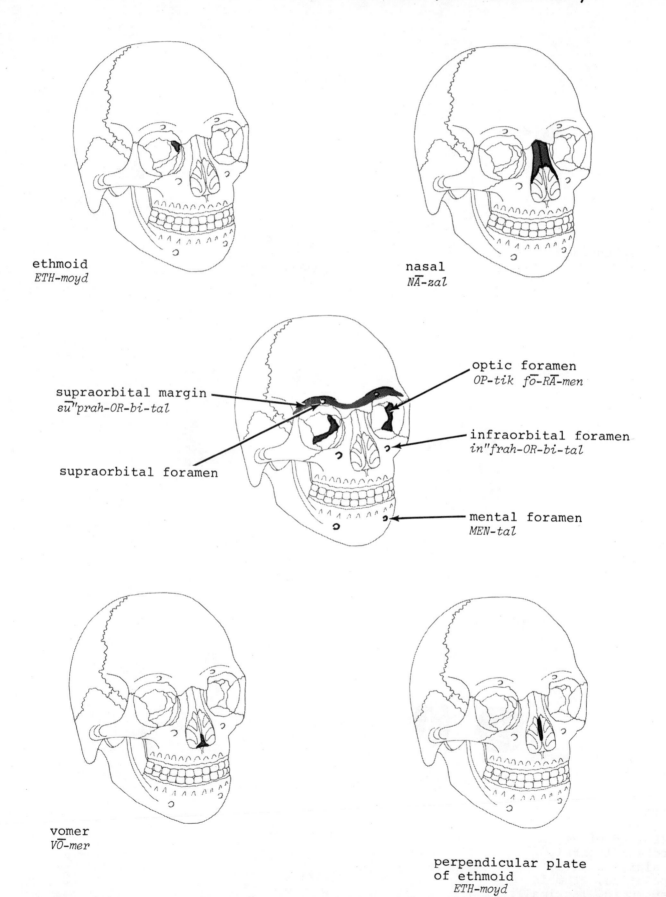

ethmoid
ETH-moyd

nasal
NĀ-zal

supraorbital margin
su"prah-OR-bi-tal

supraorbital foramen

optic foramen
OP-tik fō-RĀ-men

infraorbital foramen
in"frah-OR-bi-tal

mental foramen
MEN-tal

vomer
VŌ-mer

perpendicular plate
of ethmoid
ETH-moyd

CRANIAL AND FACIAL BONES (Lateral View)

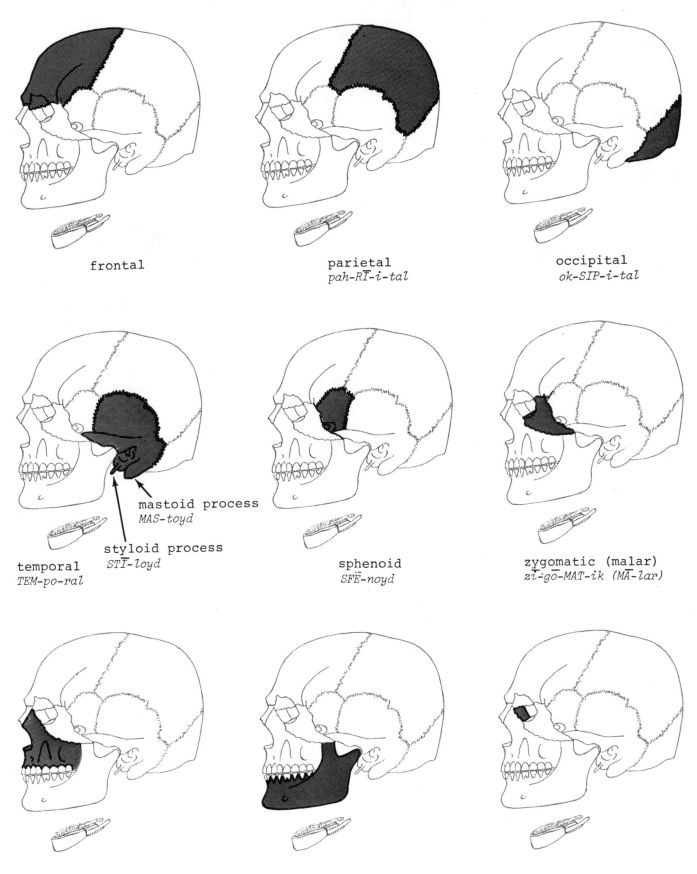

frontal

parietal
pah-RĪ-i-tal

occipital
ok-SĪP-i-tal

temporal
TEM-po-ral

mastoid process
MAS-toyd

styloid process
STĪ-loyd

sphenoid
SFĒ-noyd

zygomatic (malar)
zī-gō-MAT-ik (MĀ-lar)

maxilla

mandible

lacrimal

CRANIAL AND FACIAL BONES (Lateral View)

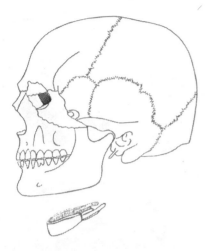

ethmoid
ETH-moyd

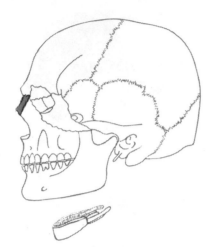

nasal
NĀ-zal

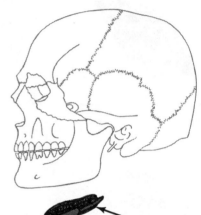

hyoid
HĪ-oyd

greater cornua

body

CROSS SECTION OF SKULL

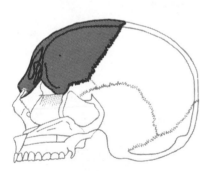

frontal
FRON-tal

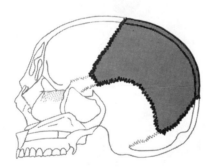

parietal
pah-RĪ-i-tal

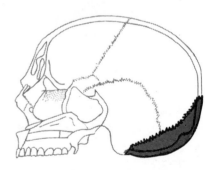

occipital
ok-SIP-i-tal

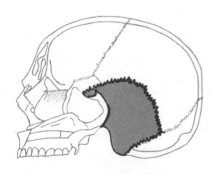

temporal
TEM-po-ral

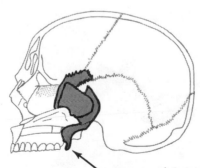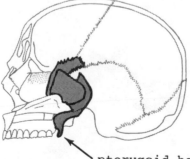

pterygoid hamulus
TER-i-goyd   HAM-ū-lus

sphenoid
SFĒ-noyd

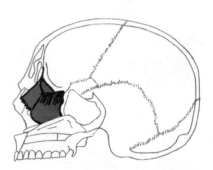

ethmoid
ETH-moyd

CROSS SECTION OF SKULL

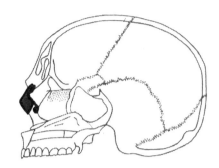

nasal
NĀ-zal

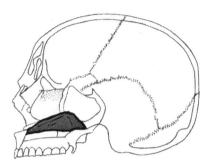

vomer
VŌ-mer

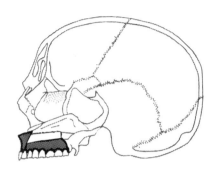

maxilla
mak-SIL-ah

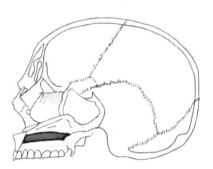

palatine
PAL-ah-tin or
PAL-ah-tin

FLOOR OF CRANIAL CAVITY

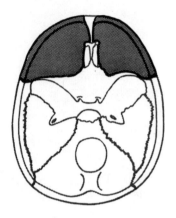

frontal
FRON-tal

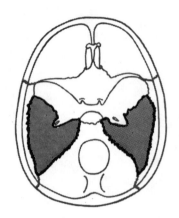

temporal
TEM-po-ral

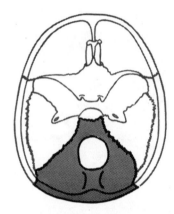

occipital
ok-SIP-i-tal

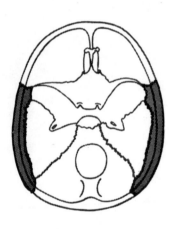

parietal
pah-RĪ-i-tal

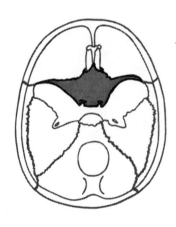

lesser wing of sphenoid
SFĒ-noyd

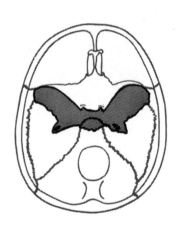

greater wing of sphenoid

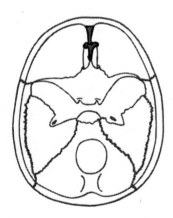

crista galli
KRIS-tah GAL-lē

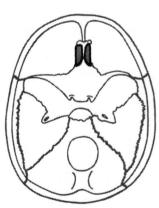

cribriform plate
KRIB-re-form
of ethmoid
ETH-moyd

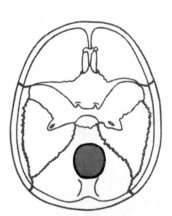

foramen magnum
fo-RĀ-men MAG-num
(where spinal
cord enters)

OSSICLES OF THE EAR

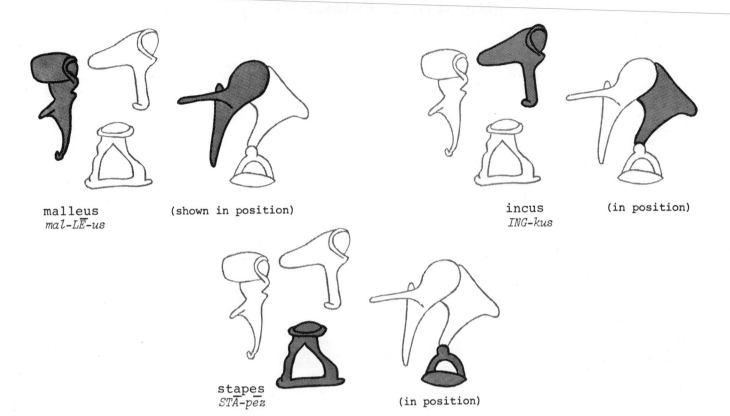

malleus
mal-LĒ-us

(shown in position)

incus
ING-kus

(in position)

stapes
STĀ-pēz

(in position)

RIGHT SIDE OF MANDIBLE

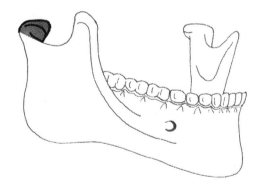

condyloid
KON-di-loyd

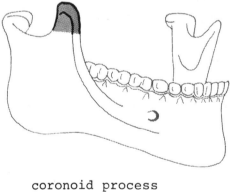

coronoid process
KŌR-ō-noyd

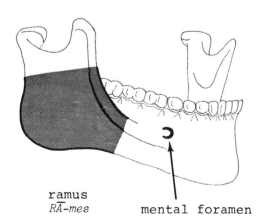

ramus
RĀ-mes

mental foramen

UPPER EXTREMITY
(Right Side, Palmer View)

clavicle.................................. 2
scapula.................................. 2
humerus.................................. 2
ulna.................................... 2
radius................................... 2
carpus
    navicular (scaphoid).................. 2
    lunate (semilunar)................... 2
    triangular (cuneiform)............... 2
    pisiform............................. 2
    greater multangular (trapezium)....... 2
    lesser multangular (trapezoid)....... 2
    capitate (os magnum)................. 2
    hamate (unciform).................... 2
metacarpus..............................10
phalanges...............................28

TOTAL BONES OF UPPER EXTREMITIES $\overline{64}$

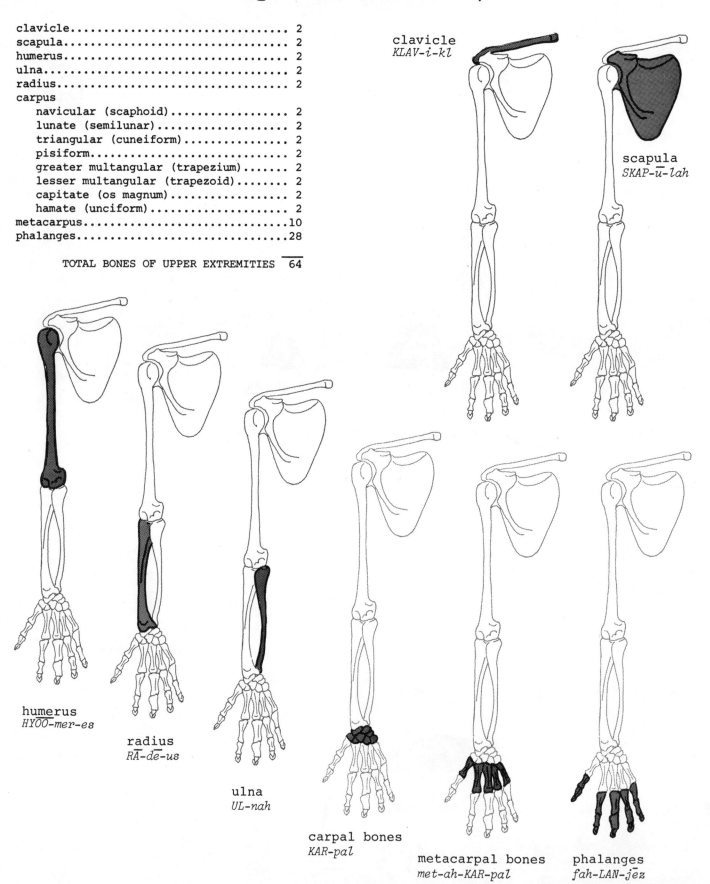

clavicle
KLAV-i-kl

scapula
SKAP-$\overline{u}$-lah

humerus
HY$\overline{OO}$-mer-es

radius
R$\overline{A}$-d$\overline{e}$-us

ulna
UL-nah

carpal bones
KAR-pal

metacarpal bones
met-ah-KAR-pal

phalanges
fah-LAN-j$\overline{e}$z

RIGHT CLAVICLE
(Superior or Upper Surface)

acromial extremity
ah-KRŌ-mē-al

coracoid tuberosity
KOR-ah-koyd too-be-ROS-i-tē

sternal extremity
STER-nal

RIGHT SCAPULA
(Posterior or Dorsal Surface)

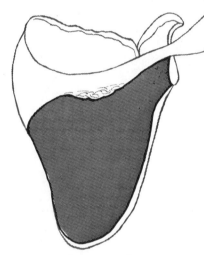  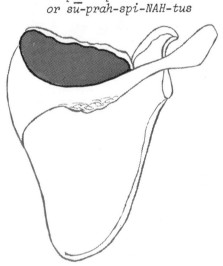

supraspinatus fossa
*sū"prah-spi-NĀ-tus FOS-ah
or sū-prah-spi-NAH-tus*

coracoid process
KOR-ah-koyd

infraspinatus fossa
*in"frah-spi-NĀ-tus
or in"frah-spi-NAH-tus*

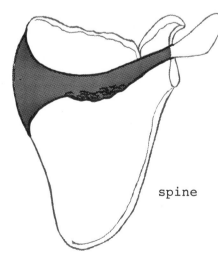 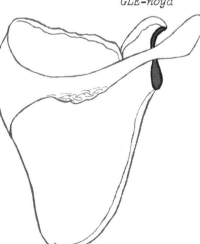 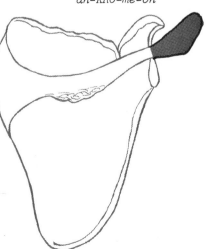

glenoid cavity
*GLEN-oyd or
GLĒ-noyd*

acromion process
ah-KRŌ-mē-on

spine

RIGHT HUMERUS (Anterior View)

head

greater
tubercle
TOO-ber-kl

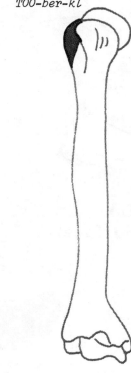

intertubercular sulcus
in"ter-too-BER-kyoo-ler SUL-kus

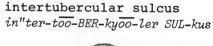
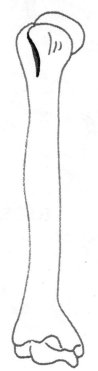

lesser
tubercle

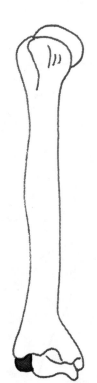

lateral
epicondyle
ep-i-KON-dīl

capitulum
ka-PET-ū-lum

coronoid fossa
KOR-o-noyd FOS-ah

trochlea
TRO-klē-ah

medial
epicondyle

RIGHT HUMERUS (Posterior View)

head

anatomical neck
an-ah-TOM-i-kl

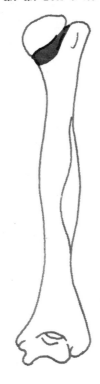

surgical neck
SER-ji-kl

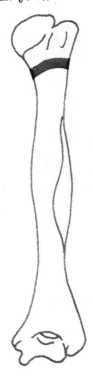

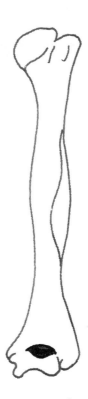

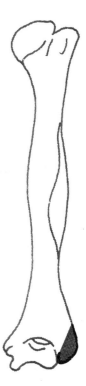

trochlea
TRŌ-klē-ah

medial epicondyle

olecranon fossa
ō-LEK-rah-non

lateral epicondyle

RADIUS AND ULNA (Posterior View)

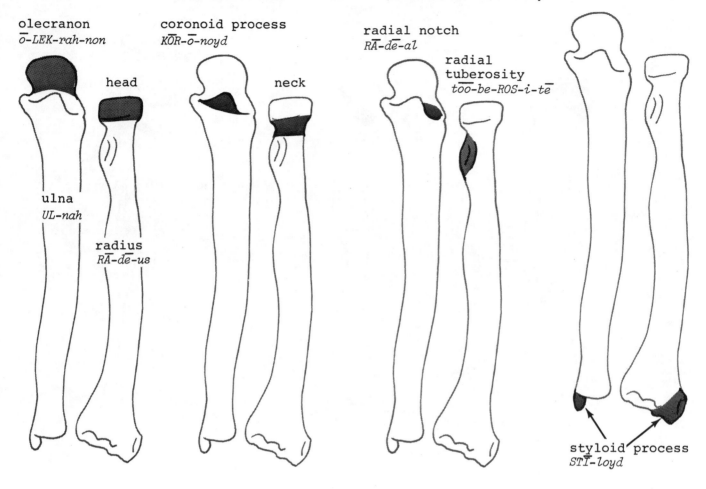

olecranon
ō-LEK-rah-non

head

coronoid process
KŌR-ō-noyd

neck

radial notch
RĀ-dē-al

radial tuberosity
too-be-ROS-i-tē

ulna
UL-nah

radius
RĀ-dē-us

styloid process
STĪ-loyd

UPPER PART OF ULNA (Side View)

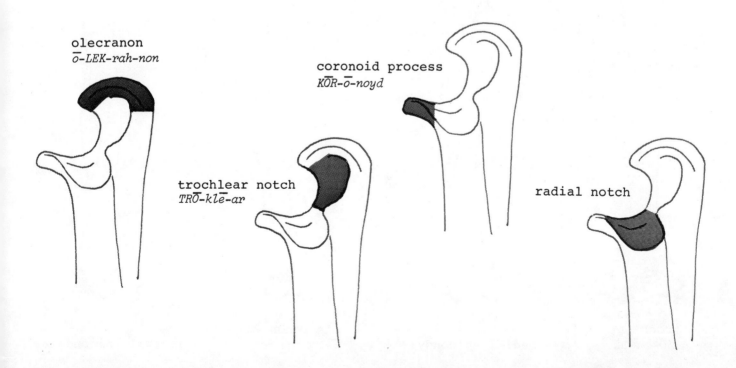

olecranon
ō-LEK-rah-non

coronoid process
KŌR-ō-noyd

trochlear notch
TRŌ-klē-ar

radial notch

RIGHT HAND (Palmer View)
The Eight Carpal Bones of the Wrist

greater multangular
mul-TANG-gu̅-lar
(trapezium)
(trah-PĒ-ze̅-ahm)

navicular
nah-VIK-ye-lar

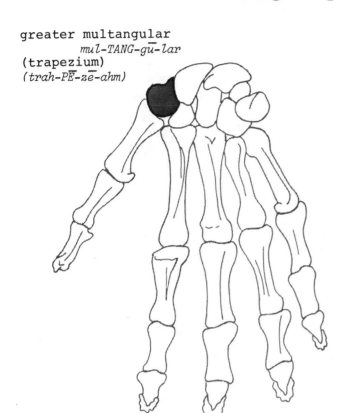

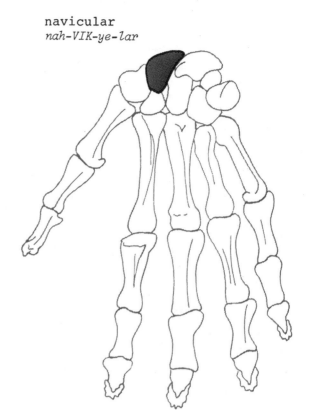

Some of the bones of the
wrist are known by either
of two names. Both are
given when this occurs.

lesser multangular
(trapezoid)
(TRAP-i-zoyd)

capitate
KAP-i-ta̅t

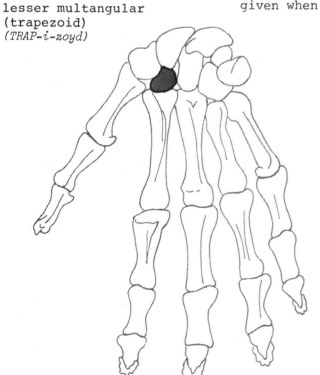

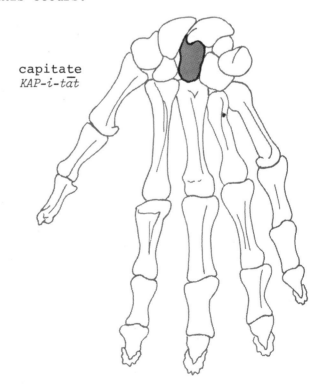

RIGHT HAND (Palmer View)

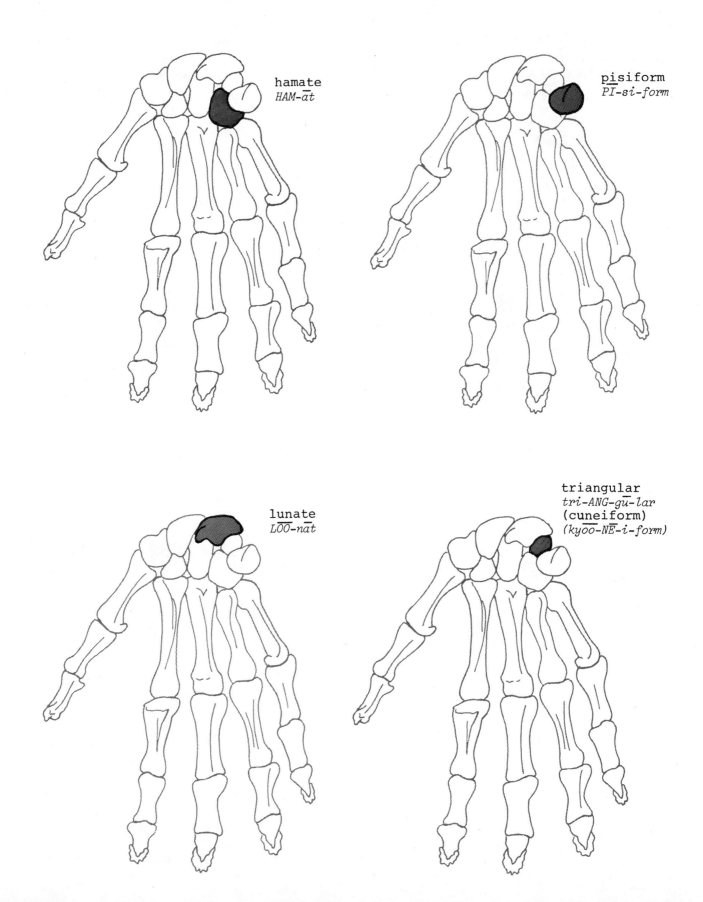

hamate
HAM-āt

pisiform
PĪ-si-form

lunate
LŌŌ-nāt

triangular
tri-ANG-gū-lar
(cuneiform)
(kyōō-NĒ-i-form)

RIGHT HAND (Palmer View)

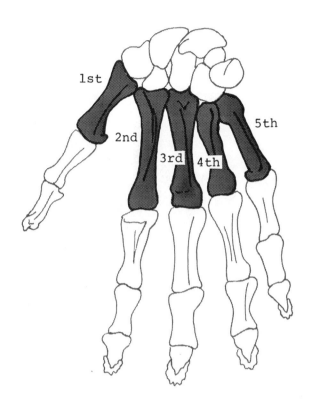

1st
2nd
3rd 4th
5th

metacarpals
met-ah-KAR-pals

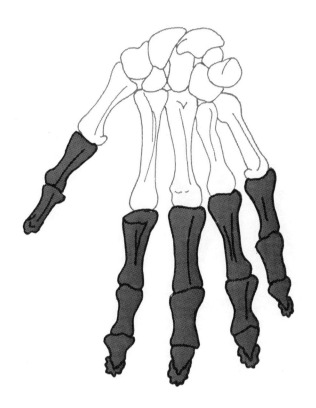

phalanges
fah-LAN-jēz

TRUNK (Anterior View)

cervical vertebra.................. 7
thoracic vertebra..................12
lumbar vertebra.................... 5
sacral vertebra................... 1
coccygeal vertebra............... 1

TOTAL VERTEBRAE IN THE ADULT   26

ribs..............................24
sternum........................... 1

TOTAL BONES IN THE TRUNK $\overline{51}$

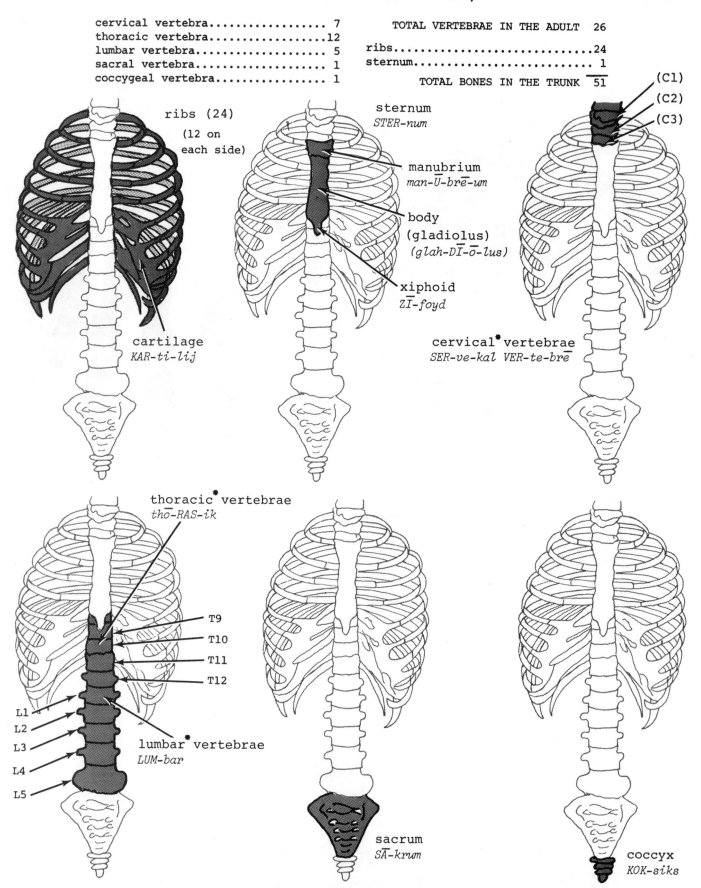

ribs (24)
(12 on
each side)

cartilage
KAR-ti-lij

sternum
STER-num

manubrium
man-U̅-bre̅-um

body
(gladiolus)
(glah-DI̅-o̅-lus)

xiphoid
ZI̅-foyd

(C1)
(C2)
(C3)

cervical* vertebrae
SER-ve-kal VER-te-bre̅

thoracic* vertebrae
tho̅-RAS-ik

T9
T10
T11
T12

L1
L2
L3
L4
L5

lumbar* vertebrae
LUM-bar

sacrum
SA̅-krum

coccyx
KOK-siks

*The various divisions of the vertebrae are abbreviated to C for cervical, T for thoracic, L for lumba:

TRUNK (Dorsal View)

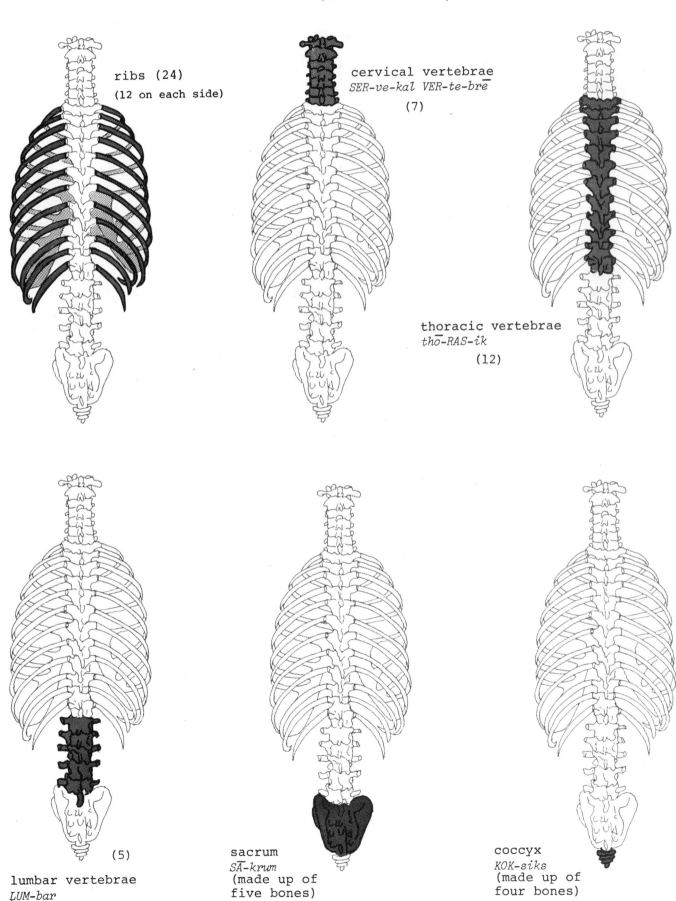

ribs (24)
(12 on each side)

cervical vertebrae
SER-ve-kal VER-te-brē
(7)

thoracic vertebrae
thō-RAS-ik
(12)

(5)

lumbar vertebrae
LUM-bar

sacrum
SĀ-krum
(made up of
five bones)

coccyx
KOK-siks
(made up of
four bones)

LUMBAR VERTEBRAE (Lateral View)

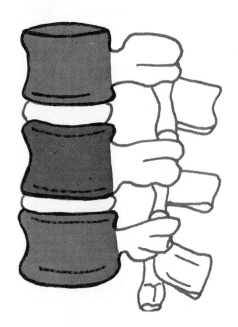 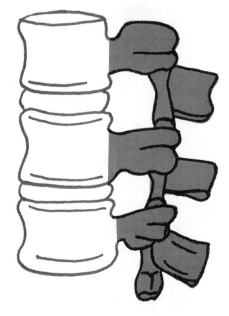 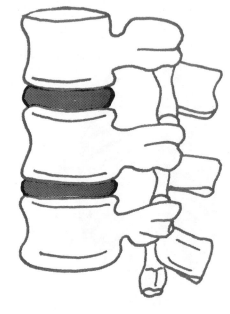

body of vertebrae
VER-te-brē

spinous processes
SPĪ-nus

intervertebral cartilaginous
*in-ter-VER-te-bral kar-ti-LAJ-i-nus
or in-ter-ver-TĒ-bral*

disks (or discs)

THORACIC VERTEBRA

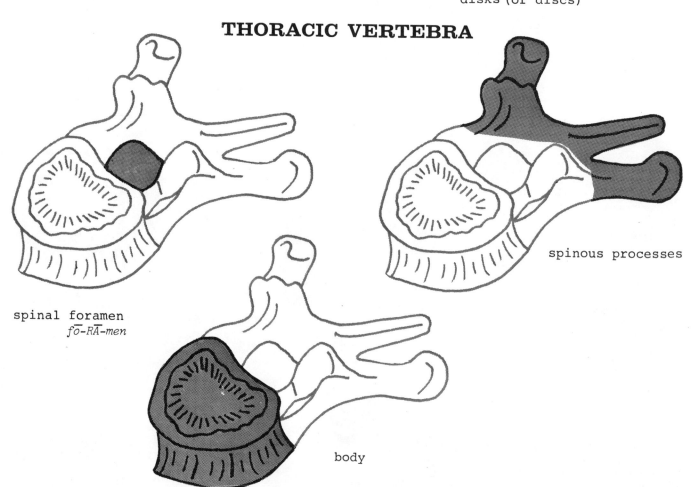

spinal foramen
fō-RĀ-men

spinous processes

body

LUMBAR VERTEBRA (Viewed from Above)

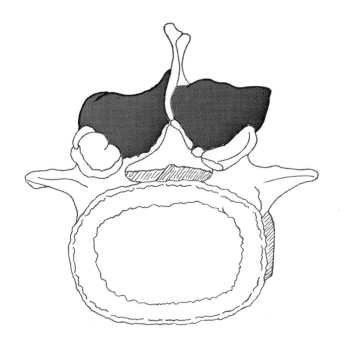

lamina
LAM-i-nah

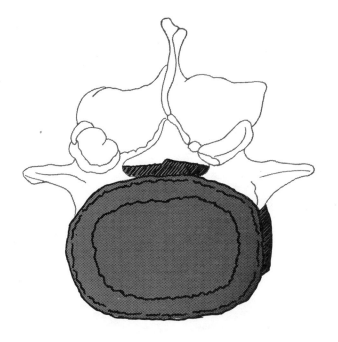

body

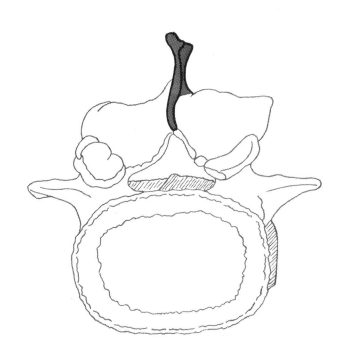

spinous process
SPĪ-nus

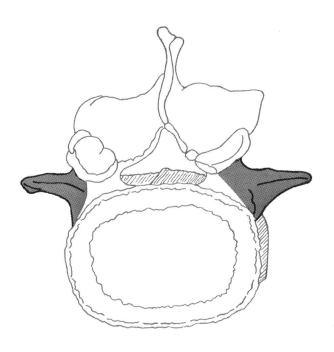

transverse process

LUMBAR VERTEBRA (Viewed from Above)

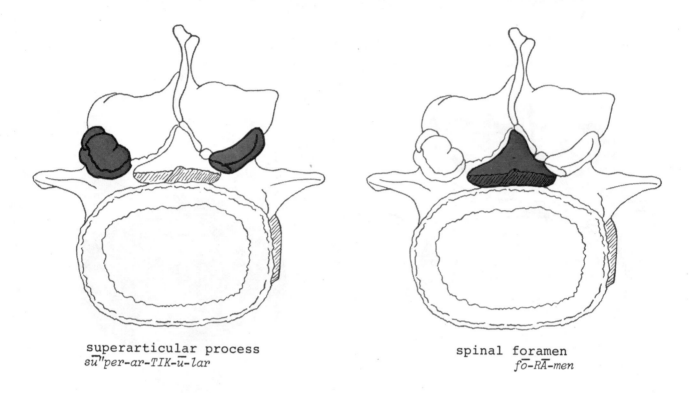

superarticular process
sū″per-ar-TIK-ū-lar

spinal foramen
fō-RĀ-men

LUMBAR VERTEBRA (Lateral View)

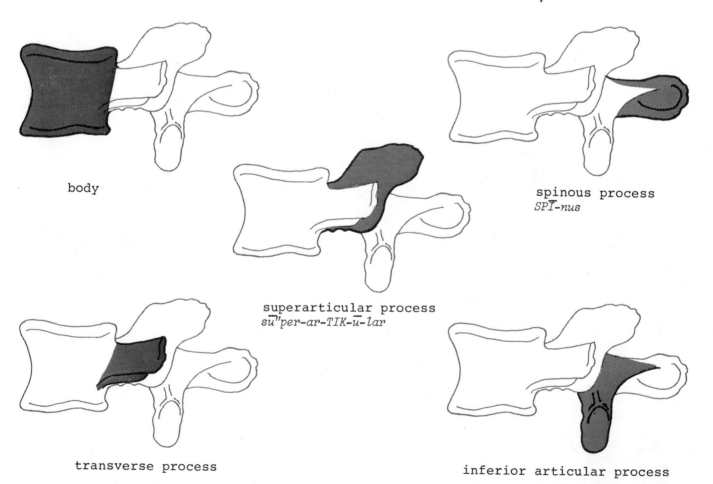

body

superarticular process
sū″per-ar-TIK-ū-lar

spinous process
SPĪ-nus

transverse process

inferior articular process

LOWER EXTREMITY (Anterior View)

hip bone (os coxae)...........2
femur........................2
patella......................2
tibia........................2
fibula.......................2

tarsus
    calcaneus (calcaneum).....2
    talus.....................2
    cuboid....................2
    navicular (scaphoid)......2
    third cuneiform...........2
    second cuneiform..........2
    first cuneiform...........2

metatarsus..................10
phalanges...................28

TOTAL BONES OF
LOWER EXTREMITIES   62

ilium
IL-ē-um

ischium
IS-kē-em

femur
FĒ-mer

patella
pah-TEL-lah

fibula
FIB-yoo-lah

tibia
TIB-ē-ah

tarsal bones
TAR-sal

proximal
PROK-si-mal
phalanx
FĀ-lanx

metatarsals
met-ah-TAR-sals

phalanges
fah-LAN-jēz

MALE PELVIS

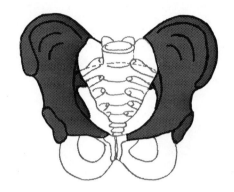

ilium
IL-ē-um

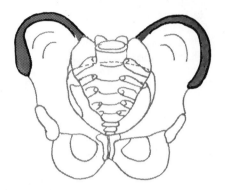

iliac crest
IL-ē-ak

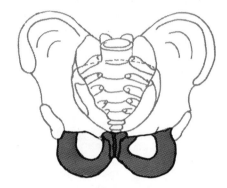

ischium
IS-kē-em

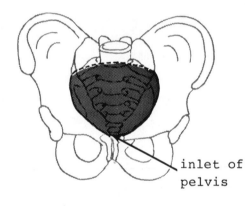

inlet of
pelvis

lesser or true pelvis

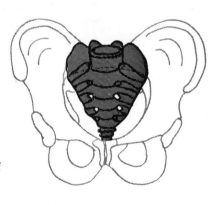

sacrum and coccyx

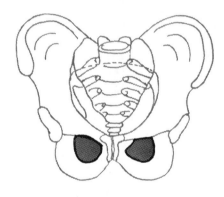

obturator foramen
OB-too-rā-ter fō-RĀ-men

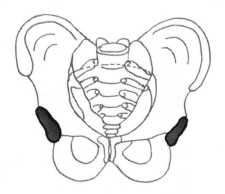

acetabulum
as"i-TAB-ye-lem

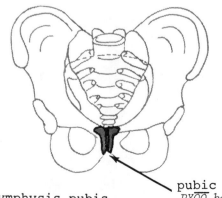

symphysis pubis
SIM-fi-sis PYOO-bis

pubic arch
PYOO-bik

FEMALE PELVIS

ilium
IL-ē-um

iliac crest
IL-ē-ak

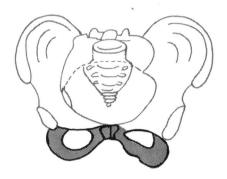

ischium
IS-kē-em

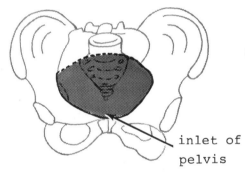

inlet of
pelvis

lesser or true pelvis

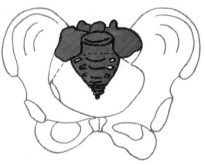

sacrum and coccyx

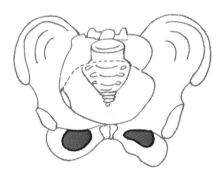

obturator foramen
OB-too-rā-ter fō-RA-men

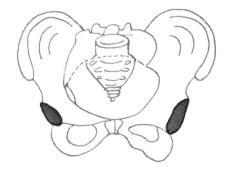

acetabulum
as"i-TAB-ye-lem

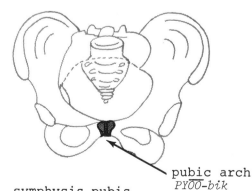

pubic arch
PYOO-bik

symphysis pubis
SIM-fi-sis PYOO-bis

THE FEMUR

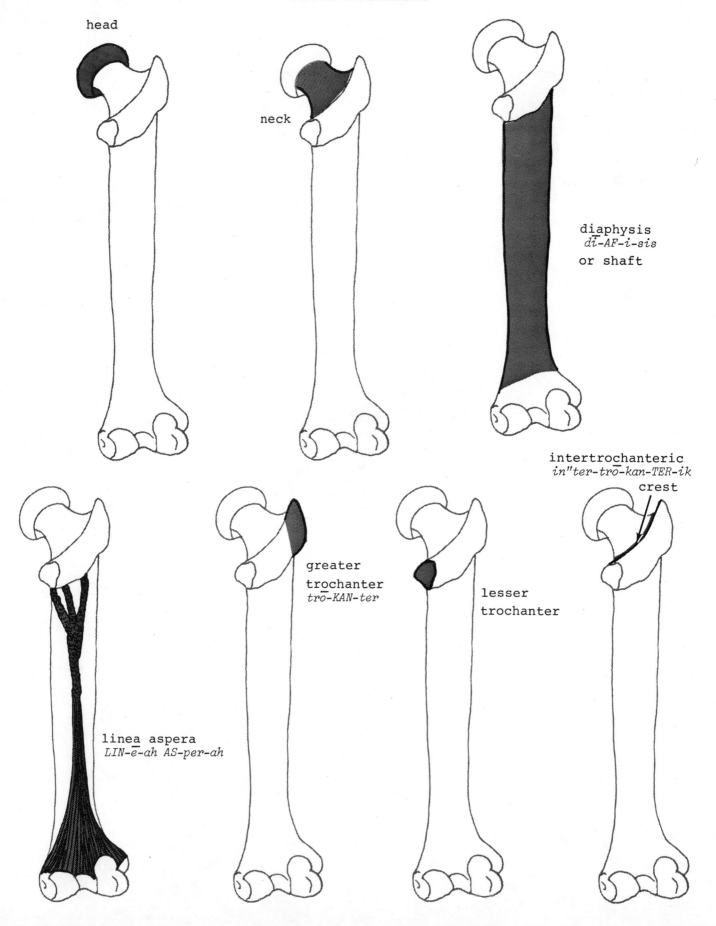

head

neck

diaphysis
dī-AF-i-sis
or shaft

intertrochanteric
in"ter-trō-kan-TER-ik
crest

greater
trochanter
trō-KAN-ter

lesser
trochanter

linea aspera
LIN-ē-ah AS-per-ah

THE FEMUR

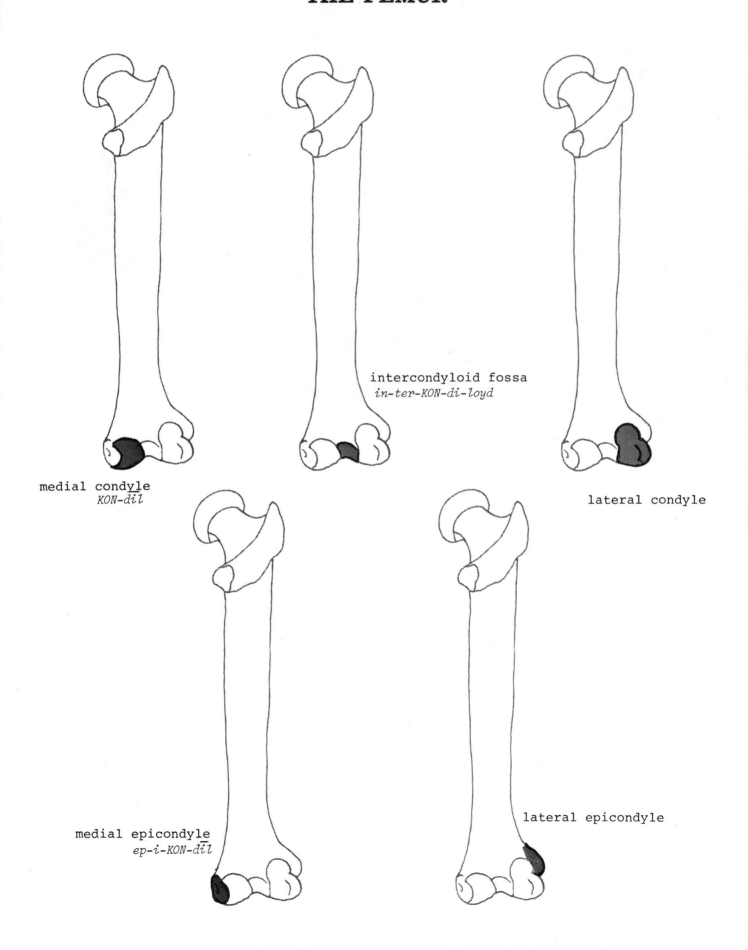

intercondyloid fossa
in-ter-KON-di-loyd

medial condyle
KON-dil

lateral condyle

medial epicondyle
ep-i-KON-dil

lateral epicondyle

GROWTH AREAS OF A LONG BONE

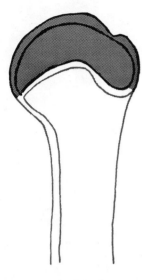 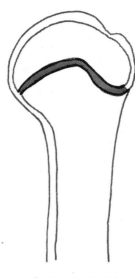 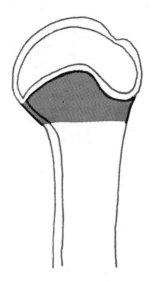 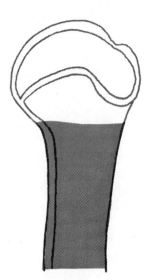

epiphysis
ē-PIF-i-sis

epiphyseal disc
ep-i-FIZ-ē-al

metaphysis
met-AF-i-sis

diaphysis or shaft
dī-AF-i-sis

The metaphysis is the extremity of the diaphysis or shaft of a long bone where it joins the epiphysis.

THE TIBIA AND FIBULA (Posterior View)

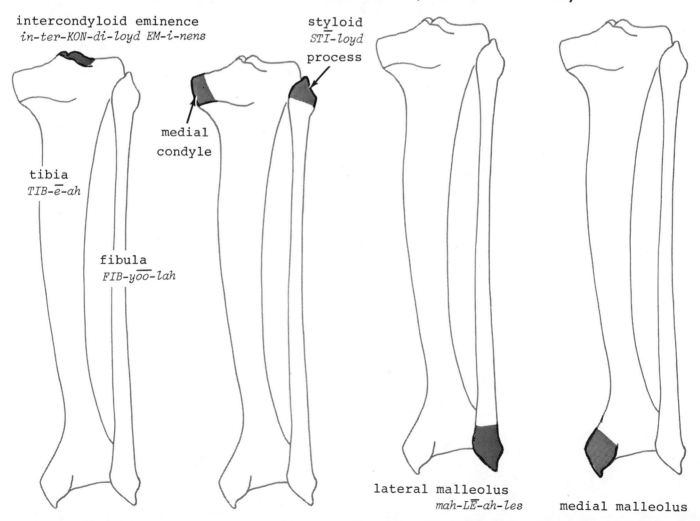

intercondyloid eminence
in-ter-KON-di-loyd EM-i-nens

styloid
STĪ-loyd
process

medial
condyle

tibia
TIB-ē-ah

fibula
FIB-yoo-lah

lateral malleolus
mah-LĒ-ah-les

medial malleolus

THE TIBIA AND FIBULA (Anterior View)

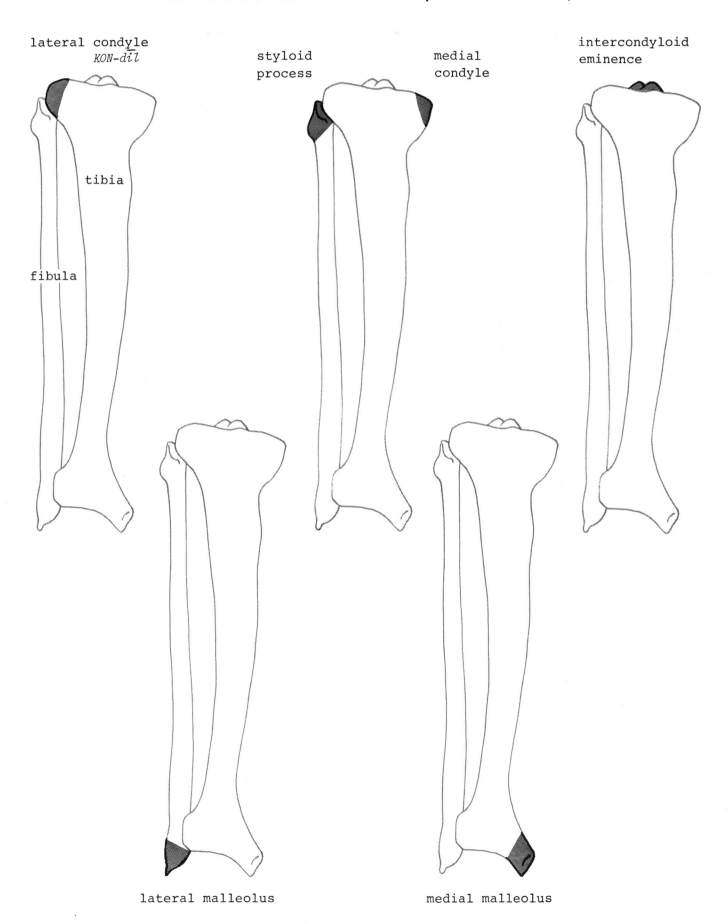

lateral condyle
KON-dil

styloid
process

medial
condyle

intercondyloid
eminence

tibia

fibula

lateral malleolus

medial malleolus

RIGHT FOOT (From Above)

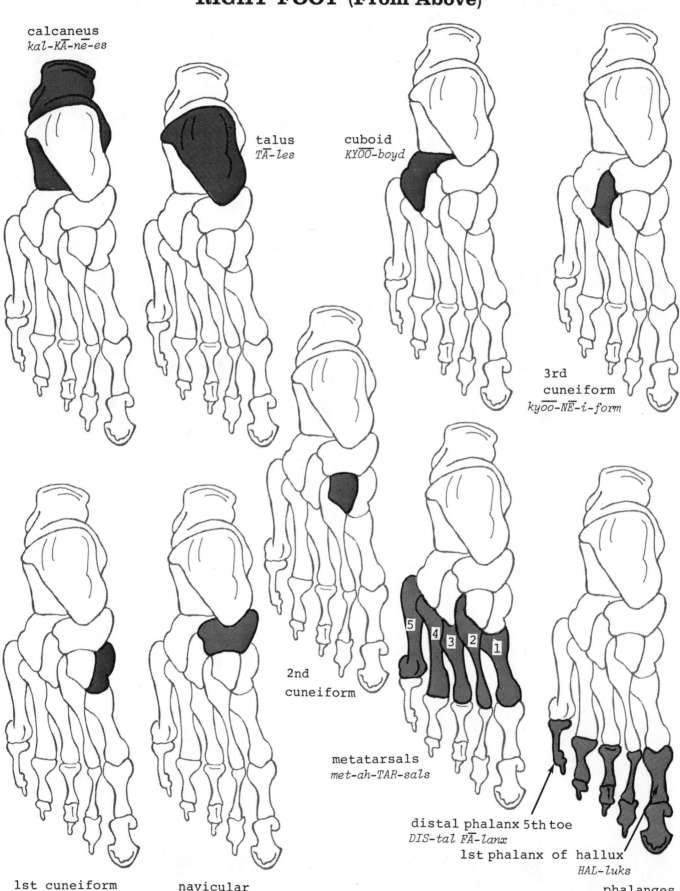

calcaneus
kal-KA̅-ne̅-es

talus
TA̅-les

cuboid
KYOO̅-boyd

3rd
cuneiform
kyoo̅-NE̅-i-form

2nd
cuneiform

5 4 3 2 1

metatarsals
met-ah-TAR-sals

distal phalanx 5th toe
DIS-tal FA̅-lanx

1st phalanx of hallux
HAL-luks

1st cuneiform

navicular
nah-VIK-ye-lar

phalanges
fah-LAN-je̅z

JOINTS AND ARTICULATIONS

SYNARTHROSES
sin-ar-THRŌ-sēz
(Immovable Joints)

AMPHIARTHROSES
am"phē-ar-THRŌ-sēz
(Slightly Movable Joints)

SKULL

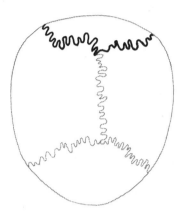

lambdoidal suture
lam-DOYD-al

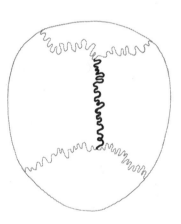

sagittal suture
SAJ-i-tal

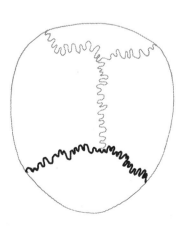

coronal suture
kō-RŌ-nal

MALE PELVIS

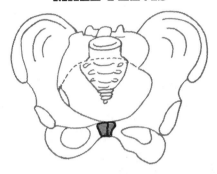

symphysis pubis joint
SIM-fi-sis   PYOO-bis

FEMALE PELVIS

sacroiliac joint
sā-krō-IL-ē-ak

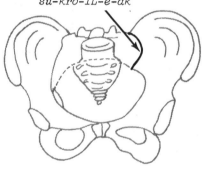

SYNDESMOSIS
sin-dez-MŌ-sis

TIBIOFIBULAR JOINT
tib-ē-ō-FIB-ū-lar

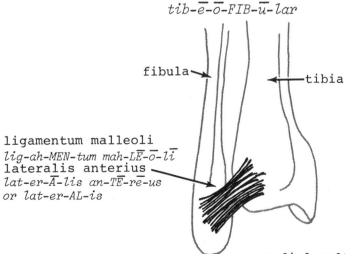

fibula

tibia

ligamentum malleoli
lig-ah-MEN-tum mah-LĒ-ō-lī
lateralis anterius
lat-er-Ā-lis an-TĒ-rē-us
or *lat-er-AL-is*

lateral malleolus
mah-LĒ-ah-les

medial malleolus

JOINTS AND ARTICULATIONS
DIARTHROSES (Freely Movable Joints)
dī-ar-THRŌ-sez

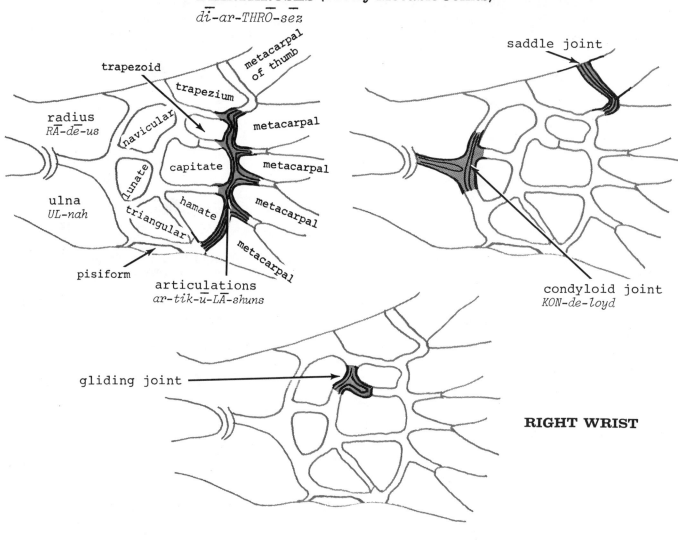

RIGHT WRIST

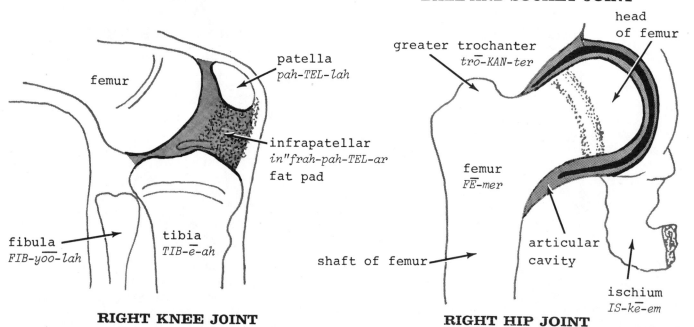

HINGE JOINT

BALL-AND-SOCKET JOINT

RIGHT KNEE JOINT

RIGHT HIP JOINT

JOINTS AND ARTICULATIONS
RIGHT KNEE JOINT CAPSULE

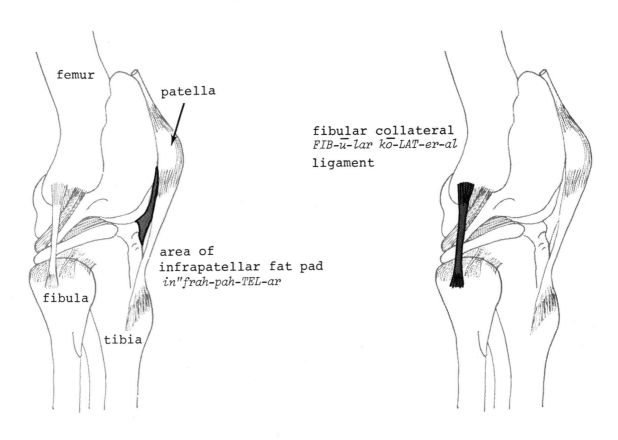

femur

patella

fibular collateral
FIB-ū-lar kō-LAT-er-al
ligament

area of
infrapatellar fat pad
in"frah-pah-TEL-ar

fibula

tibia

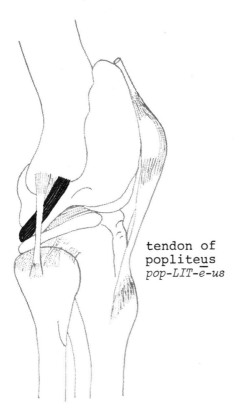

tendon of
popliteus
pop-LIT-ē-us

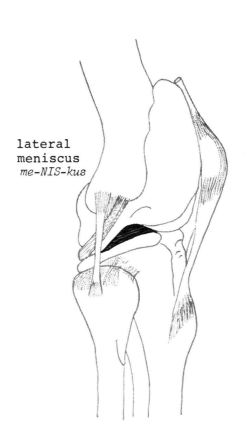

lateral
meniscus
me-NIS-kus

JOINTS AND ARTICULATIONS
RIGHT KNEE JOINT CAPSULE (Lateral View)

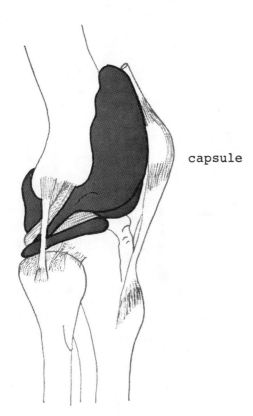

capsule

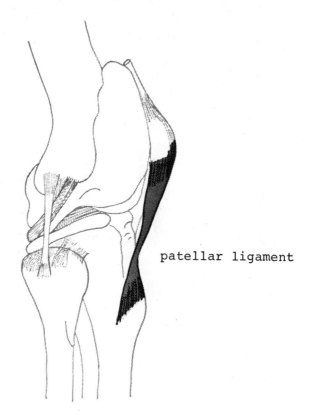

patellar ligament

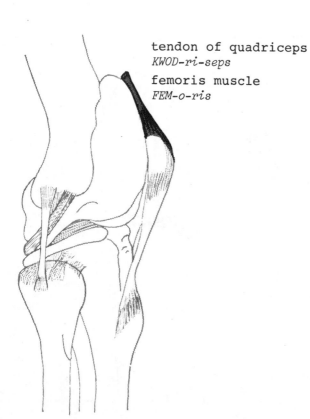

tendon of quadriceps
KWOD-ri-seps
femoris muscle
FEM-o-ris

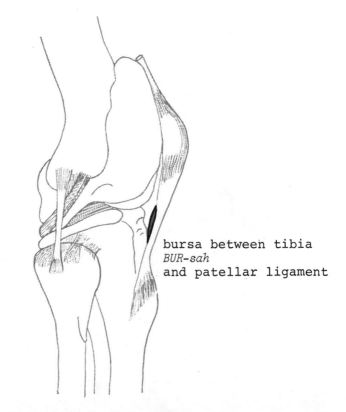

bursa between tibia
BUR-sah
and patellar ligament

JOINTS AND ARTICULATIONS
LIGAMENTS OF LEFT SHOULDER

coracoacromial ligament
kor"ah-kō-ah-KRŌ-me-al

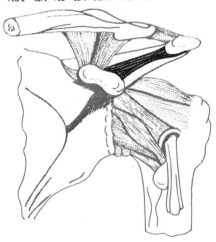

capsular ligament
KAP-sū-lar

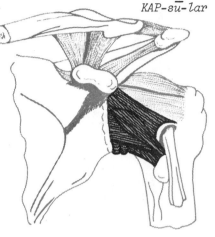

coracoclavicular ligament
kor"ah-kō-klah-VIK-u-lar

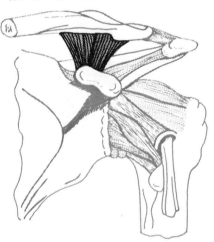

coracohumeral ligament
kor"ah-kō-HYOO-mer-al

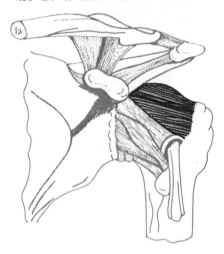

superior acromioclavicular ligament
ah-krō"me-ō-klah-VIK-u-lar

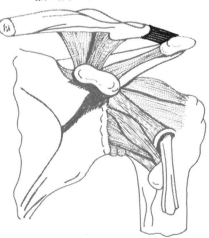

superior transverse ligament

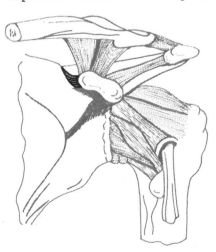

LIGAMENTS OF LEFT SHOULDER

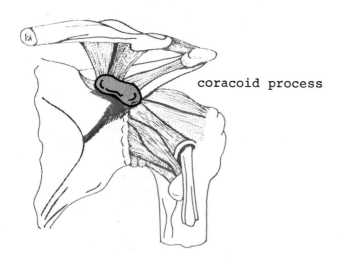

coracoid process

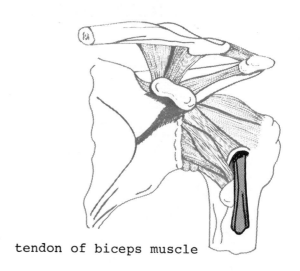

tendon of biceps muscle

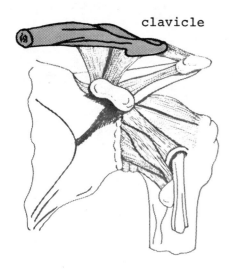

clavicle

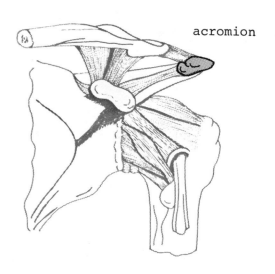

acromion

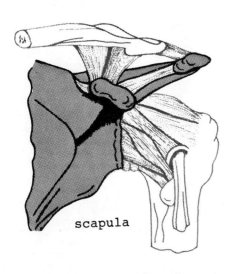

scapula

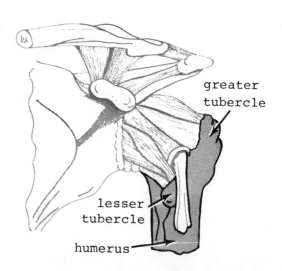

greater
tubercle

lesser
tubercle

humerus

JOINTS AND ARTICULATIONS

RIGHT SHOULDER JOINT CAPSULE (Anterior View)

synovial capsule
si-NŌ-vē-al

bursa under subscapularis
BUR-sah   *sub"skap-ū-LAR-is*

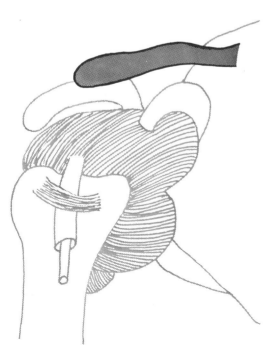

clavicle

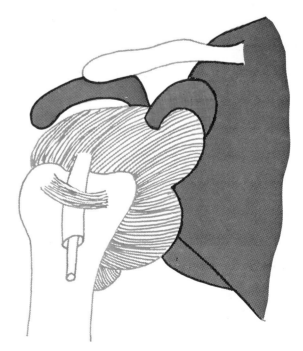

scapula

JOINTS AND ARTICULATIONS

RIGHT SHOULDER JOINT CAPSULE (Anterior View)

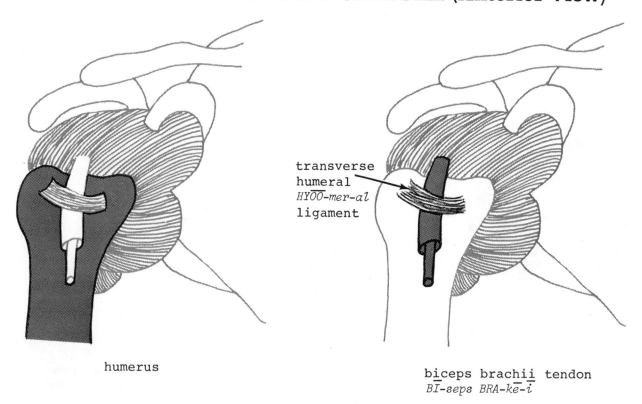

transverse
humeral
HYOO-mer-al
ligament

humerus

biceps brachii tendon
BI-seps BRA-ke-i

THE COMPOSITE PARTS OF A JOINT

<u>Cartilage</u> furnishes elastic, supporting connective tissues which protect bones at joints from shock and give the skeleton more flexibility.

The <u>articular capsule</u> is composed of the fibrous capsule (external) and the synovial lining (internal) known as the synovial membrane.

The <u>fibrous capsule</u> is attached to the whole circumference of the articular end of each bone entering the joint, and, thus, entirely surrounds the articulation.

The <u>synovial membrane</u> covers the inner surface of the fibrous capsule, forming a closed sac called the synovial cavity.

<u>Synovial tendon sheaths</u> facilitate the gliding of tendons which pass through fibrous and bone tunnels such as those under the flexor retinaculum of the wrist. These sheaths are closed sacs, one layer of the synovial membrane lining the tunnel, the other reflected over the surface of the tendon.

<u>Synovial bursae</u> are clefts in connective tissue between muscles, tendons, ligaments and bones. They are made into closed sacs by a synovial lining, similar to that of a true joint, which may in some cases be continuous through an opening in the wall with the lining of a joint cavity. They facilitate the gliding of muscles or tendons over bony or ligamentous prominences, and are named according to their location: subcutaneous, submuscular and subtendinous.

<u>Ligaments</u> are composed mainly of bundles of collagenous fibers placed parallel with, or closely inter-laced with one another, and present a white, shining, silvery appearance. They are pliant and flexible, so as to allow perfect freedom of movement, but are also strong, tough, and inextensible, so as not to yield readily to applied force. The main function of ligaments is to hold the bones together in the joints.

INDEX OF ENGLISH TRANSLATION

The material presented in each index is to aid in spelling and better identification of medical terms. To understand the language of medicine it is imperative to learn its English translation. Many terms are repeated in one or more of the indices which makes their translation easier to remember.

acetabulum
a vinegar cruet
Resemblance to a vinegar vessel
or a cup-shaped vessel.

col/later/al
together   side

dist/al
farthest

acromi/al
acromion

concha (sing.)
shell

eminence
projection

acromio/clavicul/ar
acromion   clavicle

conchae (pl.)
shells

epi/condyle
on, upon   knuckle

acro/mion
tip   shoulder

condyle
knuckle

epi/physe/al
on, upon   to grow

amphi/arthroses
double   joints

condyl/oid
knuckle

epi/physis
on, upon   to grow

ana/tom/ical
up   to cut

coraco/acromi/al
coracoid process   acromion

ethm/oid
sieve

anterius
in front

coraco/clavicul/ar
coracoid process   clavicle

femoris
femur

bi/ceps brachii
two   head   arm

coraco/humer/al
coracoid process humerus

femur
thigh

bursa
a purse

corac/oid
crow's beak

fibula
buckle
Its position to the tibia makes it look like
the pin of a brooch.

calcaneus
heel bone

cornua
horn

fibul/ar
fibula

capit/ate
head

coron/al
crown

foramen
opening

capit/ulum
head   small

coron/oid
crow

fossa
ditch

carp/al
wrist

cribri/form
sieve

front/al
forehead

cartilage
gristle

crista galli
crest   cock's comb

gladiolus
little sword (Latin)

cartilagin/ous
gristle

cub/oid
cube

glen/oid
socket

cervic/al
neck

cunei/form
wedge

ham/ate
hooked

clav/icle
key   small

dia/physis
through   to grow

ham/ulus
hook   little

coccyx
a cuckoo
Because it resembles the beak of the cuckoo

di/arthroses
through   joints

humer/al
shoulder

See Page 8
for Suffix
Endings

humerus shoulder	lumb/ar loin	obturator to occlude, stop up
hy/oid Greek letter "U"	lun/ate moon	occipit/al back of head
ili/ac hip bone	magnum great, large	olecranon head or point of elbow
ilium groin (hip bone)	malar cheek	opt/ic eye
incus anvil	malleoli hammer	orbit circle
infra/orbit/al beneath orbit	malleolus hammer	orbit/al orbit
infra/patell/ar beneath patella	malleus hammer	palatine palate
infra/spin/atus beneath spine adjective	mandible a jaw	pariet/al wall
inter/condyl/oid between knuckle	manubrium handle	patella pan (kneepan)
inter/trochanter/ic between runner	mast/oid breast Named by Galen (an early Italian anatomist) because of its resemblance to a breast.	patell/ar patella
inter/tuber/cul/ar between node small	maxilla jaw bone	pelvis a washbasin
inter/vertebr/al between vertebra	meniscus crescent (interarticular fibrocartilage)	phalanges (pl.) a line of soldiers Because bones of toes and fingers are arranged in rows or "ranks."
ischium hip, haunch	ment/al chin	phalanx (sing.) a line of soldiers
lacrim/al tear	meta/carp/als beyond wrist	pisi/form pea
lambd/oid/al Greek letter "L"	meta/physis beyond to grow	popliteus back of knee
lamina thin plate	meta/tars/als beyond a flat surface	proxim/al nearest
later/alis side adjective	mult/angul/ar many angle	pteryg/oid wing
ligamentum ligament, to bind	nas/al nose	pub/ic genitals
linea aspera line rough	navi/cul/ar ship small	pubis genitals

quadri/ceps
four   head

radi/al
rod, ray

radius
rod, ray

ramus
a branch

sacro/ili/ac
sacrum   ilium adjective

sacrum
sacred
Probably named because of the shape of the
bone resembling certain sacrificial vessels.

sagitt/al
arrow, straight

scapula
shoulder blade

sphen/oid
wedge

stapes
stirrup

stern/al
sternum

sternum
the chest

styl/oid
stake, pole

sub/scapul/aris
beneath   scapula   adjective

sulcus
trench

super/articul/ar
above   joint

supra/orbit/al
above   orbit

supra/spin/atus
above   spine adjective

surgic/al
handwork

sym/physis
together   to grow

syn/arthroses
together   joints

syn/desm/osis
together to bind

syn/ovi/al
with   egg
The fluid is something like raw egg
white in consistency.

talus
ankle

tars/al
a flat surface

tempor/al
time, temple

thorac/ic
chest

tibia
a pipe, a flute
The shin bone from which flutes or pipes were made.

tibio/fibul/ar
tibia   fibula

trapezium
table, counter

trapez/oid
table, counter

tri/angul/ar
three   angle

trochanter
runner
The ball on which the hip bone turns in the socket.

trochlea
pulley

trochle/ar
pulley

tuber/cle
node   small

tuber/osity
node   condition

turbinates
shaped like a top

ulna
elbow

vertebra (sing.)
to turn

vertebrae (pl.)
to turn

vomer
ploughshare

xiph/oid
sword (Greek)

zygomat/ic
yoke

CHAPTER III
MUSCULAR SYSTEM

There are three kinds of muscle tissue in the body.

1. Striated or voluntary muscles which are concerned with voluntary action such as raising the arms, walking, writing and speaking. Their function is to operate the bones of the body to produce motion.

2. Nonstriated or smooth muscles are known as involuntary muscles. They occur in the stomach, intestines and blood vessels and are controlled by the medulla oblongata and the cranial nerves through the autonomic nervous system.

3. Indistinctly striated muscle is known as cardiac muscle and is found only in the heart.

The muscles featured in this chapter are the striated or voluntary muscles. They are also the ones most frequently encountered in the paramedical field. There are many more muscles in the human body but, in most cases, they do not present any obstacles to paramedical personnel.

PLEASE NOTE: The first time a muscle is shown in an illustration the Origin, Insertion and Action of that muscle will be given as well as its pronunciation. When the muscle appears again in different views of that particular anatomical area reference will be made to the page on which the pronunciation, Origin, Insertion and Action were first given. Further reference will also appear in the main Index following Chapter X.

INTERESTING FACTS ABOUT MUSCLES: The derivation of naming muscles comes from various
sources:

    1. The situation of the muscle as the <u>brach</u>ialis, <u>pectoralis</u>, <u>supraspinatus</u>.
                                             arm   **breast bone**   **above** **spine**

    2. The direction of the muscle as the <u>rectus</u>, <u>obliquus</u> (external and internal
                                            straight   **slanting**
       oblique muscles), and the <u>transversus</u> abdominis (<u>transversus</u>).
                                     across **to turn**

    3. The action of the muscle as <u>flexors</u>, <u>extensors</u>.
                                  to bend   **out to stretch**

    4. The shape of the muscle as the <u>deltoid</u>, <u>trapezius</u>, <u>rhomboideus</u>.
                            Greek "delta" △   **table, counter**   **a lozenge shaped figure**

    5. The number of divisions of a muscle as the <u>biceps</u>, <u>triceps</u>, <u>quadriceps</u>.
                                       two head **three**   **four**

    6. The points of attachment of a muscle as the <u>sternocleidomastoid</u>, <u>omohyoid</u>.
                                   sternum   **clavicle** **mastoid process** **shoulder** **hyoid bone**

Practically every muscle acting upon a joint is matched by another muscle which
has opposite action. The opposing muscles to the flexors are the extensors. In the arm
the flexors cause the arm to bend and the extensors extend or straighten the arm.

The direction of muscle fibers depends on the function of the muscle. The fibers
of the transversus muscle are horizontal while those of the rectus muscle are vertical.
Wherever the direction of the fibers appears important they are so indicated on the drawings. Surgeons make incisions in the direction of the fibers. Thus, they would cut ver-
tically through the rectus muscle and then horizontally through the transversus muscle.

ORIGIN AND INSERTION OF MUSCLES

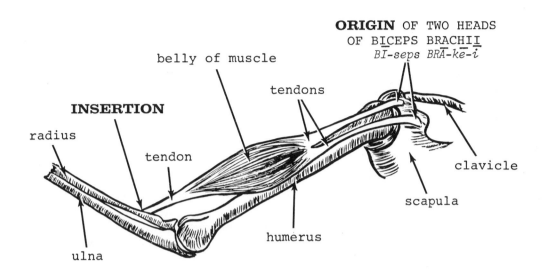

Most skeletal muscles pass over joints. Some of these joints are movable, while others are immovable. When a joint is movable the terms *origin* and *insertion* of the muscle are used.

The *origin* is the end attached to the relatively less movable bone. The *insertion* is the attachment to a bone which moves in the ordinary activity of the body. As a muscle contracts it increases in diameter and pulls the attachments at each end near to each other.

The attachment nearer the center of the body is usually described as the *origin* and the attachment which is farther out in the body is termed the *insertion.*

A muscle usually ends in a tendon or tendons or an aponeurosis by which it is attached to the bone.

DESCRIPTION OF FASCIA, LIGAMENT, APONEUROSIS AND TENDON

FASCIA—A sheet or band of fibrous tissue which covers the body under the skin and invests the muscles and certain organs.

LIGAMENT—Any tough, fibrous band which connects bones or supports viscera.

APONEUROSIS—Fibrous membrane of a pearly white color, iridescent, and glistening which represents a very flattened tendon.

TENDON—The fibrous cord of connective tissue in which the fibers of a muscle end and by which a muscle is attached to a bone or other structure. It somewhat resembles a cable.

SUPERFICIAL MUSCLES OF HEAD AND NECK

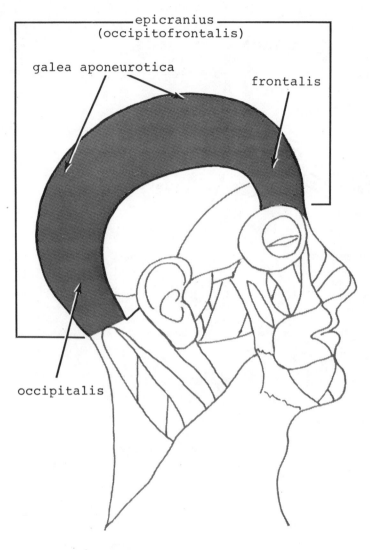

epicranius
(occipitofrontalis)

galea aponeurotica

frontalis

occipitalis

epicranius (occipitofrontalis)
ep-i-KRĀ-nē-us (ok-sip"i-tō-fron-TAL-is) *

occipital portion (occipitalis)
ok-SIP-i-tal   (ok-sip-i-TAL-is)

ORIGIN: occipital bone and mastoid process
         of temporal bone
INSERTION: galea aponeurotica
           GĀ-lē-ah ap"o-nū-ROT-ik-ah
ACTION: draws scalp backward

frontal portion (frontalis)
FRON-tal   (fron-TAL-is)

ORIGIN: continuous with corrugator super-
         cilia, procerus, orbicularis oculi
INSERTION: galea aponeurotica
ACTION: elevates eyebrows, pulls scalp
         forward

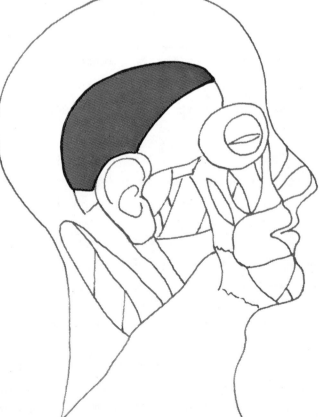

auricularis superior
*aw"rik-ū-LAR-is***

ORIGIN: galea aponeurotica
INSERTION: upper margin of root of auricle
ACTION: raises pinna of ear

*Optional pronunciation for the *alis* suffix
 is *Ā-lis.* (*fron-TĀ-lis*)
**Optional pronunciation for the *aris* suffix
  is *Ā-ris.* (*aw"rik-ū-LĀ-ris*)

SUPERFICIAL MUSCLES OF HEAD AND NECK

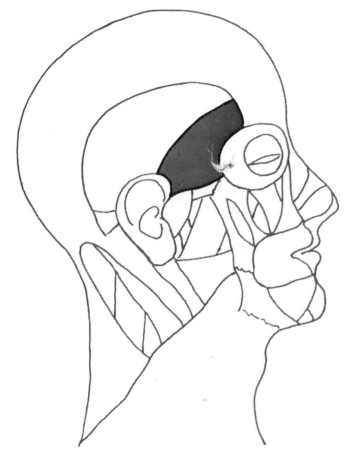

auricularis anterior

O: superficial temporal fascia
I: cartilage of auricle
A: draws pinna of ear forward

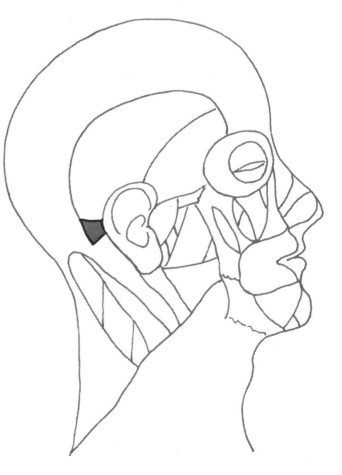

auricularis posterior

O: mastoid process
I: posterior portion of root of auricle
A: draws pinna of ear backward

From hereon the ORIGIN will be designated
by O, INSERTION by I and ACTION by A.

SUPERFICIAL MUSCLES OF HEAD AND NECK

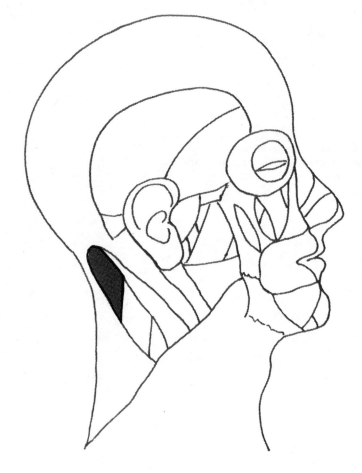

splenius capitis
SPLĒ-nē-us KAP-i-tis

O: lower half of ligamentum nuchae; spines of
   C7, T1, T2 and T3 vertebrae
I: mastoid process and occipital bone
A: rotates and extends head and neck and
   flexes sidewise

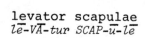

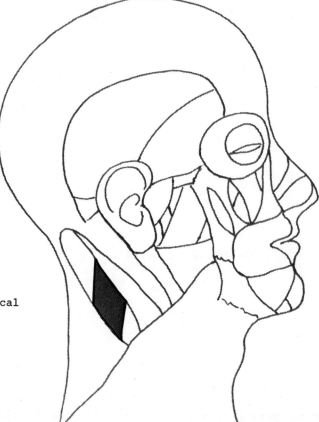

levator scapulae
lē-VĀ-tur SCAP-ū-lē

O: transverse process of four upper cervical
   vertebrae
I: pcsterior edge of scapula
A: raises upper angle of scapula; aids in
   rotating head

SUPERFICIAL MUSCLES OF HEAD AND NECK

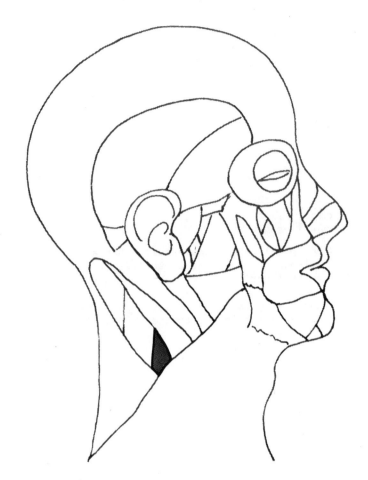

scalenus medius
skah-LĒ-nus MĒ-dē-us

O: transverse processes of C2 to C6 vertebrae
I: upper surface at first rib
A: helps raise first and second ribs, bends
   vertebral column

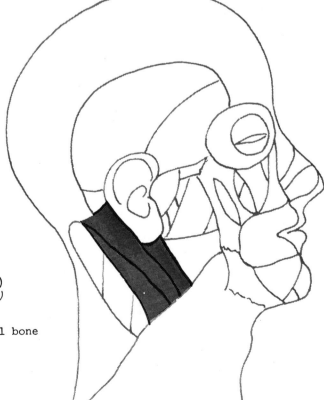

sternocleidomastoid (sternomastoid)
stur"nō-klī"dō-MAS-toyd (stur"nō-MAS-toyd)

O: two heads from sternum and clavicle
I: tendon into mastoid portion of temporal bone
A: flexes vertebral column; rotates head

SUPERFICIAL MUSCLES OF HEAD AND NECK

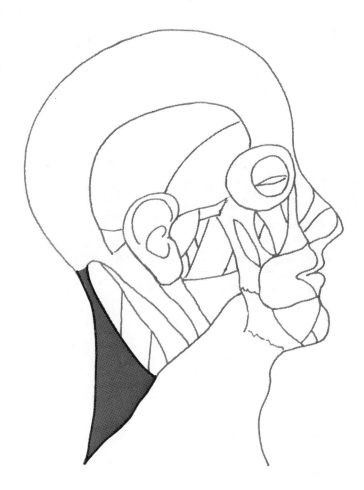

trapezius
trah-PĒ-zē-us

O: occipital bone; ligamentum nuchae and
   spinous processes of C7 to T12 vertebrae
I: ventral surface of vertebral border of
   scapula
A: moves scapula forward away from spine and
   downward and inward toward chest wall

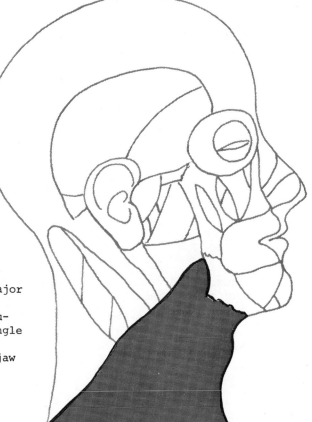

platysma
plah-TIZ-mah

O: fascia over upper parts of pectoralis major
   and deltoid
I: lower border of mandible; skin and subcutaneous tissue of lower part of face, angle
   of mouth
A: wrinkles skin of neck; depresses lower jaw
   and upper lip

SUPERFICIAL MUSCLES OF HEAD AND NECK

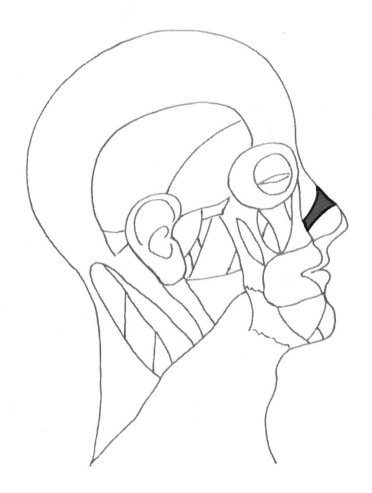

nasalis
nā-ZAL-is

O: maxilla; greater alar cartilage
I: aponeurosis over bridge of nose
A: depresses cartilaginous part of nose and
   draws ala toward septum

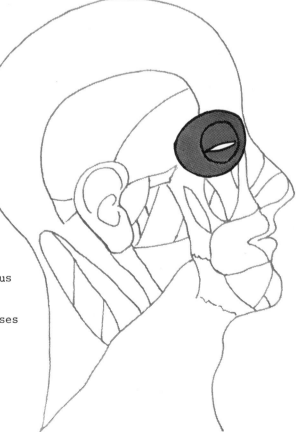

orbicularis oculi
or"bik-ū-LAR-is OK-ū-lī

O: oval sphincter muscle surrounding eye;
    frontal process of maxilla; inner canthus
    of eye; lacrimal bone
I: near origin after encircling orbit
A: closes lids, wrinkles forehead, compresses
    lacrimal sac

SUPERFICIAL MUSCLES OF HEAD AND NECK

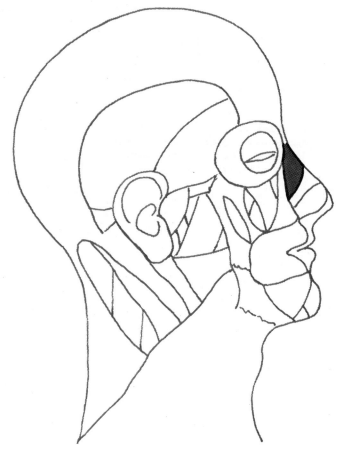

procerus
pr̄o-SER-us

O: fascia over lower part of nasal bone and
   upper part of lateral nasal cartilage
I: skin over lower part of forehead between
   eyebrows; frontalis muscle
A: draws down medial angle of eyebrows,
   causes wrinkles over bridge of nose

quadratus labii superioris
kwod-RĀ-tus LĀ-bē-ī su-PĒ-rē-or-is

THREE HEADS
   1. angular head

      O: frontal process of maxilla
      I: greater alar cartilage and skin of
         nose
      A: raises upper lip and dilates nostril

   2. infraorbital head
      in"frah-OR-bi-tal

      O: levator labii superioris
      I: upper lip
      A: gives expression of sadness

   3. zygomatic head
      zi-ḡo-MAT-ik

      O: zygomatic bone
      I: upper lip
      A: draws upper lip upward and outward

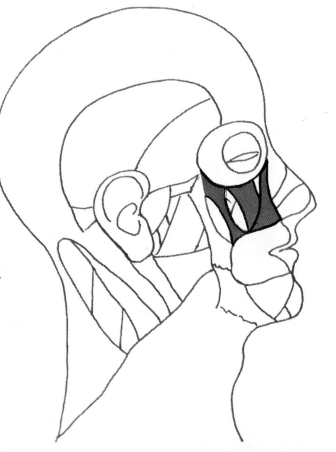

SUPERFICIAL MUSCLES OF HEAD AND NECK

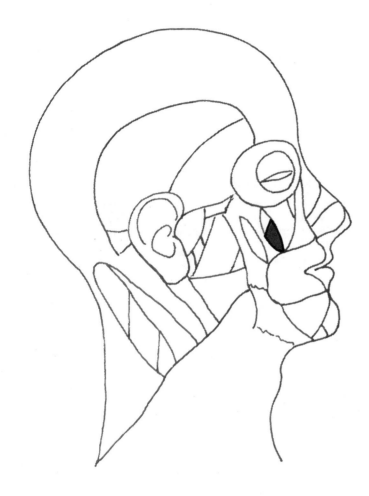

caninus
k$\bar{a}$-N$\bar{I}$-nus

O: maxilla
I: orbicularis oris and skin at angle of
   mouth
A: raises angle of mouth

zygomaticus
z$\bar{i}$-g$\bar{o}$-MAT-e-kus

O: zygomatic bone
I: angle of mouth
A: draws angle of mouth backward and upward
   as in laughing

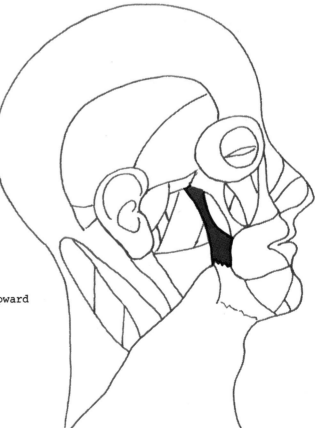

SUPERFICIAL MUSCLES OF HEAD AND NECK

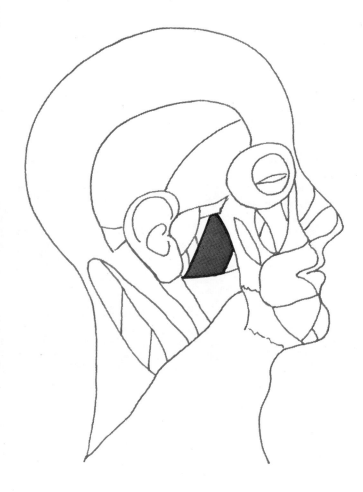

masseter
MAS-i-ter or mas-Ē-ter

O: zygomatic process and adjacent portions of
   maxilla
I: angle and lateral surface of ramus of mandible
A: closes jaw

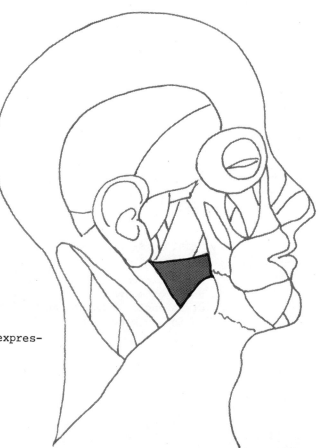

risorius
ri-SAW-rē-us

O: fascia over masseter
I: skin of angle of mouth
A: draws angle of mouth, causing an expres-
   sion of grinning

SUPERFICIAL MUSCLES OF HEAD AND NECK

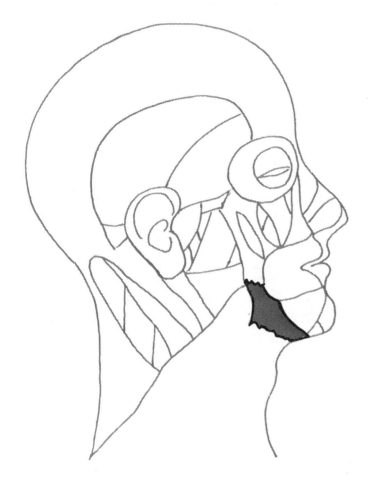

triangularis
trī-ang-gū-LAR-is

O: lower border of mandible
I: angle of mouth
A: pulls down corners of mouth

quadratus labii inferioris
              in-FĒ-rē-or-is

O: anterior portion of lower border of man-
   dible
I: skin of lower lip; orbicularis oris
A: depresses lower lip

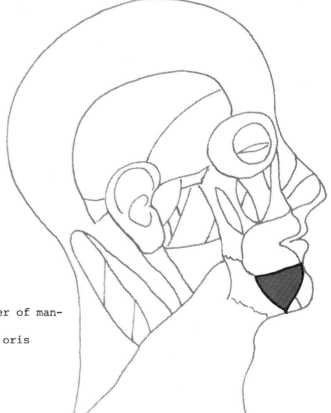

SUPERFICIAL MUSCLES OF HEAD AND NECK

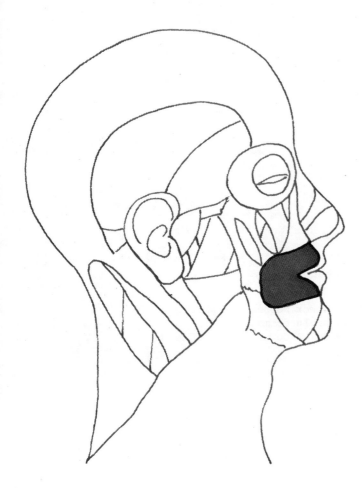

orbicularis oris
     OR-is

O: muscle fibers surrounding opening of mouth
I: angle of mouth
A: protrudes lips, pushes them forward;
   closes lips, the sphincter of the mouth

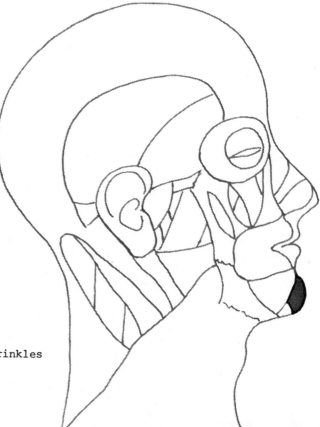

mentalis
men-TAL-is

O: incisive fossa of mandible
I: skin of chin
A: raises and protrudes lower lip; wrinkles
   skin of chin

MUSCLES OF THE EYE

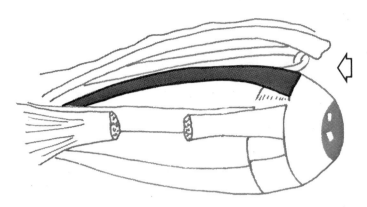

superior rectus
        REK-tus

O: apex of orbital cavity
I: upper and central portion of eyeball
A: rolls eyeball upward

superior oblique
        ob-LIK or ō-BLĒK

O: same as recti muscles
I: between superior and lateral recti of
   eyeball
A: rotates eyeball directing cornea downward and outward

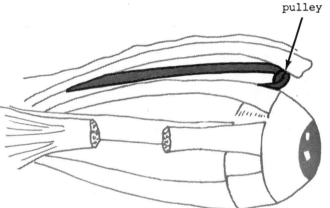

pulley

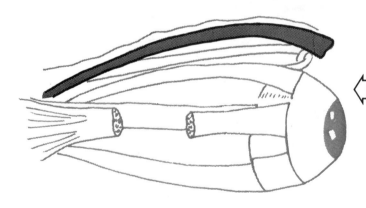

levator palpebrae superioris
lē-VĀ-tur PAL-pe-brē sū-PĒ-rē-or-is

O: lesser wing of sphenoid
I: upper tarsal plate of eye
A: raises upper lid

inferior rectus

O: same as superior and medial recti
I: lower central portion of eyeball
A: rolls eyeball downward

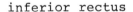

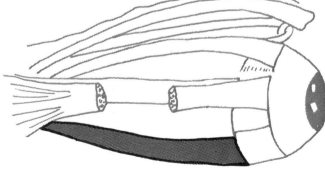

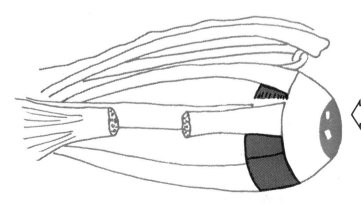

inferior oblique

O: orbital portion of maxilla
I: between superior and lateral recti of
   eyeball
A: rotates eyeball directing the cornea
   upward and outward

MUSCLES OF THE EYE

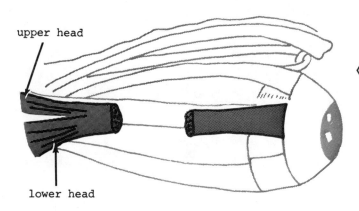

upper head

lower head

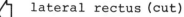

lateral rectus (cut)

O: same as superior, inferior and medial recti
I: midway on lateral portion of eyelid
A: rolls eyeball outward

medial rectus

O: same as superior rectus
I: midway on medial side of eyeball
A: rolls eyeball inward

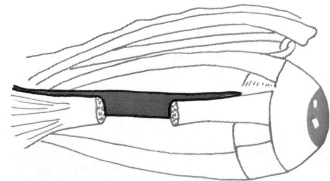

MUSCLES OF THE NECK

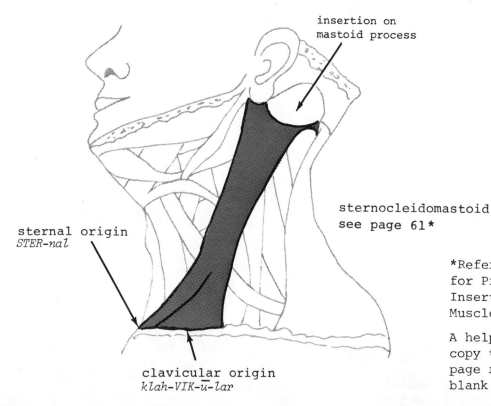

insertion on mastoid process

sternal origin
STER-nal

clavicular origin
klah-VIK-ū-lar

sternocleidomastoid
see page 61*

*Refer to the pages listed for Pronunciation, Origin, Insertion and Action of the Muscle.

A helpful study aid is to copy the material from the page referred to in the blank area provided.

MUSCLES OF THE NECK

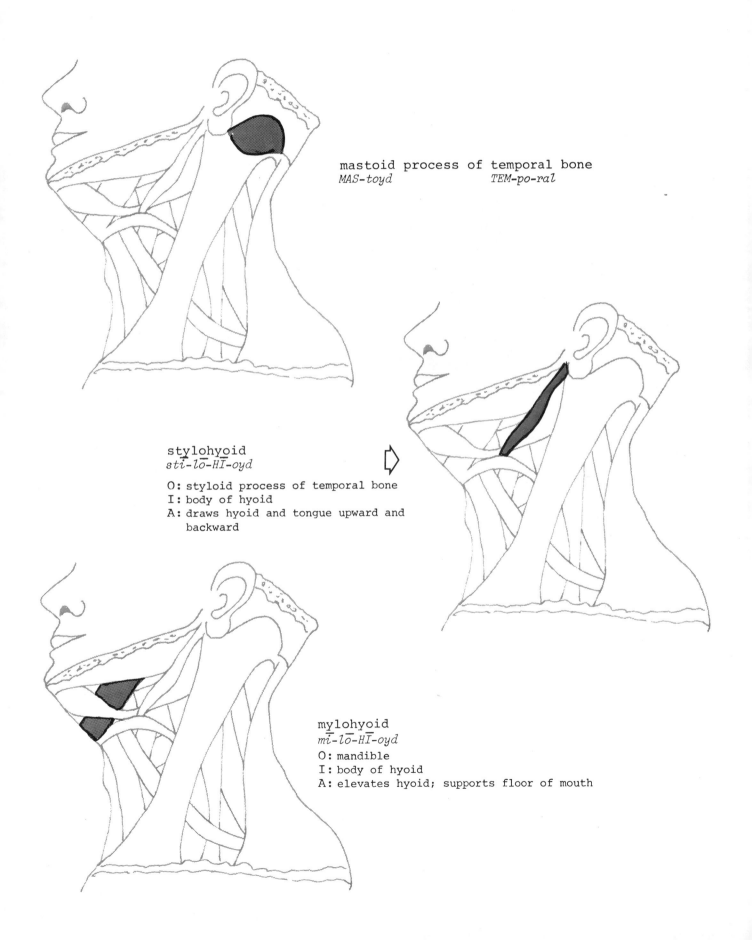

mastoid process of temporal bone
MAS-toyd   *TEM-po-ral*

stylohyoid
stī-lō-HĪ-oyd

O: styloid process of temporal bone
I: body of hyoid
A: draws hyoid and tongue upward and
   backward

mylohyoid
mī-lō-HĪ-oyd
O: mandible
I: body of hyoid
A: elevates hyoid; supports floor of mouth

MUSCLES OF THE NECK

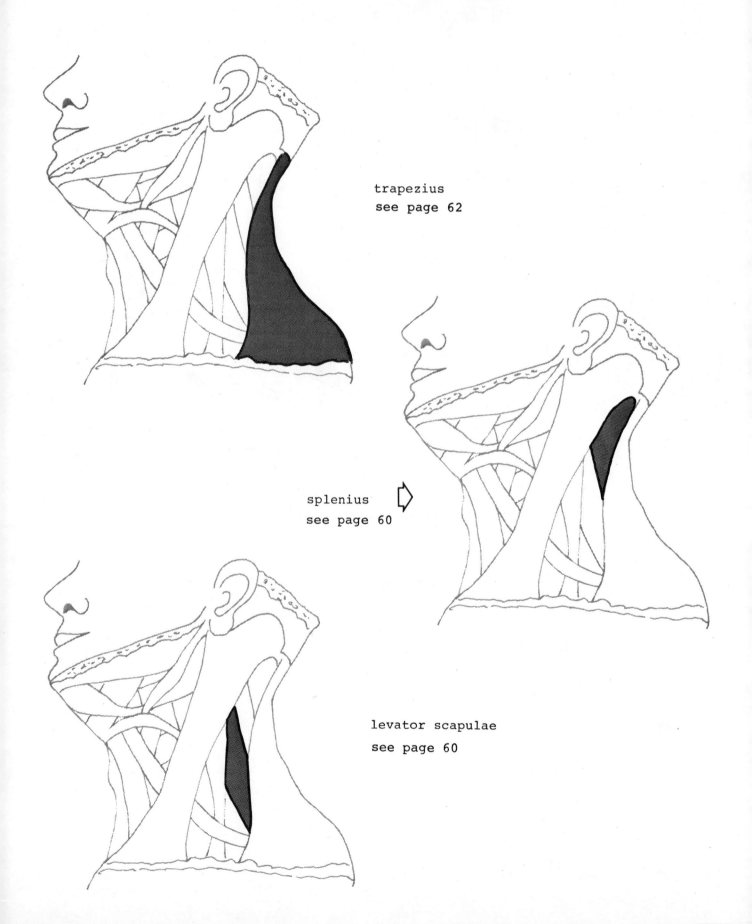

trapezius
see page 62

splenius
see page 60

levator scapulae
see page 60

MUSCLES OF THE NECK

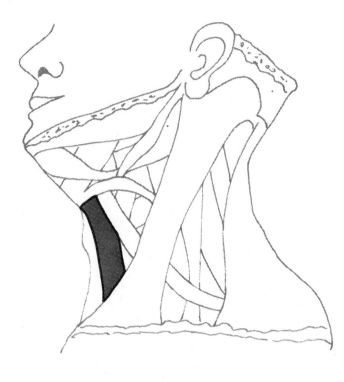

sternohyoid
ster-nō-HĪ-oyd

O: manubrium at sternum and clavicle
I: hyoid
A: depresses hyoid and larynx

omohyoid
ō-mō-HĪ-oyd

O: superior border of scapula
I: lateral border of hyoid
A: retracts and depresses hyoid; contracts
   cervical fascia

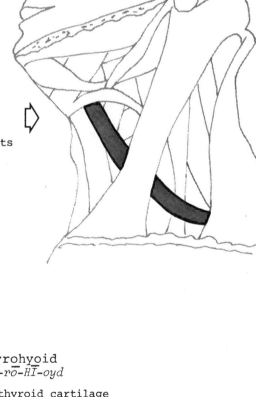

thyrohyoid
thī-rō-HĪ-oyd

O: thyroid cartilage
I: greater cornu of hyoid
A: raises and changes shape of larynx

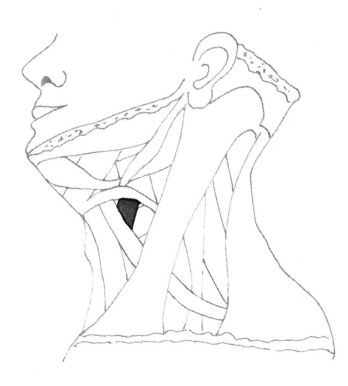

MUSCLES OF THE NECK

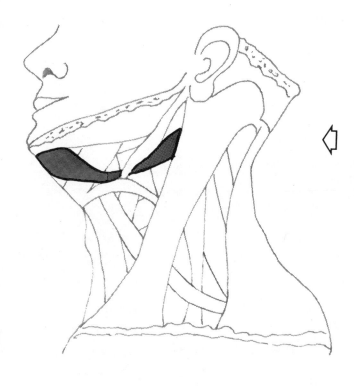

<u>di</u>gastric
di-GAS-trik

O: anterior body: mandible; posterior body:
   skull
I: hyoid bone
A: elevates and retracts hyoid and tongue

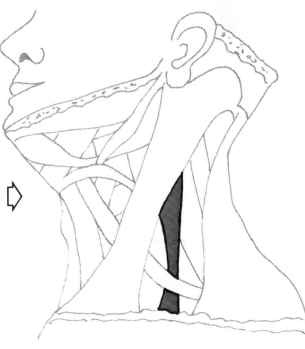

scalenus posterior
skah-LĒ-nus

O: transverse processes of C5, C6, C7 vertebrae
I: outer surfaces of second rib
A: helps raise first and second ribs, bends
   vertebral column

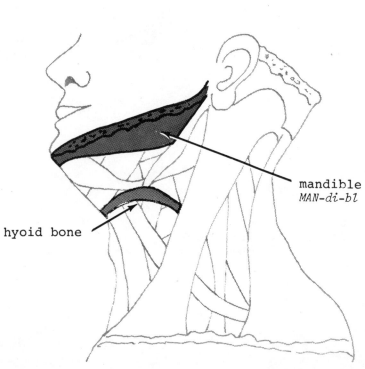

mandible
MAN-di-bl

hyoid bone

SUPERFICIAL MUSCLES OF CHEST AND UPPER ARM
(Anterior View)

deltoid
DEL-toyd

O: clavicle; acromial process and posterior
   border of scapula
I: lateral side of shaft of humerus
A: abduction, flexion, extension and rotation
   of arm

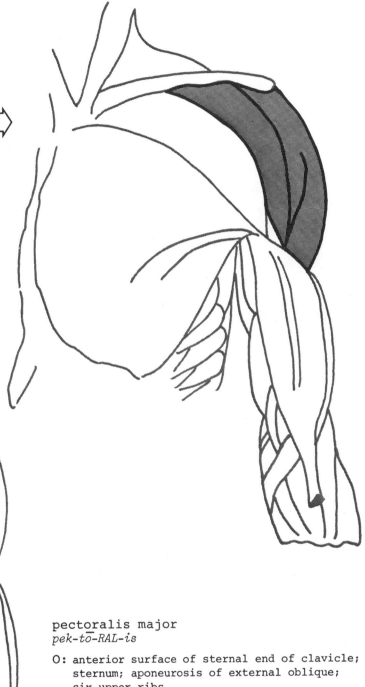

sternum

pectoralis major
pek-tō-RAL-is

O: anterior surface of sternal end of clavicle;
   sternum; aponeurosis of external oblique;
   six upper ribs
I: crest and greater tubercle of humerus
A: flexes, adducts, rotates arm medially

SUPERFICIAL MUSCLES OF CHEST AND UPPER ARM
(Anterior View)

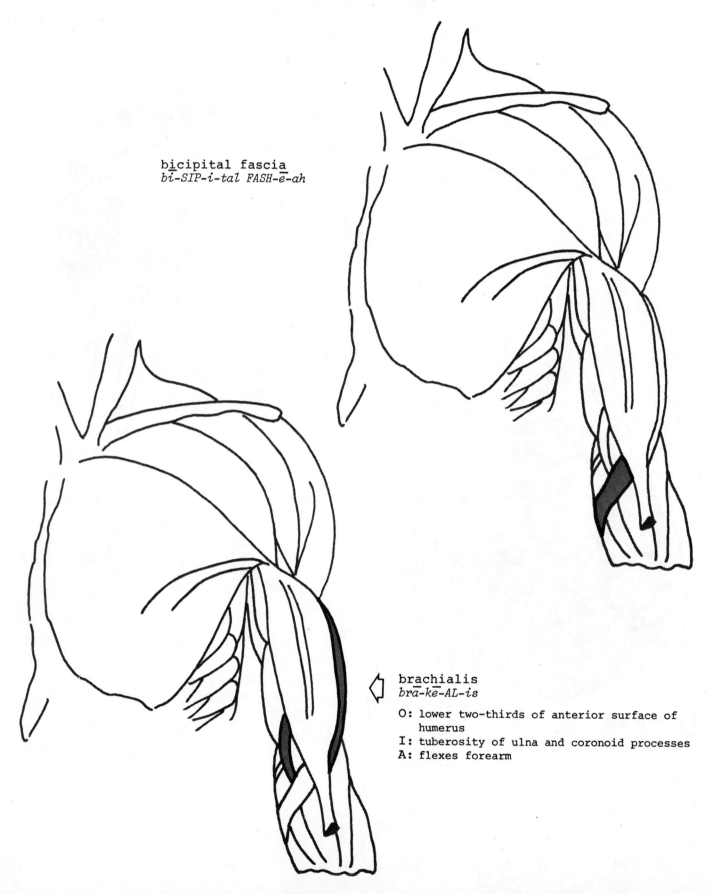

bicipital fascia
bi̅-SIP-i-tal FASH-e̅-ah

brachialis
bra̅-ke̅-AL-is

O: lower two-thirds of anterior surface of
   humerus
I: tuberosity of ulna and coronoid processes
A: flexes forearm

SUPERFICIAL MUSCLES OF CHEST AND UPPER ARM
(Anterior View)

biceps brachii
BĪ-seps BRĀ-kē-ī

O: long head: upper border of glenoid cavity
   short head: coracoid process by tendon
I: both heads: tuberosity of radius
A: flexes elbow and shoulder; supinates
   forearm

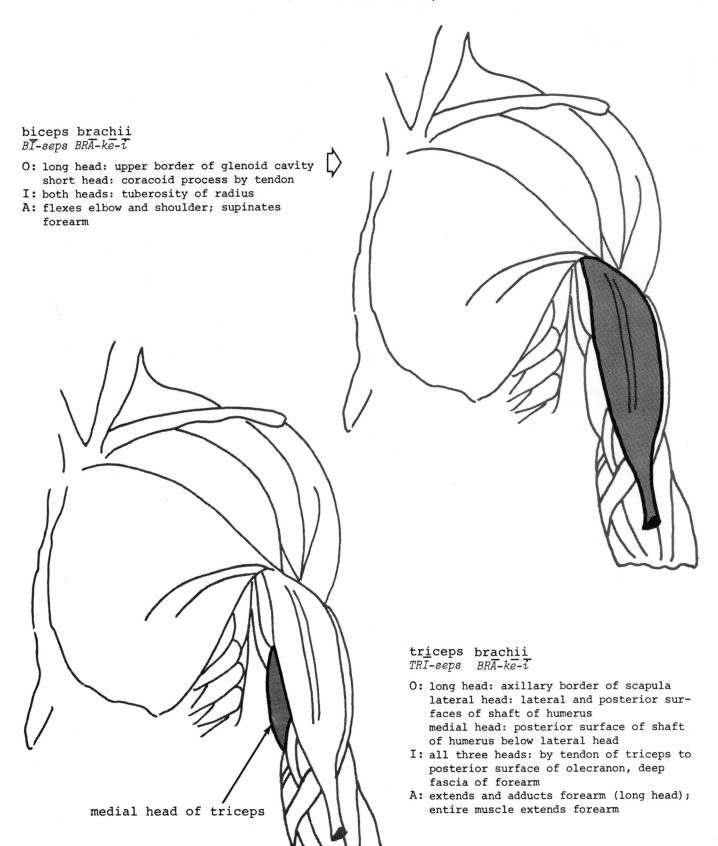

medial head of triceps

triceps brachii
TRĪ-seps BRĀ-kē-ī

O: long head: axillary border of scapula
   lateral head: lateral and posterior surfaces of shaft of humerus
   medial head: posterior surface of shaft
   of humerus below lateral head
I: all three heads: by tendon of triceps to
   posterior surface of olecranon, deep
   fascia of forearm
A: extends and adducts forearm (long head);
   entire muscle extends forearm

SUPERFICIAL MUSCLES OF CHEST AND UPPER ARM
(Anterior View)

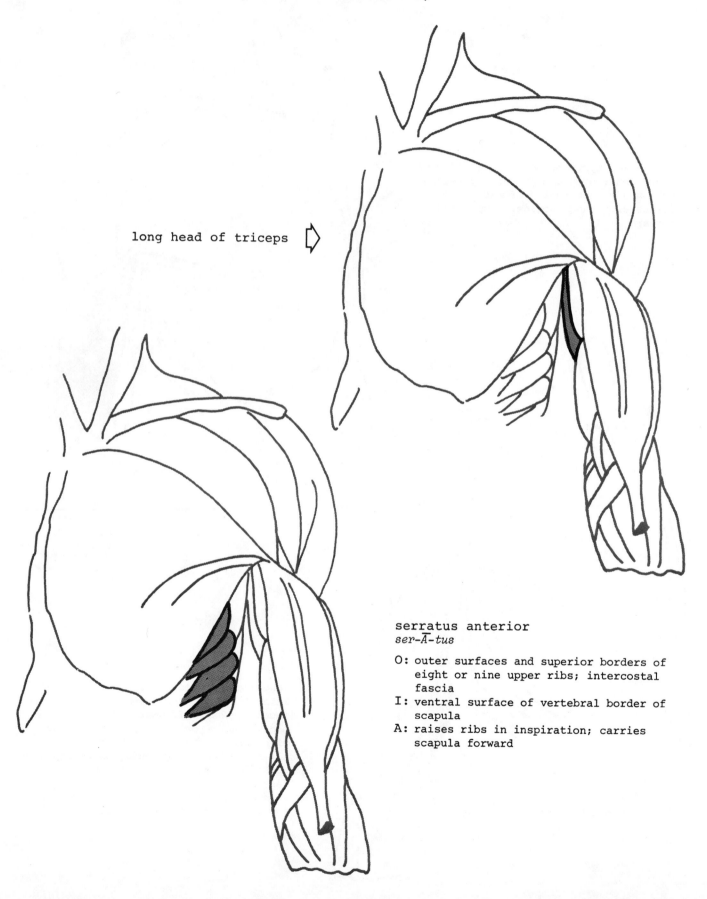

long head of triceps ▷

serratus anterior
ser-Ā-tus

O: outer surfaces and superior borders of
   eight or nine upper ribs; intercostal
   fascia
I: ventral surface of vertebral border of
   scapula
A: raises ribs in inspiration; carries
   scapula forward

SUPERFICIAL MUSCLES OF CHEST AND UPPER ARM
(Anterior View)

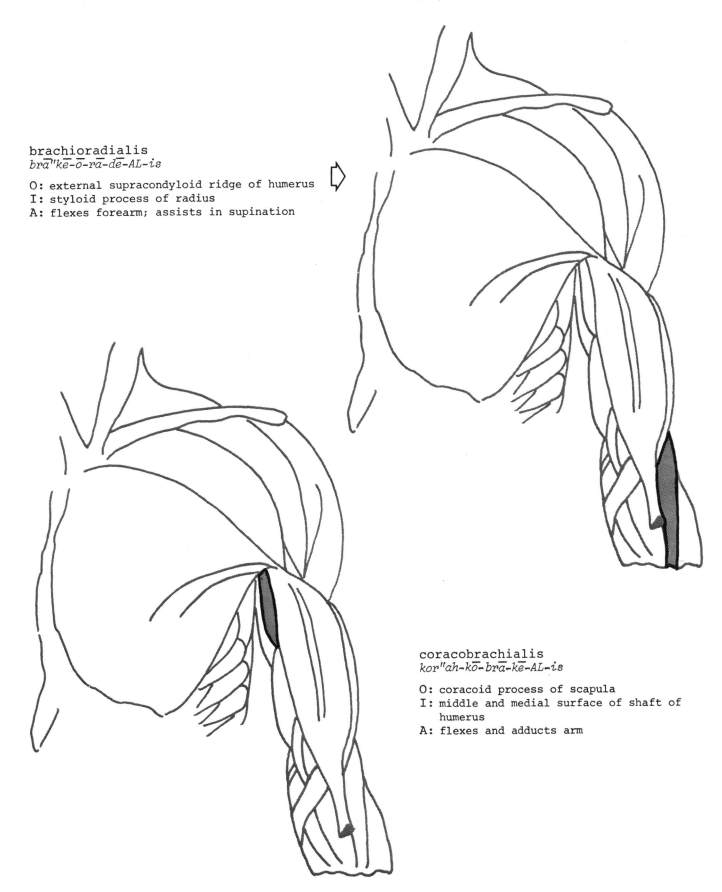

brachioradialis
brā"kē-ō-rā-dē-AL-is

O: external supracondyloid ridge of humerus
I: styloid process of radius
A: flexes forearm; assists in supination

coracobrachialis
kor"ah-kō-brā-kē-AL-is

O: coracoid process of scapula
I: middle and medial surface of shaft of
   humerus
A: flexes and adducts arm

DEEP MUSCLES OF CHEST AND UPPER ARM
(Anterior View)

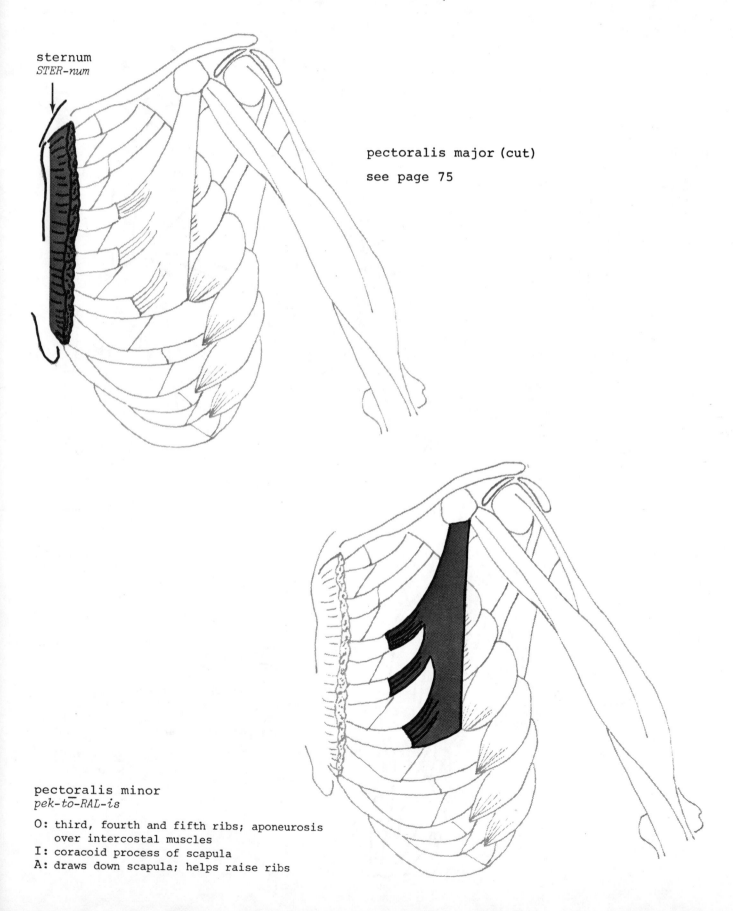

sternum
STER-num

pectoralis major (cut)

see page 75

pectoralis minor
pek-to̅-RAL-is

O: third, fourth and fifth ribs; aponeurosis
   over intercostal muscles
I: coracoid process of scapula
A: draws down scapula; helps raise ribs

DEEP MUSCLES OF CHEST AND UPPER ARM
(Anterior View)

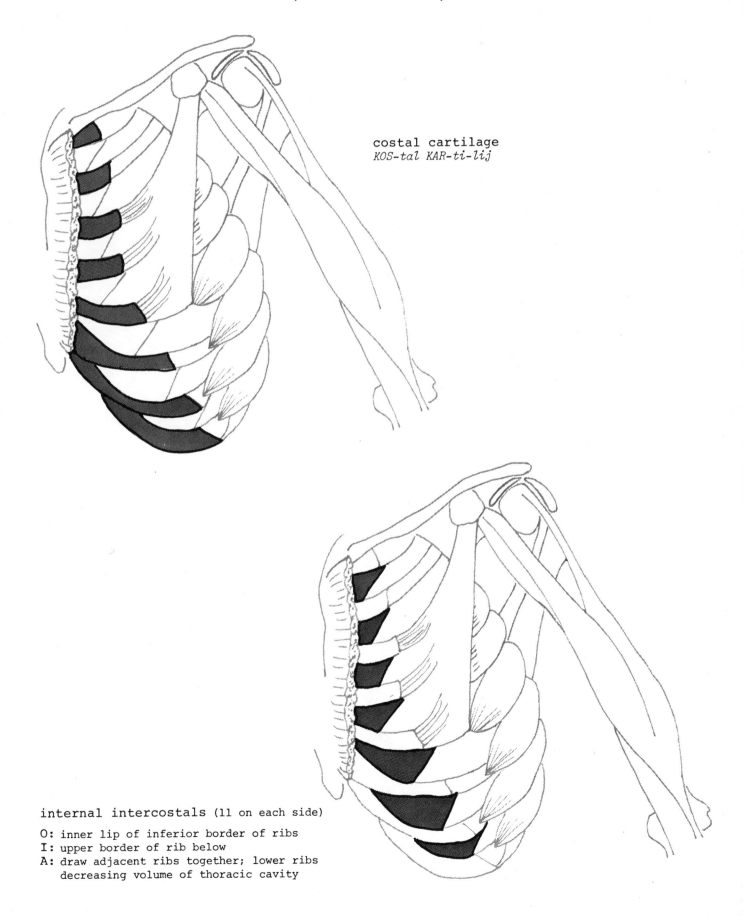

costal cartilage
KOS-tal KAR-ti-lij

internal intercostals (11 on each side)

O: inner lip of inferior border of ribs
I: upper border of rib below
A: draw adjacent ribs together; lower ribs
   decreasing volume of thoracic cavity

DEEP MUSCLES OF CHEST AND UPPER ARM
(Anterior View)

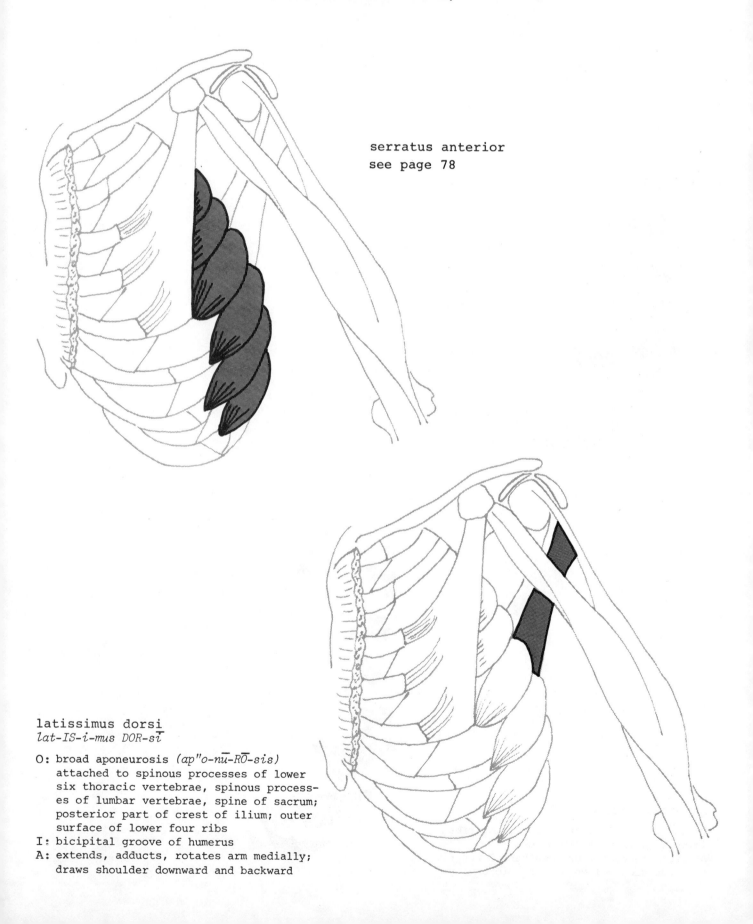

serratus anterior
see page 78

latissimus dorsi
lat-IS-i-mus DOR-sī

O: broad aponeurosis *(ap"o-nū-RŌ-sis)*
   attached to spinous processes of lower
   six thoracic vertebrae, spinous processes of lumbar vertebrae, spine of sacrum;
   posterior part of crest of ilium; outer
   surface of lower four ribs
I: bicipital groove of humerus
A: extends, adducts, rotates arm medially;
   draws shoulder downward and backward

DEEP MUSCLES OF CHEST AND UPPER ARM
(Anterior View)

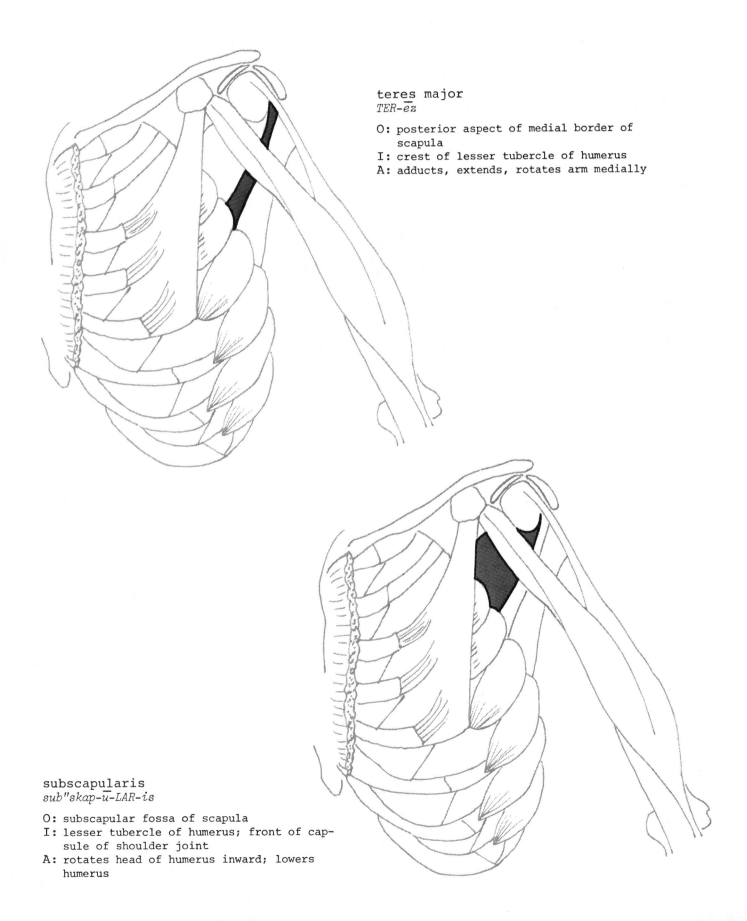

teres major
TER-ēz

O: posterior aspect of medial border of
   scapula
I: crest of lesser tubercle of humerus
A: adducts, extends, rotates arm medially

subscapularis
sub"skap-ū-LAR-is

O: subscapular fossa of scapula
I: lesser tubercle of humerus; front of capsule of shoulder joint
A: rotates head of humerus inward; lowers
   humerus

DEEP MUSCLES OF CHEST AND UPPER ARM
(Anterior View)

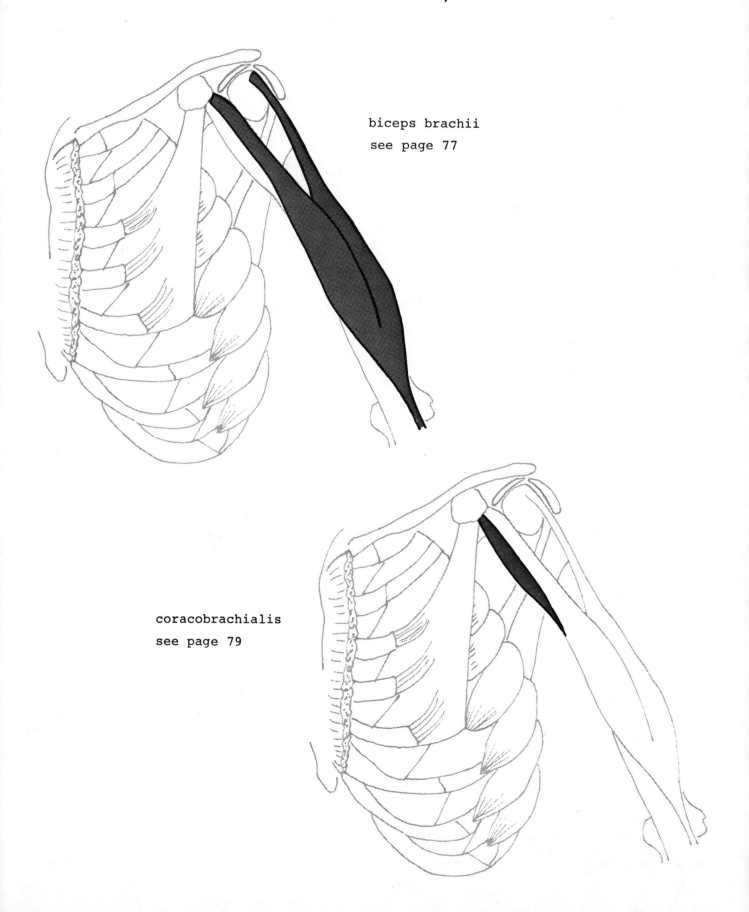

biceps brachii
see page 77

coracobrachialis
see page 79

DEEP MUSCLES OF CHEST AND UPPER ARM
(Anterior View)

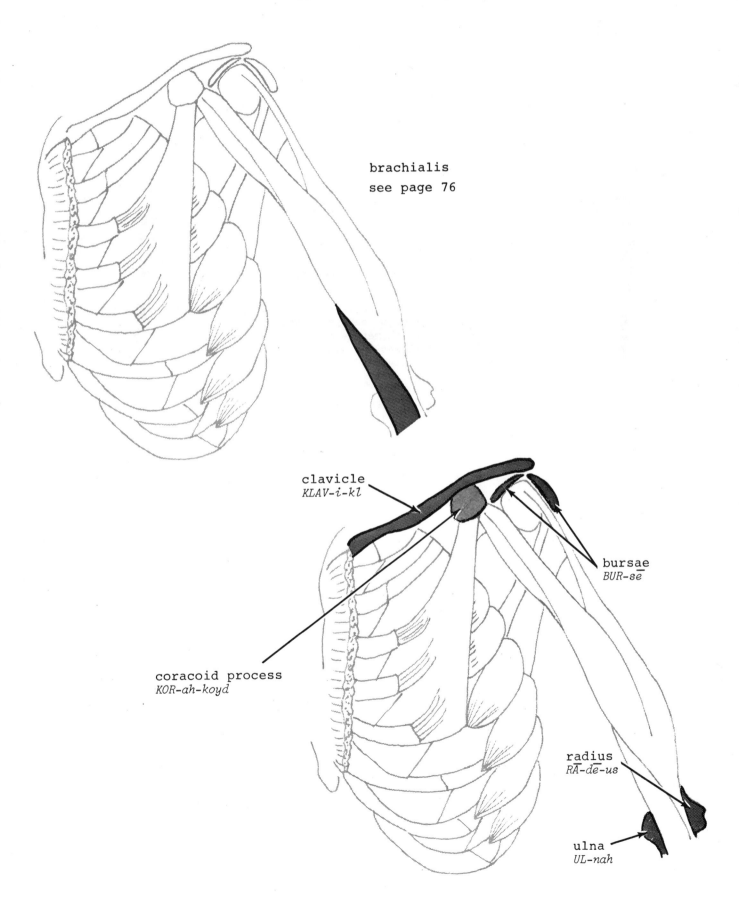

brachialis
see page 76

clavicle
KLAV-i-kl

bursae
BUR-sē

coracoid process
KOR-ah-koyd

radius
RĀ-dē-us

ulna
UL-nah

SUPERFICIAL MUSCLES, LEFT FOREARM
(Palmer View)

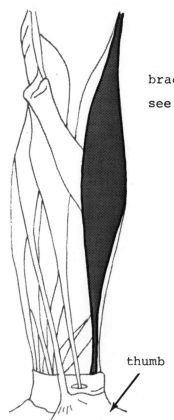

brachioradialis

see page 79

thumb

flexor carpi ulnaris
ul-NAR-is

O: first head: medial epicondyle of humerus
     second head: olecranon, ulna
I: fifth metacarpal and pisiform bone
A: flexes and adducts hand

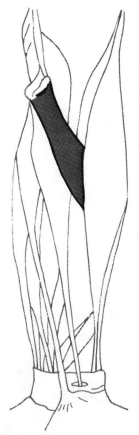

pronator teres
prō-NĀ-ter TER-ēz

O: medial epicondyle of humerus; coronoid
     process of ulna
I: middle of lateral surface of radius
A: pronates hand

palmaris longus
pal-MAR-is LONG-gus

O: medial epicondyle of humerus
I: transverse carpal ligament and palmar
     aponeurosis
A: flexes the hand

SUPERFICIAL MUSCLES, LEFT FOREARM
(Palmer View)

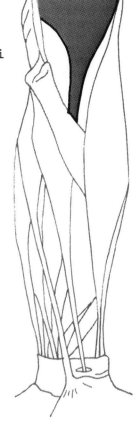

biceps brachii
see page 77

flexor carpi radialis
FLEK-ser KAR-pī rā-dē-AL-is

O: medial epicondyle of humerus
I: bases of second and third metacarpal
   bones
A: flexes hand and helps to abduct it

flexor digitorum sublimis also known as
   dij-i-TOR-um SUB-li-mis
flexor digitorum superficialis
   su"per-fi-shē-AL-is

O: first head: medial epicondyle of humerus
   second head: coronoid process of ulna
   third head: oblique line of radius
I: middle phalanges of fingers
A: flexes middle and proximal phalanges

flexor pollicis longus
   POL-li-sis

O: volar (palmar) surface of body radius
I: base of distal phalanx of thumb
A: flexes second phalanx of thumb

SUPERFICIAL MUSCLES,
LEFT FOREARM (Palmer View)

medial epicondyle

flexor retinaculum (fascia rather than a
    muscle) *ret-i-NAK-ū-lum*

O: tendons and ligaments of carpus
I: palmar aponeurosis; styloid processes of
    ulna and radius
A: holds tendons of forearm close against
    wrist

SUPERFICIAL MUSCLES,
LEFT FOREARM (Dorsal View)

anconeus
an-KON-ē-us or an-KŌ-nē-us

O: back of lateral epicondyle of humerus
I: olecranon and dorsal surface of ulna
A: extends forearm

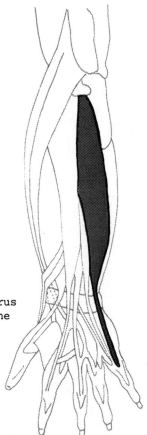

extensor carpi ulnaris
            ul-NAR-is

O: external epicondyle of humerus
I: base of fifth metacarpal bone
A: extends and adducts hand

SUPERFICIAL MUSCLES, LEFT FOREARM
(Dorsal View)

extensor digitorum communis
KOM-ū̄-nis

O: external epicondyle of humerus
I: proximal, middle and distal phalanges
   of fingers
A: extends little finger; helps extend wrist

extensor carpi radialis longus

O: external supracondylar ridge of humerus
I: base of second metacarpal bone
A: extends and abducts wrist

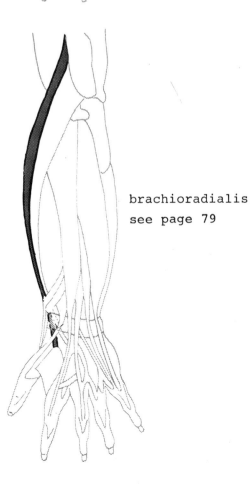

brachioradialis
see page 79

triceps brachii
see page 77

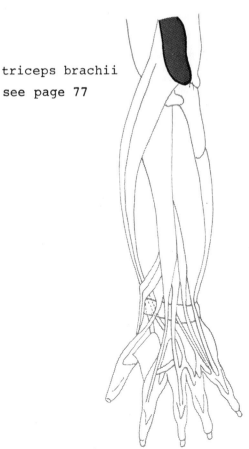

SUPERFICIAL MUSCLES, LEFT FOREARM
(Dorsal View)

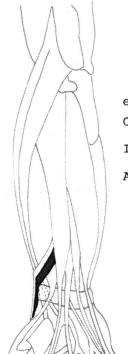

extensor carpi radialis brevis
eks-TEN-ser   rā-dē-AL-is BREV-is

O: external epicondyle of humerus
I: base of third metacarpal bone
A: extends and abducts wrist

extensor digiti quinti proprius
DIJ-i-tī   PRŌ-prē-us
also called extensor digiti minimi

O: external epicondyle of humerus
I: back of proximal phalanx of little
   finger
A: extends little finger

extensor pollicis brevis

O: dorsal surface of radius, interosseous
   membrane
I: dorsal surface of proximal phalanx of
   thumb
A: extends thumb

abductor pollicis longus
ab-DUK-tor POL-li-sis

O: posterior surface of radius and ulna
I: radial side of base of first meta-
   carpal
A: abducts and assists in extending thumb

SUPERFICIAL MUSCLES,
LEFT FOREARM (Dorsal View)

lateral epicondyle

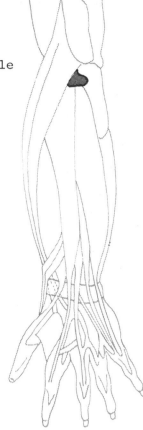

extensor pollicis longus

O: lateral side of dorsal surface of ulna
I: base of second phalanx of thumb
A: extends terminal phalanx of thumb

DEEP MUSCLES, LEFT FOREARM
(Palmer View)

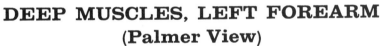

supinator
$su\text{-}pi\text{-}N\overline{A}\text{-}ter$

O: outer condyle of humerus, ligaments at
   elbow, ridge on ulna
I: anterior and lateral surfaces of body
   of radius
A: supinates hand

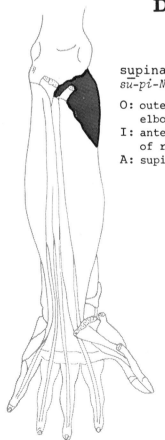

flexor pollicis longus

see page 87

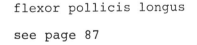

DEEP MUSCLES, LEFT FOREARM (Palmer View)

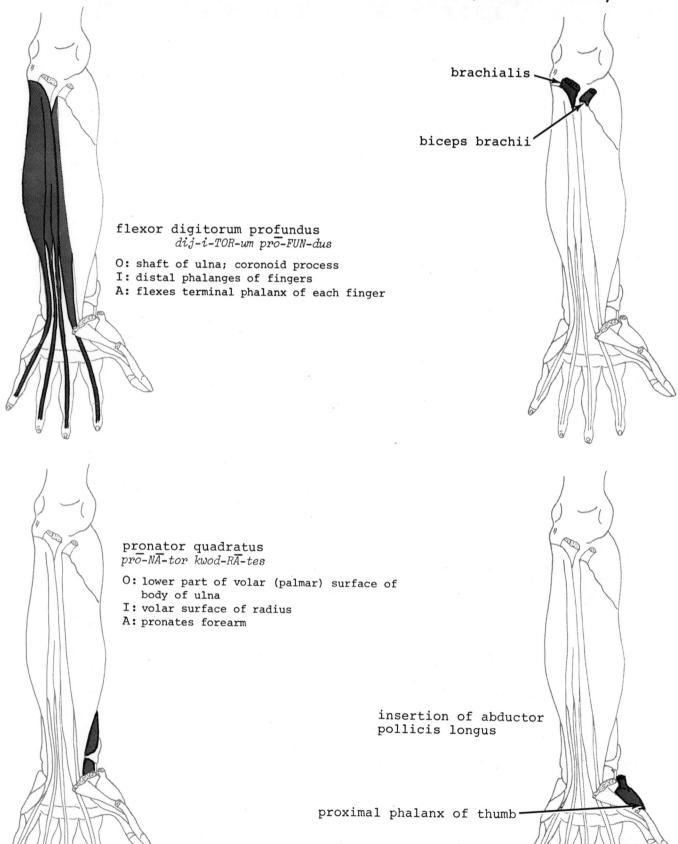

brachialis

biceps brachii

flexor digitorum profundus
dij-i-TOR-um pro-FUN-dus

O: shaft of ulna; coronoid process
I: distal phalanges of fingers
A: flexes terminal phalanx of each finger

pronator quadratus
pro-NA-tor kwod-RA-tes

O: lower part of volar (palmar) surface of
   body of ulna
I: volar surface of radius
A: pronates forearm

insertion of abductor
pollicis longus

proximal phalanx of thumb

DEEP MUSCLES, LEFT FOREARM (Palmer View)

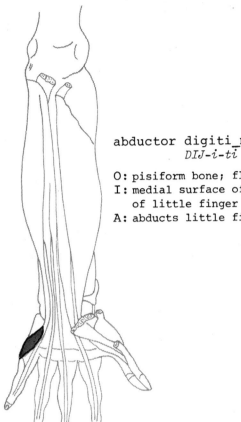

abductor digiti minimi or quinti
    DIJ-i-tī MIN-i-mī KWIN-tī

O: pisiform bone; flexor carpi ulnaris tendon
I: medial surface of base of proximal phalanx
   of little finger
A: abducts little finger

flexor pollicis brevis

O: transverse carpal ligament; tubercle of
   trapezium
I: base of proximal phalanx of thumb
A: flexes thumb

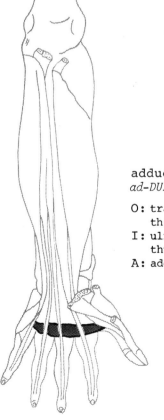

adductor pollicis
    ad-DUK-tor

O: trapezium, trapezoid, capitate, shaft of
   third metacarpal
I: ulnar side of base of first phalanx of
   thumb
A: adducts thumb

DEEP MUSCLES, LEFT FOREARM (Dorsal View)

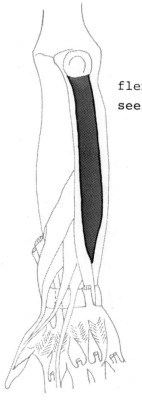

flexor digitorum profundus
see page 92

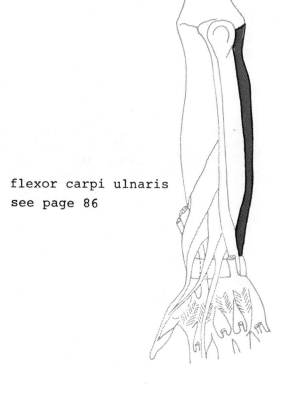

flexor carpi ulnaris
see page 86

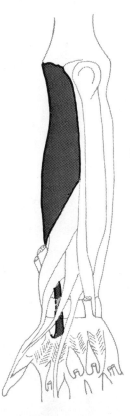

supinator
see page 91

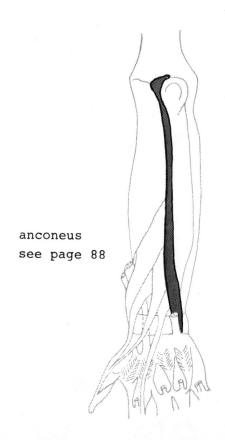

anconeus
see page 88

DEEP MUSCLES, LEFT FOREARM (Dorsal View)

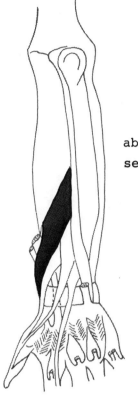

abductor pollicis longus
see page 90

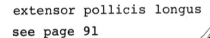

extensor pollicis longus
see page 91

extensor indicis
        IN-di-sis

O: dorsal surface of body of ulna; interosseus
   membrane
I: common extensor tendon index finger
A: extends index finger

extensor pollicis brevis
see page 90

DEEP MUSCLES, LEFT FOREARM (Dorsal View)

extensor carpi radialis longus
see page 89

extensor carpi radialis brevis
see page 90

humerus

olecranon

interossei_dorsalis
in-ter-OS-ē-ī dor-SAL-is

O: by two heads from adjacent sides of metacarpal bones
I: extensor tendons of second, third and
   fourth fingers
A: abduct, flex proximal phalanges

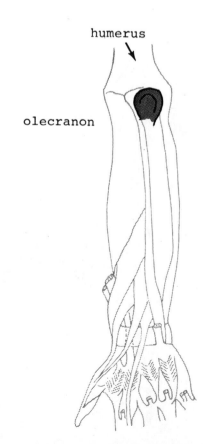

MUSCLES OF RIGHT HAND (Palmer View)

flexor pollicis brevis (cut)
see page 93

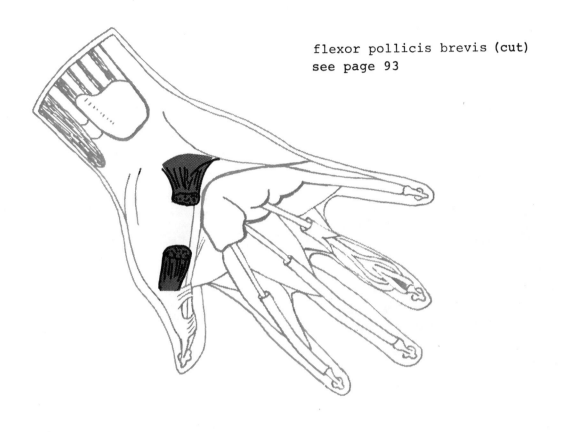

transverse carpal ligament

O: pisiform; hamate; tuberosity of scaphoid;
   ridge of trapezium
I: palmar aponeurosis; trapezium
A: holds tendons in place

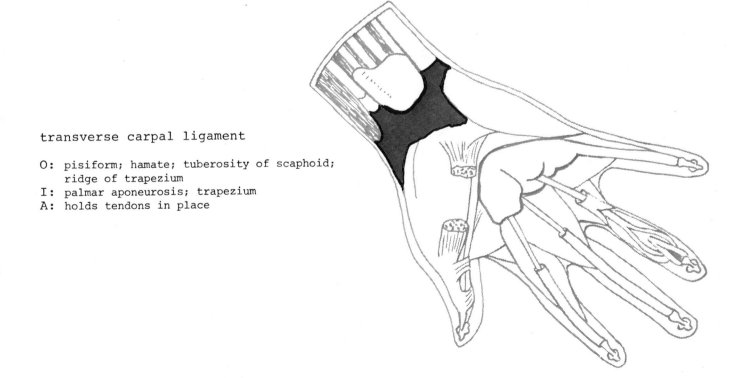

MUSCLES OF RIGHT HAND (Palmer View)

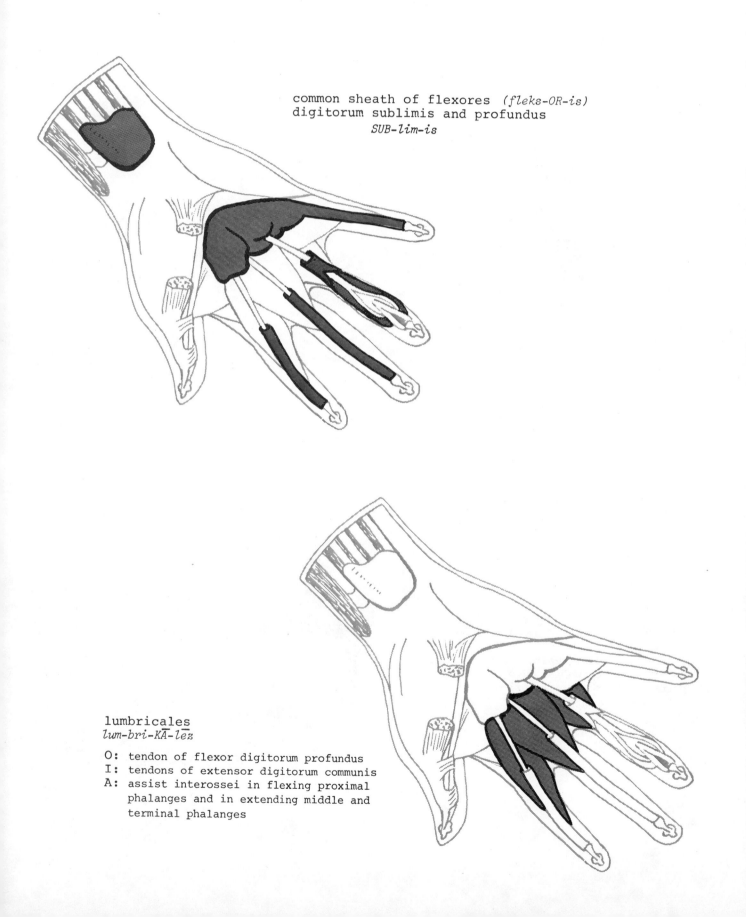

common sheath of flexores *(fleks-OR-is)*
digitorum sublimis and profundus
SUB-lim-is

lumbricales
lum-bri-KĀ-lēz

O: tendon of flexor digitorum profundus
I: tendons of extensor digitorum communis
A: assist interossei in flexing proximal
   phalanges and in extending middle and
   terminal phalanges

MUSCLES OF RIGHT HAND (Palmer View)

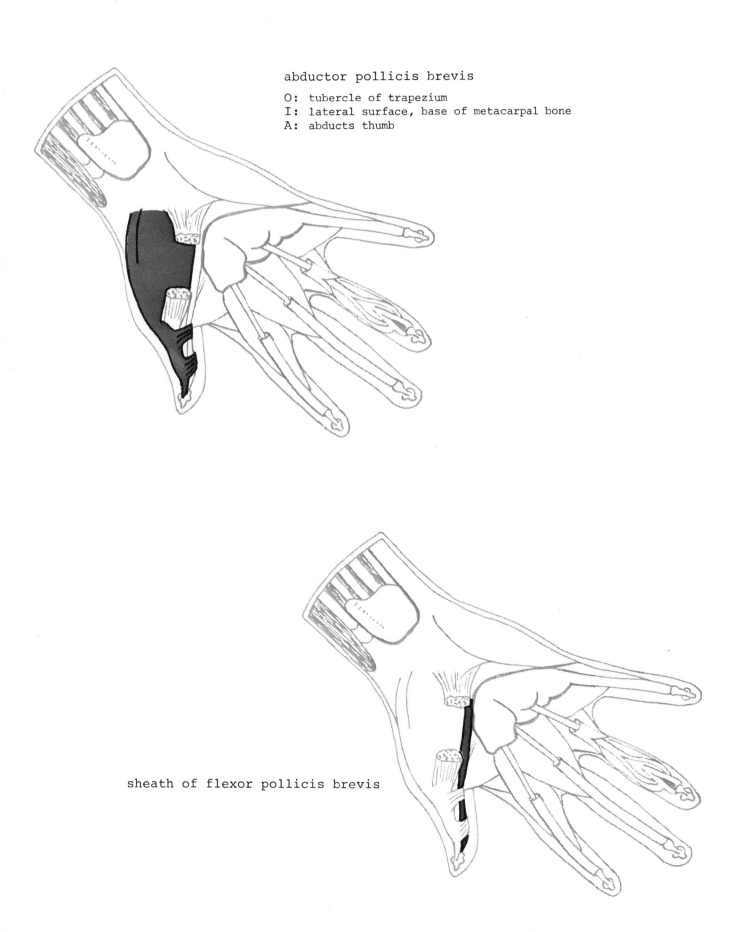

abductor pollicis brevis

O: tubercle of trapezium
I: lateral surface, base of metacarpal bone
A: abducts thumb

sheath of flexor pollicis brevis

MUSCLES OF RIGHT HAND (Palmer View)

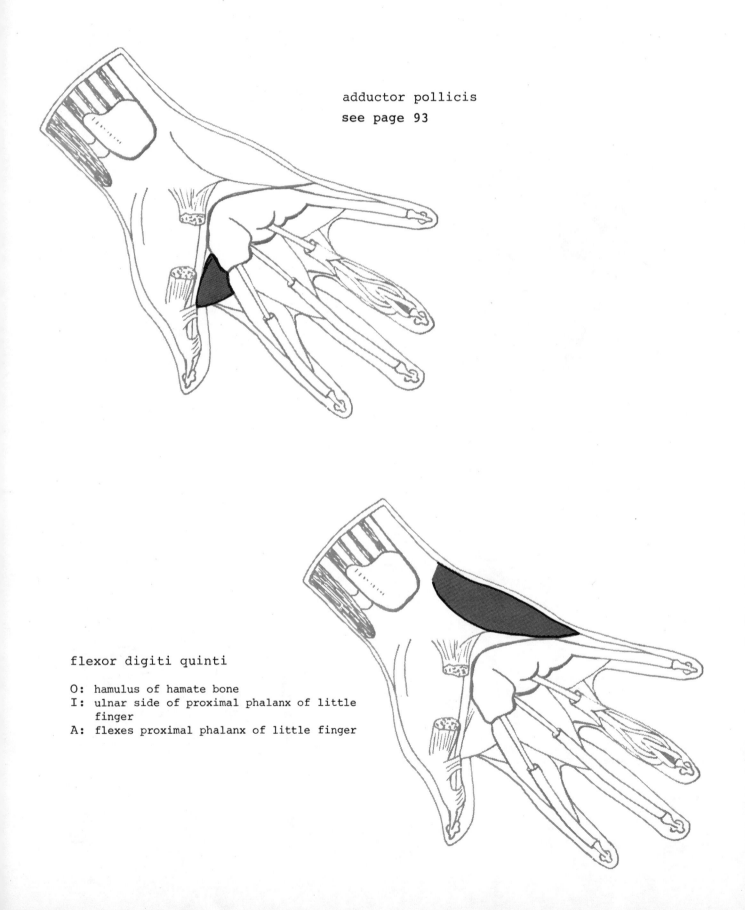

adductor pollicis
see page 93

flexor digiti quinti

O: hamulus of hamate bone
I: ulnar side of proximal phalanx of little
   finger
A: flexes proximal phalanx of little finger

MUSCLES OF RIGHT HAND (Palmer View)

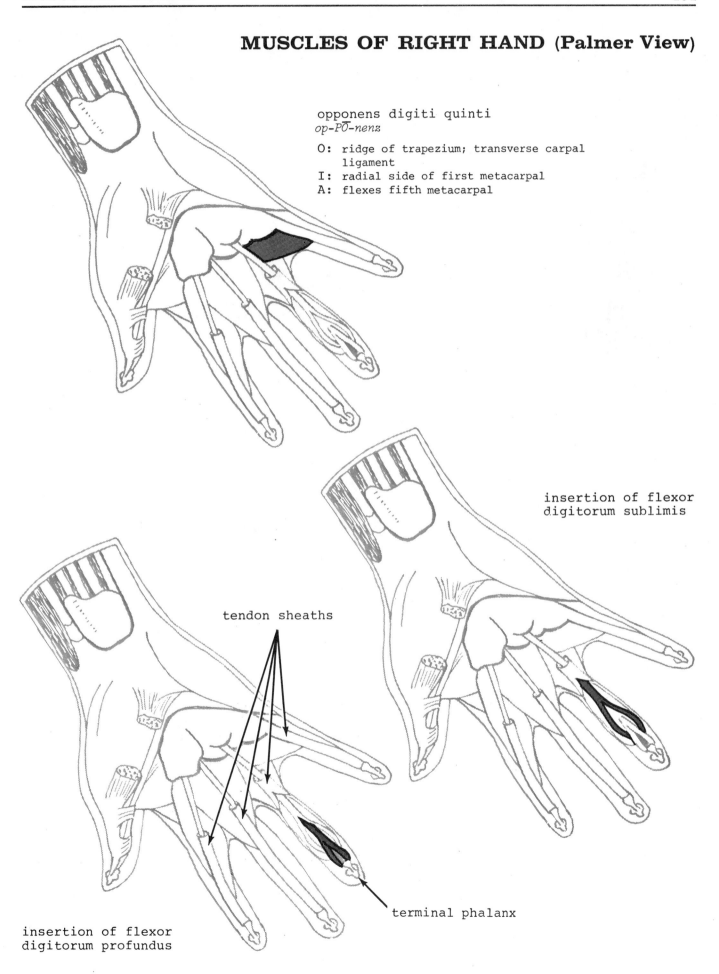

opponens digiti quinti
op-PŌ-nenz

O: ridge of trapezium; transverse carpal ligament
I: radial side of first metacarpal
A: flexes fifth metacarpal

insertion of flexor digitorum sublimis

tendon sheaths

terminal phalanx

insertion of flexor digitorum profundus

ABDOMINAL MUSCLES

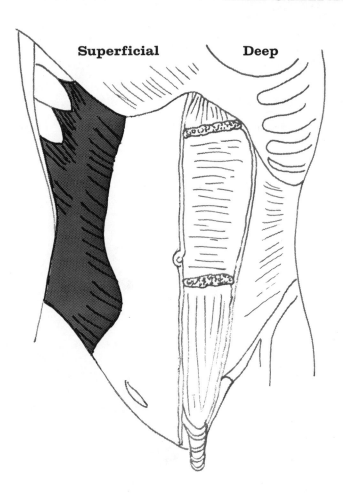

Superficial   **Deep**

external oblique

O: lower eight ribs
I: anterior half of outer lip of iliac crest;
   anterior rectus sheath
A: compresses abdominal contents

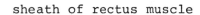

sheath of rectus muscle

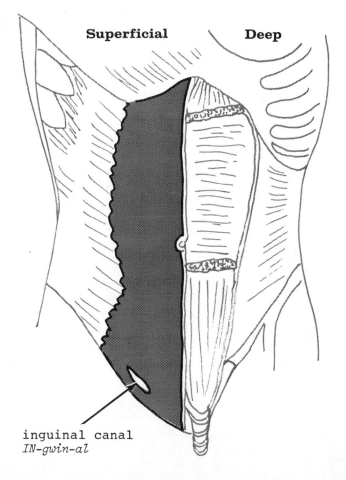

Superficial   **Deep**

inguinal canal
IN-gwin-al

ABDOMINAL MUSCLES

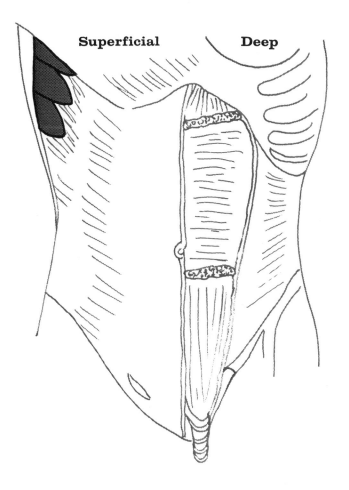

Superficial   **Deep**

serratus anterior
see page 78

latissimus dorsi
see page 82

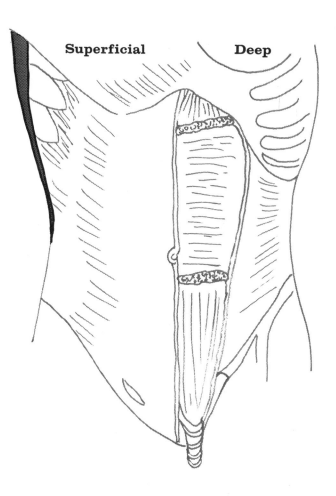

Superficial   **Deep**

ABDOMINAL MUSCLES

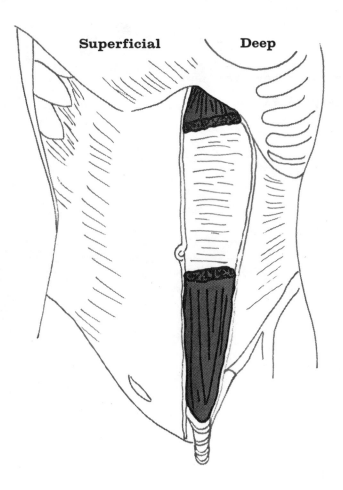

rectus abdominis (cut)
      ab-DOM-i-nis

O: crest of pubis and ligaments covering
   symphysis pubis
I: cartilages of fifth, sixth and seventh
   ribs
A: flexes vertebral column, assists in com-
   pressing abdominal wall

transversalis (transversus abdominis)
trans-ver-SAL-is (trans-VER-sus)

O: lateral third of Poupart's (inguinal) ligament; inner lip of iliac crest; inner
   surfaces of cartilages of six lower ribs;
   lumbodorsal fascia
I: crest of pubis and pectineal line; linea
   alba
A: compresses viscera and flexes thorax

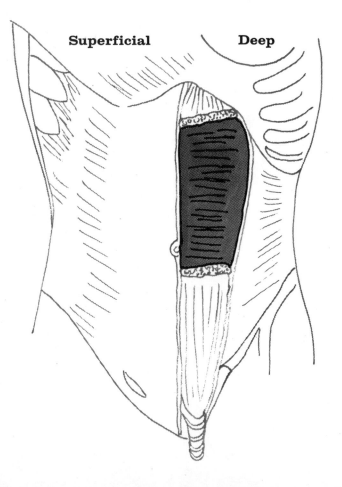

ABDOMINAL MUSCLES

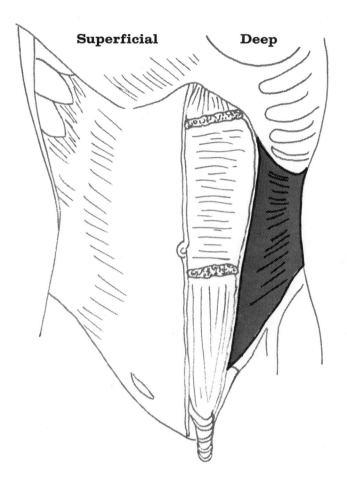

Superficial   **Deep**

internal oblique

O: crest of ilium; Poupart's ligament;
   lumbar fascia
I: six lowest ribs; linea alba; crest of
   pubes
A: compresses abdominal contents

cremaster
krē-MAS-ter

O: internal oblique and Poupart's ligament
I: cremasteric *(krē-mas-TER-ik)* fascia
   and spine of pubis
A: retracts testicle

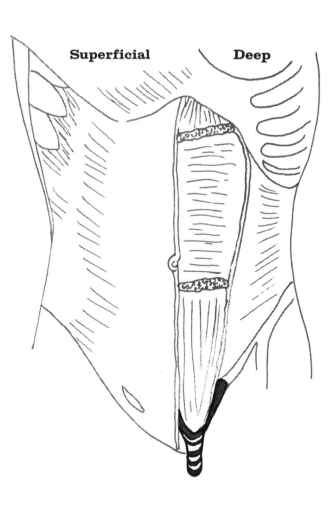

Superficial   **Deep**

ABDOMINAL MUSCLES

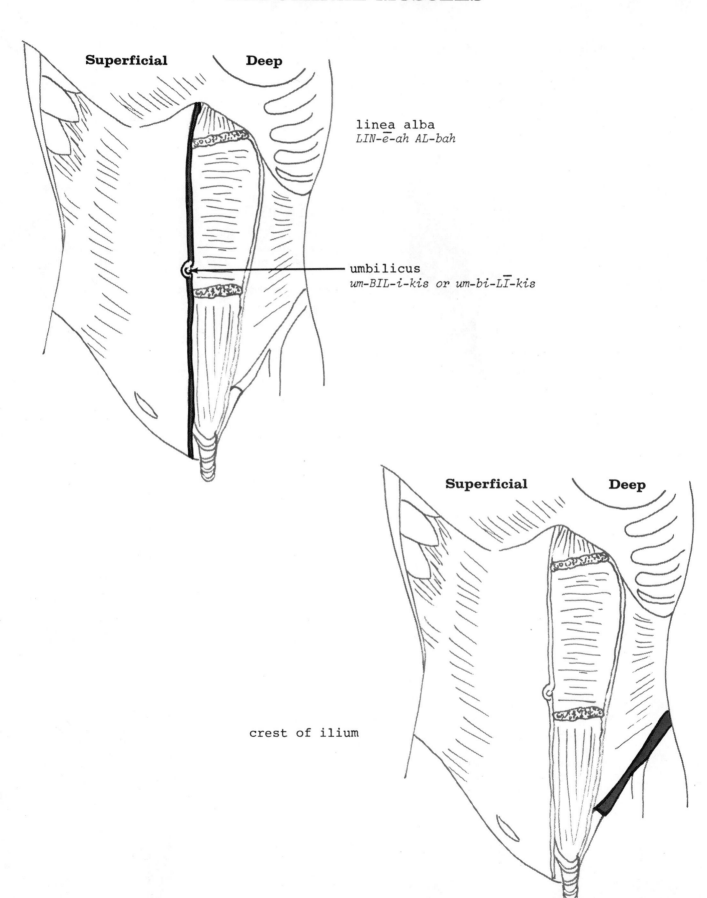

Superficial   **Deep**

linea alba
LIN-ē-ah AL-bah

umbilicus
um-BIL-i-kis or um-bi-LĪ-kis

crest of ilium

Superficial   **Deep**

MUSCLES OF BACK

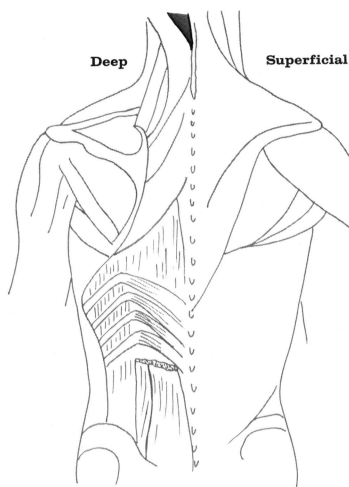

Deep   **Superficial**

semispinalis capitis
sem"ē-spi-NAH-lis* KAP-i-tis

O: transverse processes of five or six upper
   thoracic and four lower cervical vertebrae
I: occipital bone
A: rotates head and draws it backward

*Optional pronunciation of *alis* suffix
  is *Ā-lis*. *sem"ē-spi-NĀ-lis*

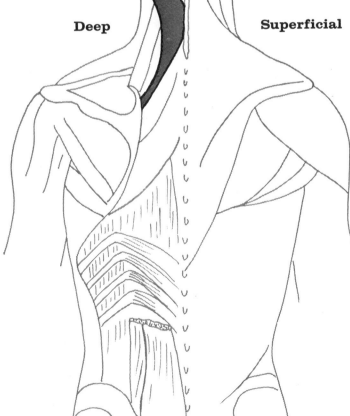

Deep   **Superficial**

splenius capitis
SPLĒ-nē-us

O: lower half of ligamentum nuchae, spines
   of seventh cervical and first three thoracic vertebrae
I: mastoid process; occipital bone
A: rotates and extends head and neck and
   flexes sidewise

splenius cervicis
           SER-vi-sis

O: spinous processes of third to sixth thoracic vertebrae
I: transverse process of two or three upper
   cervical vertebrae
A: extends, flexes sidewise, and rotates neck
   and head

MUSCLES OF BACK

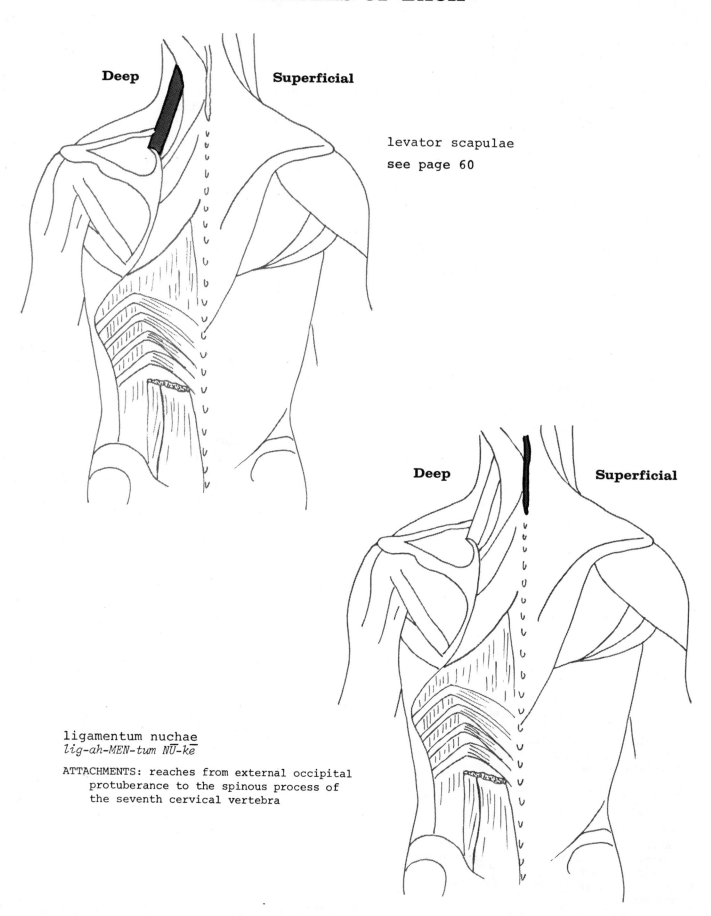

Deep   **Superficial**

levator scapulae
see page 60

Deep   **Superficial**

ligamentum nuchae
lig-ah-MEN-tum NŪ-kē

ATTACHMENTS: reaches from external occipital
    protuberance to the spinous process of
    the seventh cervical vertebra

MUSCLES OF BACK

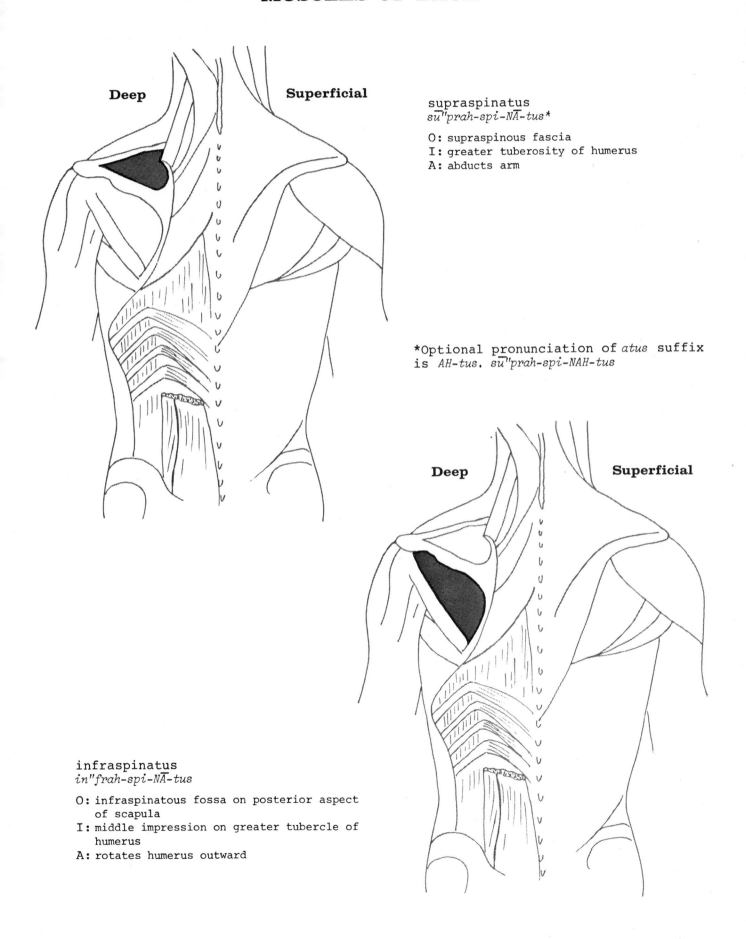

Deep   **Superficial**

supraspinatus
sū"prah-spi-NĀ-tus *

O: supraspinous fascia
I: greater tuberosity of humerus
A: abducts arm

*Optional pronunciation of *atus* suffix
is *AH-tus*. *sū"prah-spi-NAH-tus*

Deep   **Superficial**

infraspinatus
in"frah-spi-NĀ-tus

O: infraspinatous fossa on posterior aspect
   of scapula
I: middle impression on greater tubercle of
   humerus
A: rotates humerus outward

MUSCLES OF BACK

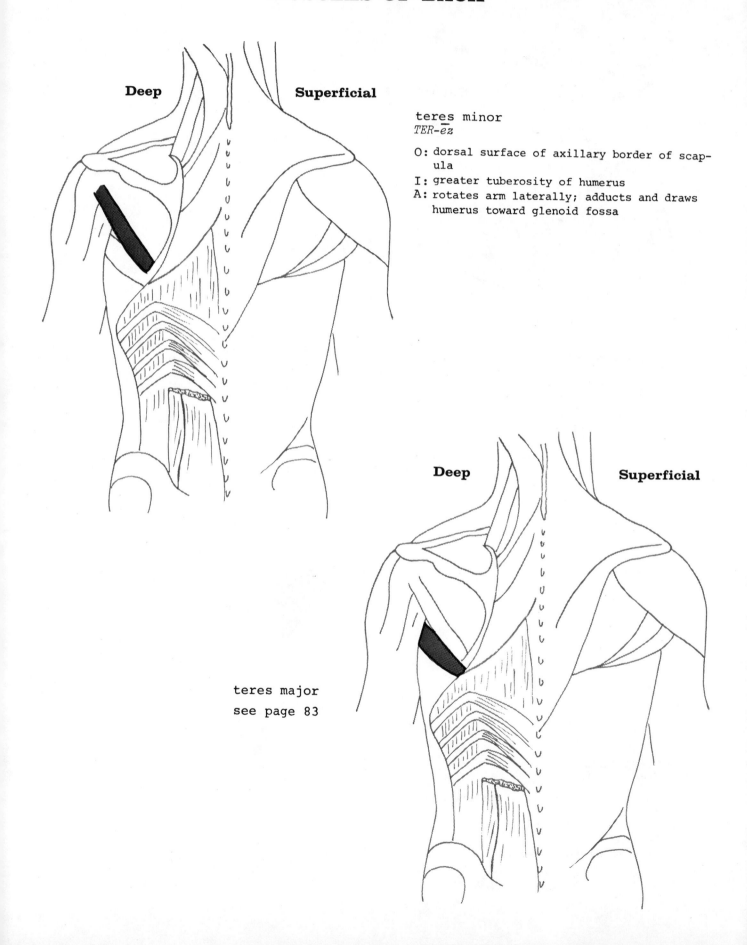

Deep

Superficial

teres minor
TER-ēz

O: dorsal surface of axillary border of scapula

I: greater tuberosity of humerus

A: rotates arm laterally; adducts and draws
humerus toward glenoid fossa

Deep

Superficial

teres major
see page 83

MUSCLES OF BACK

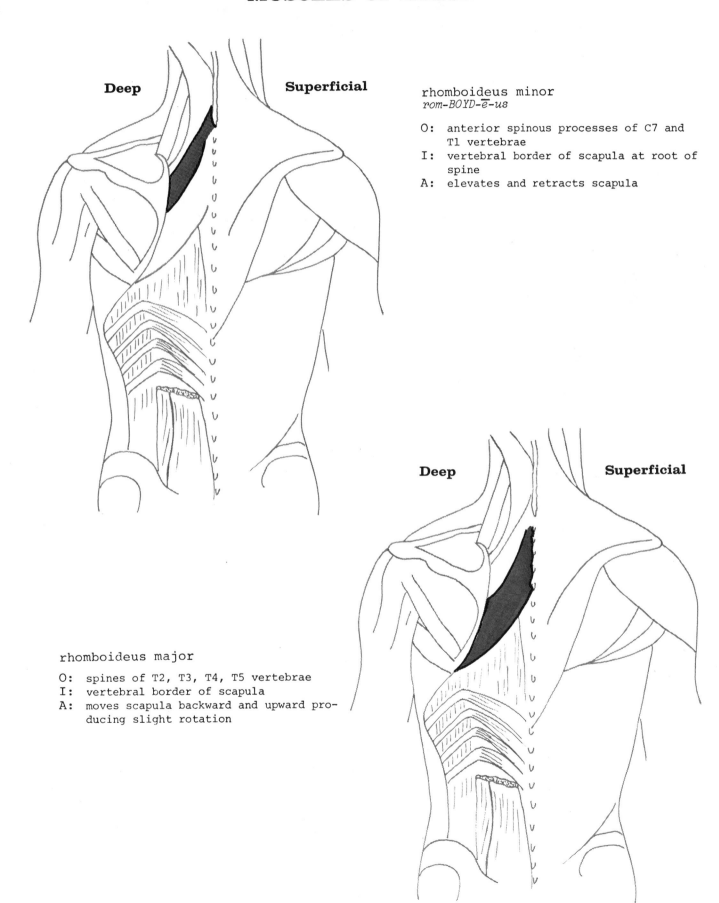

Deep   **Superficial**

rhomboideus minor
rom-BOYD-ē-us

O: anterior spinous processes of C7 and
   T1 vertebrae
I: vertebral border of scapula at root of
   spine
A: elevates and retracts scapula

rhomboideus major

O: spines of T2, T3, T4, T5 vertebrae
I: vertebral border of scapula
A: moves scapula backward and upward pro-
   ducing slight rotation

Deep   **Superficial**

MUSCLES OF BACK

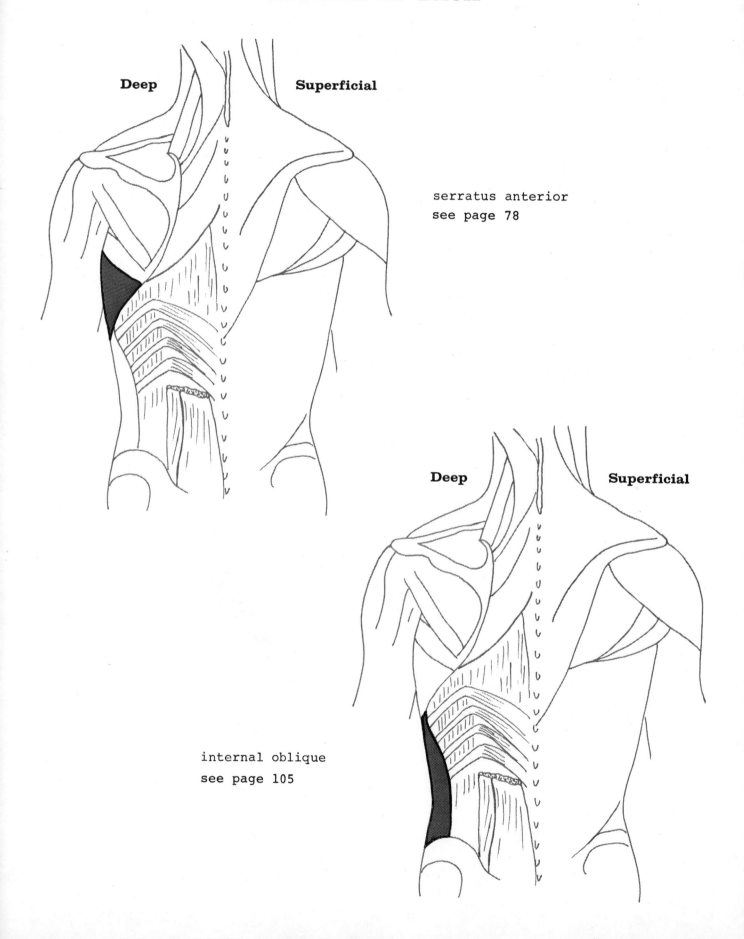

Deep   **Superficial**

serratus anterior
see page 78

Deep   **Superficial**

internal oblique
see page 105

MUSCLES OF BACK

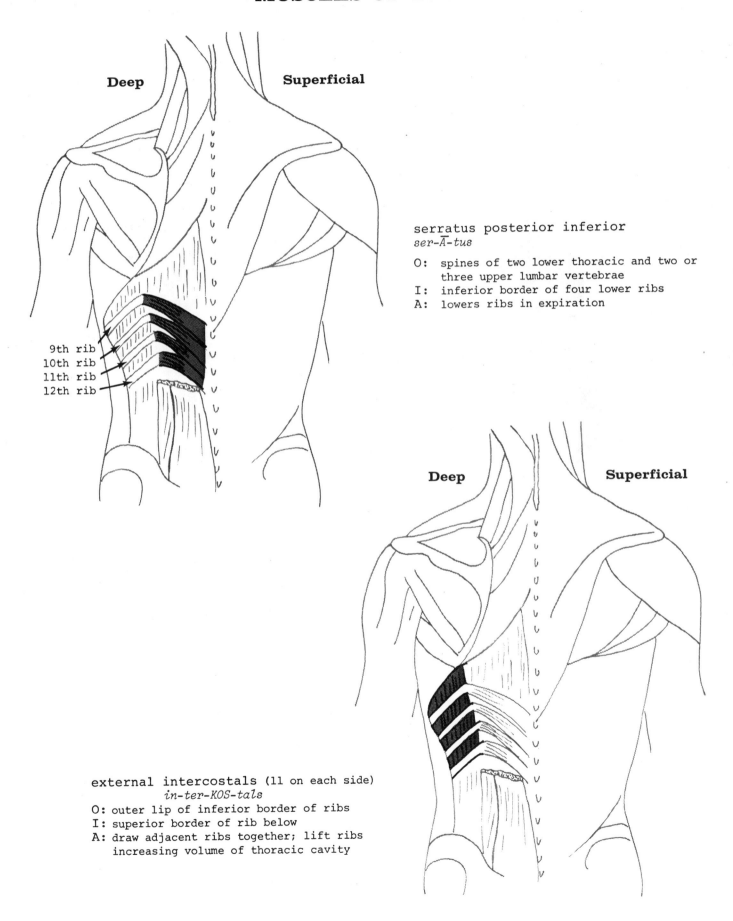

Deep   **Superficial**

serratus posterior inferior
ser-$\overline{A}$-tus

O: spines of two lower thoracic and two or
   three upper lumbar vertebrae
I: inferior border of four lower ribs
A: lowers ribs in expiration

9th rib
10th rib
11th rib
12th rib

Deep   **Superficial**

external intercostals (11 on each side)
in-ter-KOS-tals

O: outer lip of inferior border of ribs
I: superior border of rib below
A: draw adjacent ribs together; lift ribs
   increasing volume of thoracic cavity

MUSCLES OF BACK

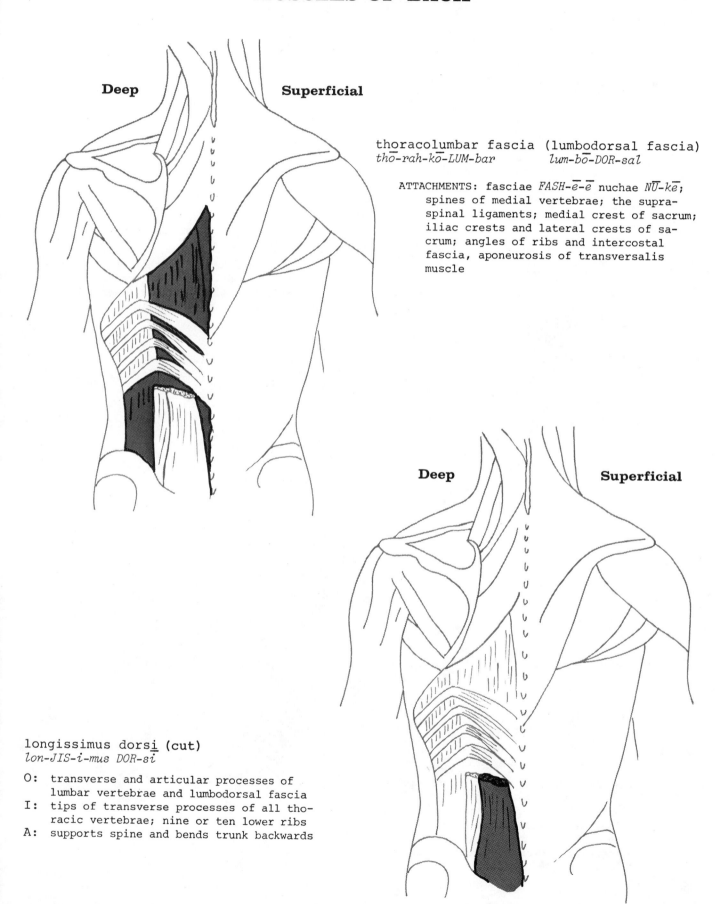

Deep   **Superficial**

thoracolumbar fascia (lumbodorsal fascia)
thō-rah-kō-LUM-bar   lum-bō-DOR-sal

ATTACHMENTS: fasciae *FASH-ē-ē* nuchae *NŪ-kē*;
spines of medial vertebrae; the supra-
spinal ligaments; medial crest of sacrum;
iliac crests and lateral crests of sacrum; angles of ribs and intercostal
fascia, aponeurosis of transversalis
muscle

Deep   **Superficial**

longissimus dors<u>i</u> (cut)
lon-JIS-i-mus DOR-sī

O: transverse and articular processes of
    lumbar vertebrae and lumbodorsal fascia
I: tips of transverse processes of all tho-
    racic vertebrae; nine or ten lower ribs
A: supports spine and bends trunk backwards

MUSCLES OF BACK

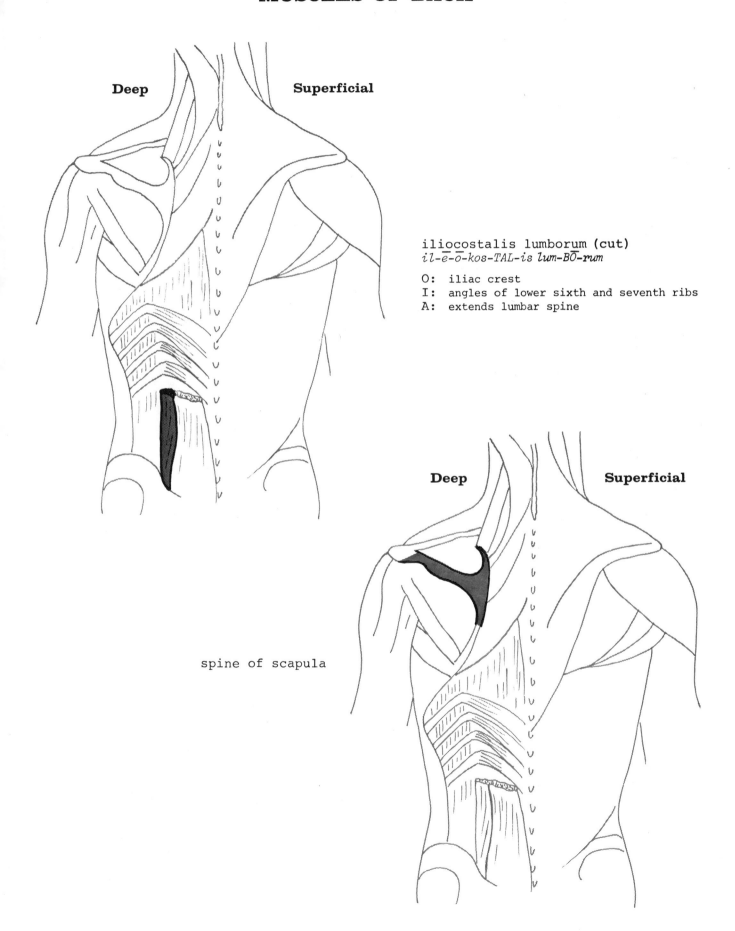

Deep   **Superficial**

iliocostalis lumborum (cut)
il-ē-ō-kos-TAL-is lum-BŌ-rum

O: iliac crest
I: angles of lower sixth and seventh ribs
A: extends lumbar spine

Deep   **Superficial**

spine of scapula

MUSCLES OF BACK

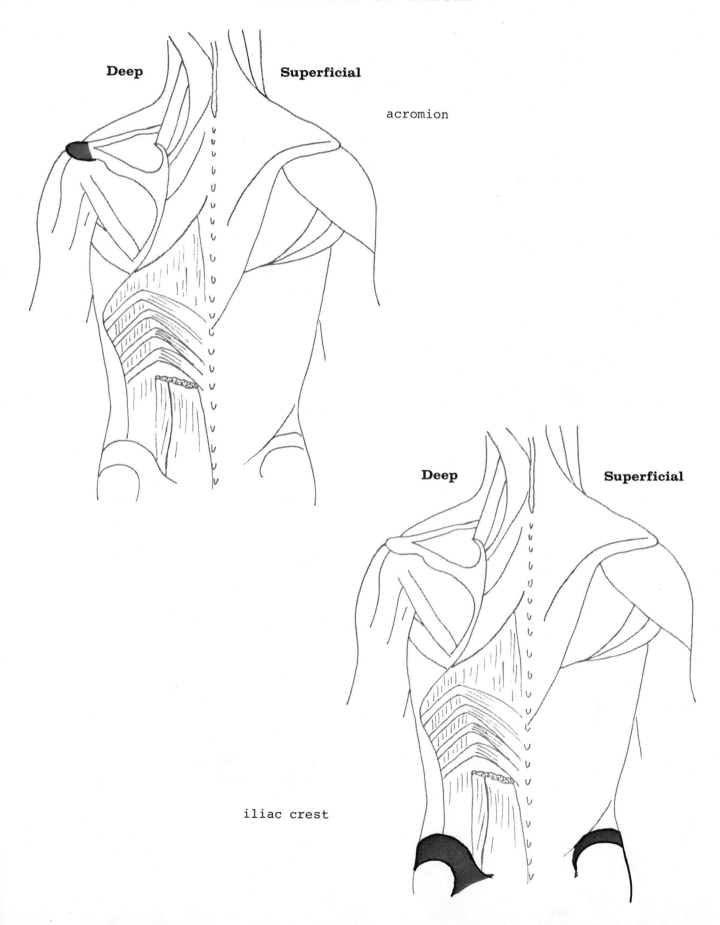

Deep   **Superficial**

acromion

Deep   **Superficial**

iliac crest

MUSCLES OF BACK

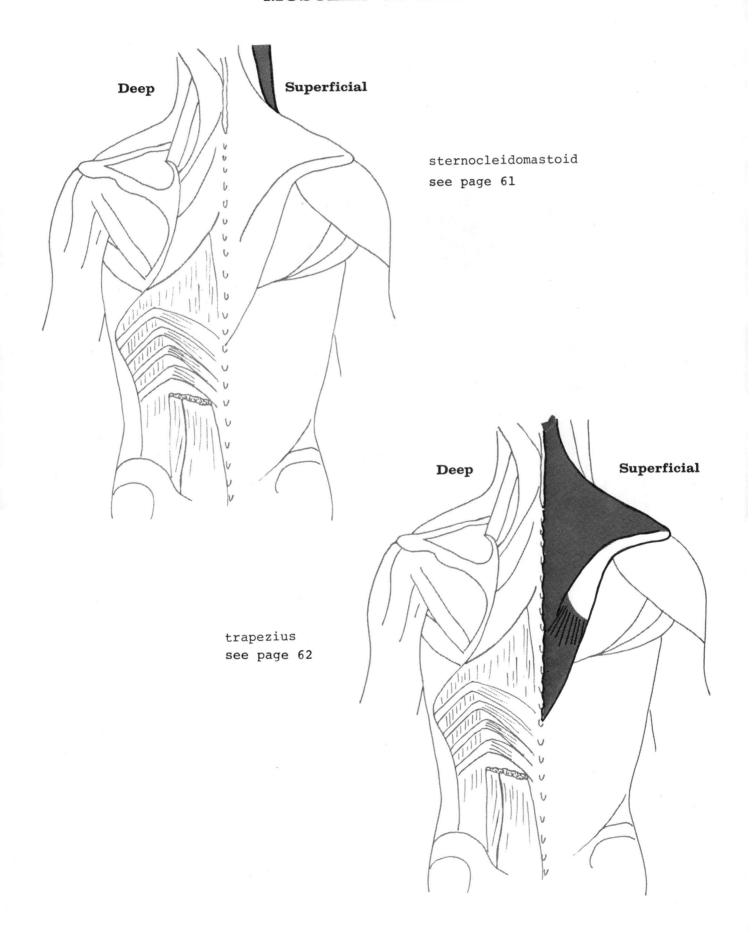

Deep

Superficial

sternocleidomastoid
see page 61

Deep

Superficial

trapezius
see page 62

MUSCLES OF BACK

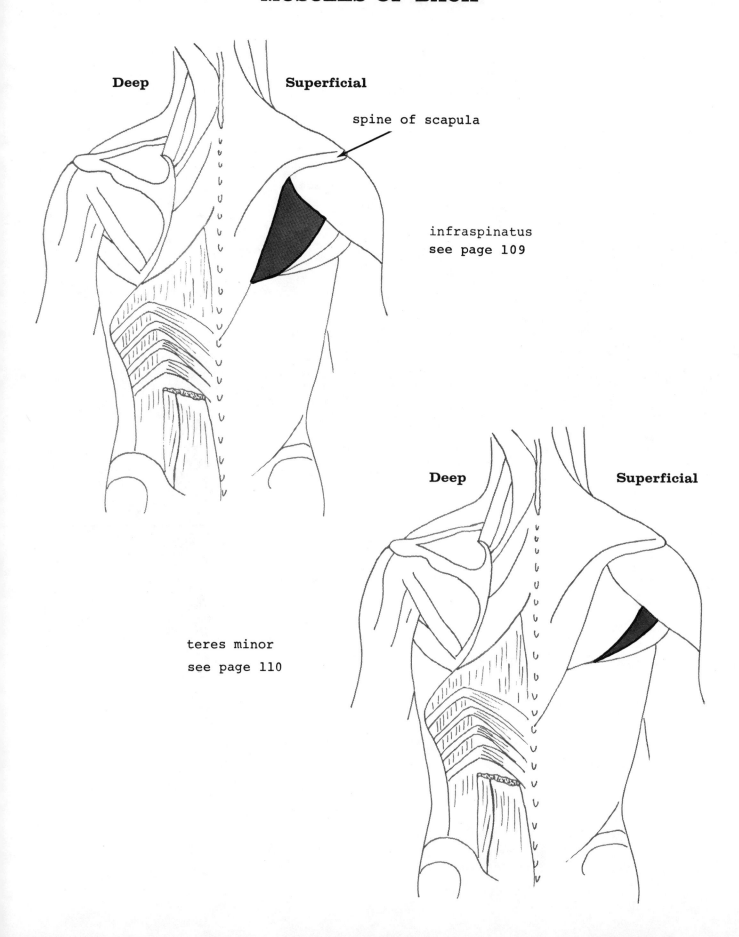

Deep

Superficial

spine of scapula

infraspinatus
see page 109

teres minor
see page 110

Deep

Superficial

MUSCLES OF BACK

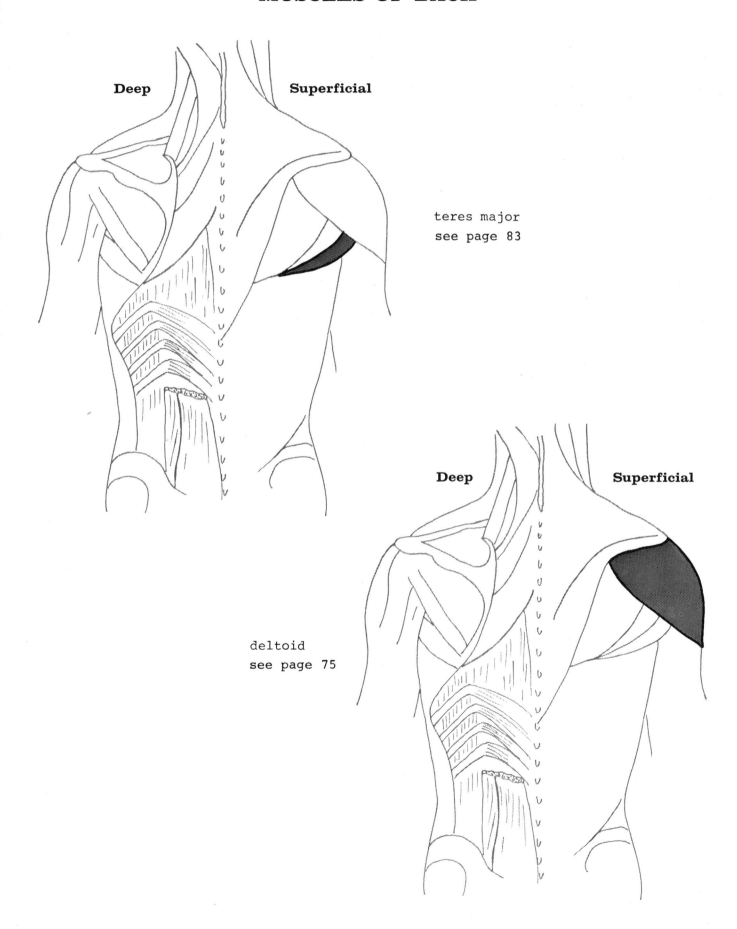

Deep   **Superficial**

teres major
see page 83

Deep   **Superficial**

deltoid
see page 75

MUSCLES OF BACK

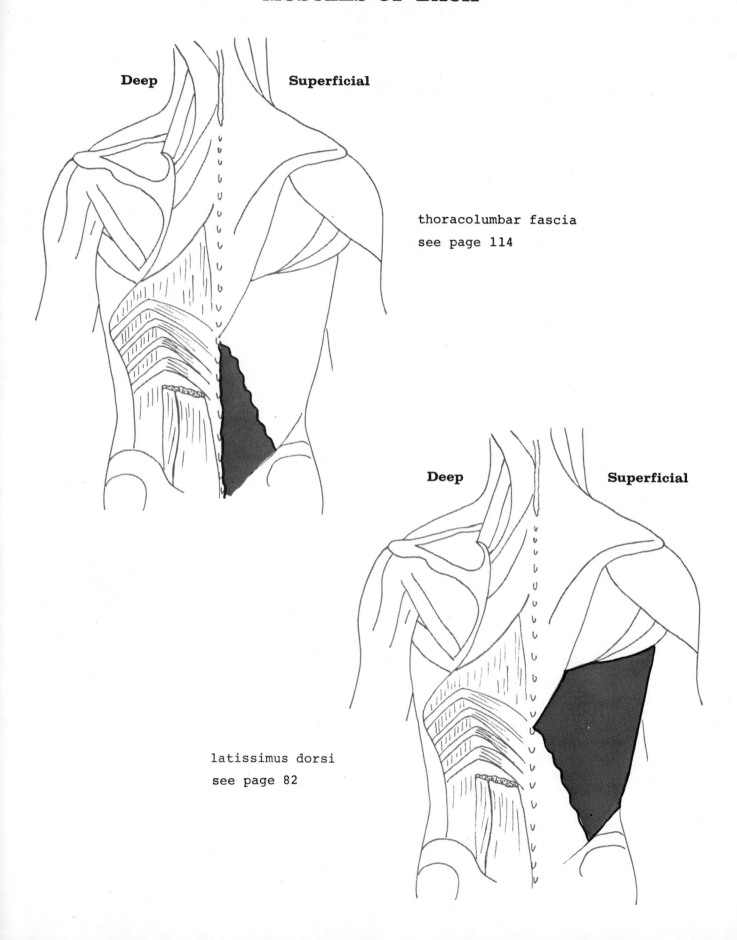

Deep   **Superficial**

thoracolumbar fascia
see page 114

Deep   **Superficial**

latissimus dorsi
see page 82

DEEP MUSCLES OF BACK

longissimus capitis
KAP-i-tis

O: transverse processes fourth and fifth
upper thoracic vertebrae
I: mastoid process of temporal bone
A: draws head backward; rotates head

iliocostalis cervicis

O: angles of first six ribs
I: transverse processes of fourth and sixth
cervical vertebrae
A: extends cervical spine

iliocostalis thoracis
il-ē-ō-kos-TAL-is THOR-ah-sis

O: upper border of angles of six
lower ribs
I: angles of six upper ribs
A: extends cervical spine

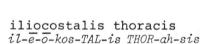

longissimus cervicis
SER-vi-sis

O: transverse processes fourth and fifth
upper thoracic vertebrae
I: transverse processes of second and
sixth cervical vertebrae
A: extends cervical vertebrae

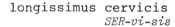

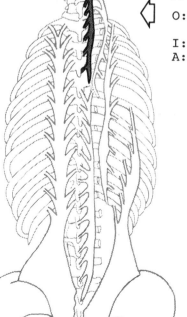

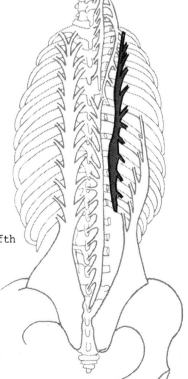

DEEP MUSCLES OF BACK

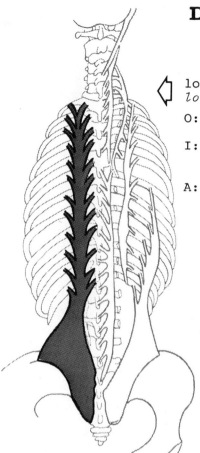

longissimus thoracis
lon-JIS-i-mus

O: transverse processes lumbar vertebrae
and lumbosacral fascia
I: transverse processes of all thoracic
and upper lumbar vertebrae; ninth and
tenth lower ribs
A: extends thoracic vertebrae

spinalis thoracis
spi-NAH-lis

O: spinous processes of upper lumbar
and lower thoracic vertebrae
I: spines of upper vertebrae
A: extends vertebral column

iliocostalis lumborum
see page 115

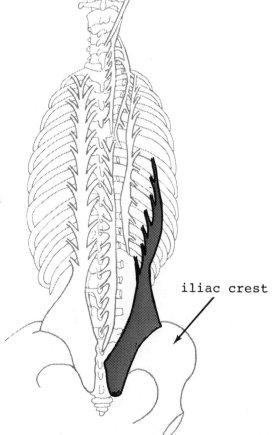

iliac crest

SUPERFICIAL MUSCLES, RIGHT THIGH (Lateral View)

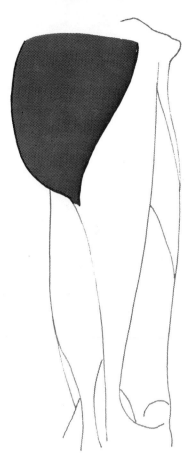

gluteus maximus
glū-TĒ-us MAX-i-mus

O: posterior gluteal
line of ilium and
posterior surface
of sacrum and
coccyx

I: fascia lata; glu-
teal tuberosity
of femur

A: extends, abducts
and rotates thigh
outward

tensor fascia lata
TEN-ser FASH-ē-ah LAH-tah

O: anterior part of iliac
spine

I: iliotibial band of
fascia

A: tenses fascia lata;
flexes thigh

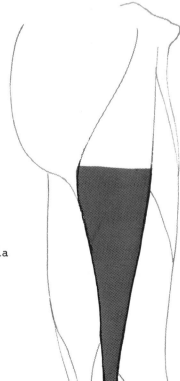

iliotibial tract
il"ē-ō-TIB-ē-al

O: lower end of fascia
lata

I: tibia

A: tendinous tract
from fascia lata

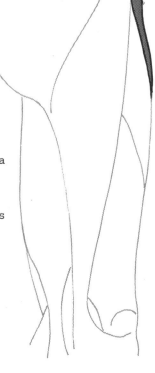

sartorius
sar-TŌ-rē-us

O: anterior super-
ior spine of
ilium

I: medial surface
of shaft of tibia

A: flexes leg on
thigh and thigh
on pelvis; ab-
ducts and rotates
thigh outward

SUPERFICIAL MUSCLES, RIGHT THIGH (Lateral View)

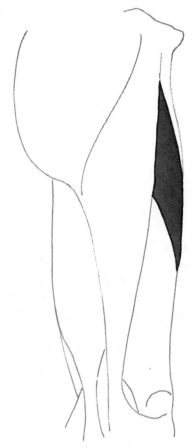

rectus femoris
REK-tus FEM-o-ris
(one of four heads
of quadriceps fem-
oris)

O: anterior infer-
   ior iliac spine;
   brim of acetab-
   ulum
I: proximal border
   of patella
A: extends leg;
   flexes thigh

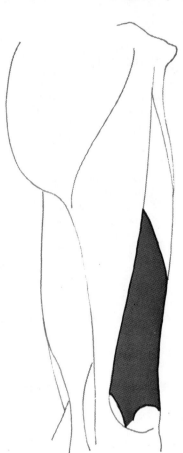

vastus lateralis
VAS-tus lat-er-AL-is
(one of four heads
of quadriceps fem-
oris)

O: broad aponeur-
   osis from great-
   er trochanter
   and linea aspera
   of femur
I: inserts into
   base of patella
   and condyles and
   tuberosity of
   tibia
A: extends knee

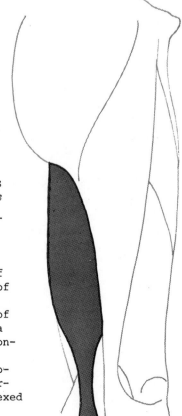

biceps femoris
BI-seps FEM-o-ris

O: long head: tu-
   berosity of
   ischium
   short head:
   lateral lip of
   linea aspera of
   femur
I: lateral side of
   head of fibula
   and lateral con-
   dyle of tibia
A: flexes leg, ro-
   tates it later-
   ally after flexed

gastrocnemius
gas-trok-NĒ-mē-us

O: two heads from
   lateral and me-
   dial condyles of
   femur; adjacent
   part of capsule
   of knee
I: aponeurosis
   unites with ten-
   don of soleus to
   form tendo cal-
   caneus (Achilles'
   tendon)
A: extends foot at
   ankle; flexes
   femur upon tibia

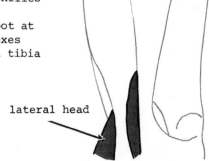

lateral head

SUPERFICIAL MUSCLES, RIGHT THIGH (Lateral View)

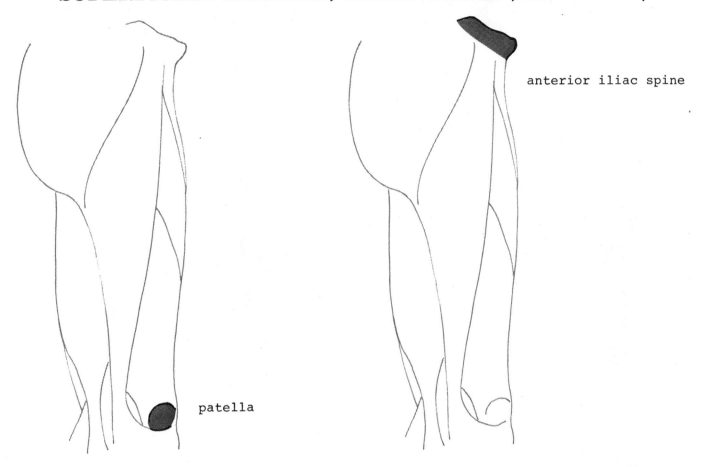

anterior iliac spine

patella

SUPERFICIAL MUSCLES, RIGHT THIGH (Anterior View)

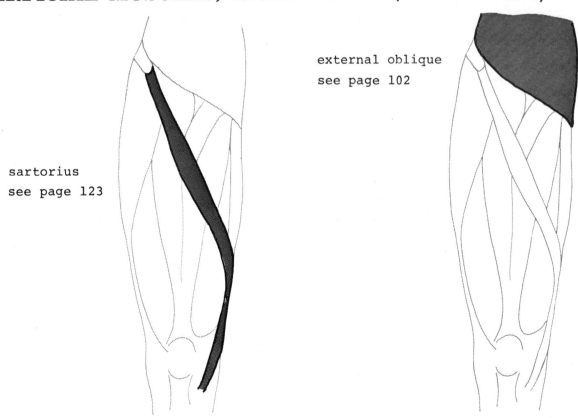

external oblique
see page 102

sartorius
see page 123

SUPERFICIAL MUSCLES, RIGHT THIGH (Anterior View)

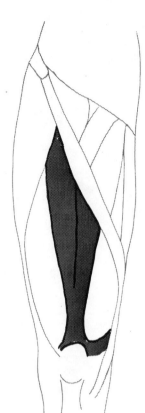

rectus femoris
see page 124

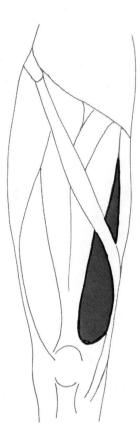

vastus medialis
    mē-dē-AL-is
(one of four heads
of quadriceps fem-
oris)

O: medial lip of
    linea aspera;
    intertrochanteric
    line
I: common tendon and
    medial side of
    patella
A: extends leg; draws
    patella inward

tensor fascia lata
see page 123

iliotibial tract
see page 123

SUPERFICIAL MUSCLES, RIGHT THIGH (Anterior View)

vastus lateralis
see page 124

vastus intermedius
in-ter-ME-de-us

(not shown)
(one of four heads of
quadriceps femoris)

O: ventral and medial
   surfaces of body
   of femur
I: common tendon of
   quadriceps femoris
A: extends leg; flex-
   es thigh

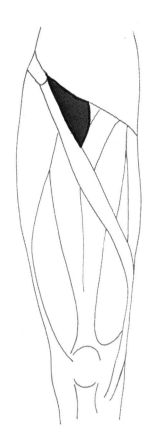

iliopsoas
il"e-o-SO-as

A compound muscle con-
sisting of the iliacus
and psoas, major and
minor

pectineus
pek-TIN-e-us or
pek-tin-E-us

O: iliopectineal line;
   spine of pubis
I: femur distal to
   lesser trochanter
A: flexes, adducts
   thigh

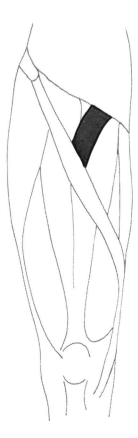

gracilis
GRAH-si-lis

O: symphysis pubis
   and pubic arch
I: medial surface
   of shaft of tibia
A: flexes and adducts
   knee and adducts
   thigh

SUPERFICIAL MUSCLES, RIGHT THIGH (Anterior View)

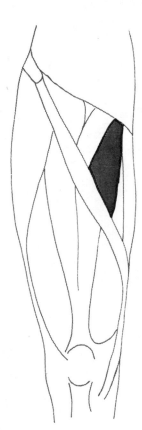

adductor longus

O: crest and sym-
   physis of pubis
I: linea aspera of
   femur
A: adducts, rotates
   and flexes thigh

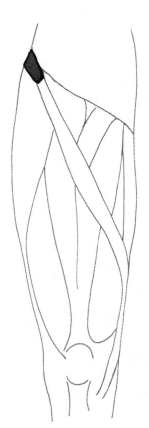

anterior superior
iliac spine

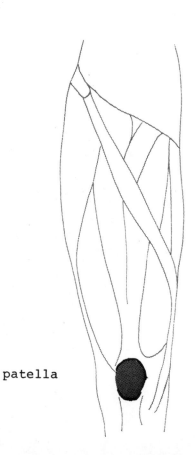

patella

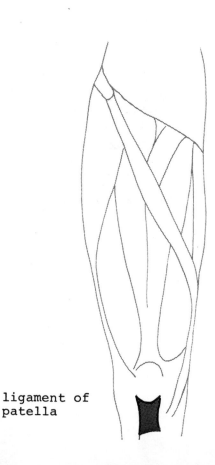

ligament of
patella

SUPERFICIAL MUSCLES, RIGHT THIGH (Dorsal View)

gluteus maximus

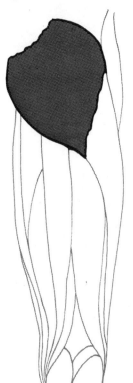

see
page
123

gluteus medius
glū-TĒ-us MĒ-dē-us

O: lateral surface of
   ilium
I: oblique ridge on
   lateral surface of
   greater trochanter
A: rotates, abducts
   and extends thigh

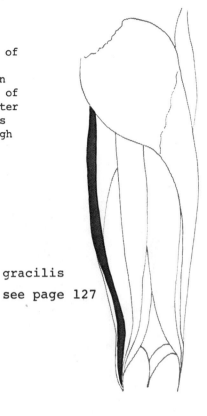

gracilis
see page 127

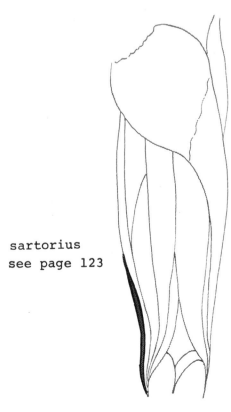

sartorius
see page 123

adductor magnus
MAG-nus

O: crest and symphysis
   of pubis
I: linea aspera of
   femur
A: adducts thigh and
   everts it

SUPERFICIAL MUSCLES, RIGHT THIGH (Dorsal View)

semitendinosus
sem-ē-ten-di-NŌ-sus

O: tuberosity of
   ischium
I: upper and medial
   surface of tibia
A: rotates leg inward

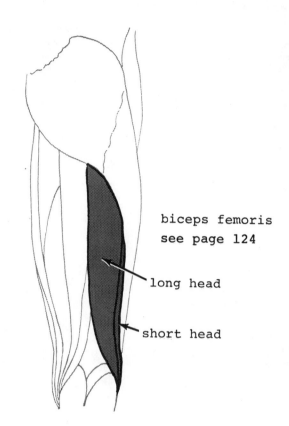

biceps femoris
see page 124

long head

short head

tensor fascia lata
see page 123

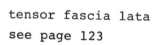
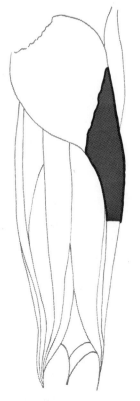

iliotibial tract
see page 123

SUPERFICIAL MUSCLES, RIGHT THIGH (Dorsal View)

semimembranosus
sem-ē-mem-brah-NŌ-sus

O: tuberosity of
   ischium
I: medial condyle of
   tibia
A: flexes leg and
   rotates it inward

plantaris
plan-TAR-is

O: distal part of the
   linea aspera, and
   from the oblique
   popliteal ligament
   of the knee joint
I: posterior part of
   calcaneus
A: plantar flexes foot;
   flexes leg

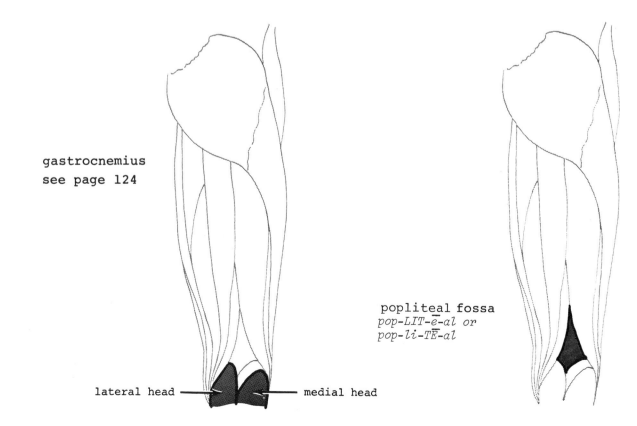

gastrocnemius
see page 124

lateral head ——→   ←—— medial head

popliteal fossa
pop-LIT-ē-al or
pop-li-TĒ-al

SUPERFICIAL MUSCLES, LOWER RIGHT LEG (Anterior View)

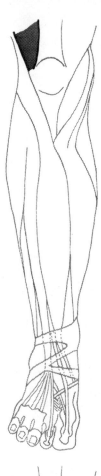

vastus lateralis
see page 124

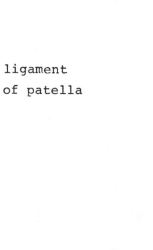

vastus medialis
see page 126

ligament
of patella

patella

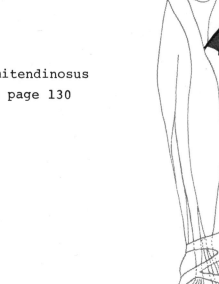

sartorius
see page 123

semitendinosus
see page 130

SUPERFICIAL MUSCLES, LOWER RIGHT LEG (Anterior View)

gastrocnemius
see page 123

soleus
sō-LĒ-us

O: posterior part
   of head of fib-
   ula and medial
   portion of tibia
I: tendo calcaneus
   (Achilles' ten-
   don) (joins with
   tendon of gas-
   trocnemius)
A: helps extend foot
   at ankle, raising
   heel; steadies
   leg upon foot

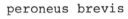

peroneus brevis

O: lower two-thirds
   of lateral surface
   of fibula
I: tuberosity at
   base of fifth
   metatarsal
A: extends and ab-
   ducts foot

peroneus longus
per-Ō-ne-us or
per-ō-NĒ-us

O: head and upper
   two-thirds of
   lateral surface
   of tibia
I: lateral side of
   first metatarsal
   and first cunei-
   form
A: extends, abducts
   and everts foot

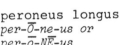

tibialis anterior
tib-ē-AL-is

O: lateral condyle
   and upper part
   of shaft of tibia
I: medial and under
   surface of first
   cuneiform and
   base of first
   metatarsal
A: flexes foot at
   ankle; elevates
   inner border of
   foot

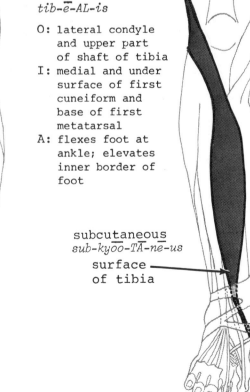

extensor digitorum
longus

O: lateral condyle of
   tibia; proximal
   three-fourths an-
   terior surface of
   fibula
I: second and third
   phalanges of sec-
   ond, third, fourth
   and fifth toes
A: extends proximal
   phalanges of the
   four small toes;
   dorsally flexes
   and pronates foot

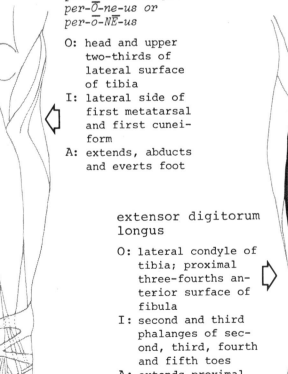

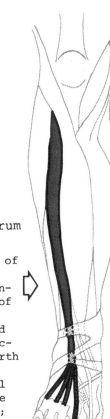

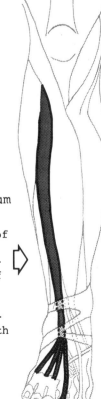

subcutaneous
sub-kyōo-TĀ-ne-us

surface
of tibia

SUPERFICIAL MUSCLES, LOWER RIGHT LEG (Anterior View)

extensor hallucis
longus   HAL-lū-sis

O: fibula and inter-
   osseous membrane
I: dorsal surface
   base of distal
   phalanx of great
   toe
A: extends great toe

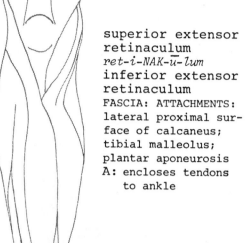

superior extensor
retinaculum
ret-i-NAK-ū-lum
inferior extensor
retinaculum
FASCIA: ATTACHMENTS:
lateral proximal sur-
   face of calcaneus;
   tibial malleolus;
   plantar aponeurosis
A: encloses tendons
   to ankle

superior

inferior

extensor hallucis
brevis

O: portion of exten-
   sor brevis digitor-
   um that goes to
   great toe
I: proximal phalanx
   of great toe
A: extends phalanges
   of great toe

interossei dorsalis
in-ter-OS-ē-ī dor-SAL-is

O: surfaces of adjacent
   metatarsal bones
I: bases of proximal
   phalanges; aponeur-
   osis of tendons of
   extensor digitorum
   longus
A: abduct, flex toes

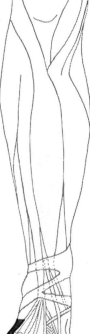

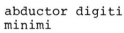

abductor digiti
minimi

O: medial and later-
   al tubercles of
   calcaneus; plan-
   tar fascia
I: fibular side of
   base of proximal
   phalanx of little
   toe
A: abducts little
   toe

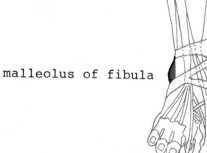

malleolus of fibula

SUPERFICIAL MUSCLES, LOWER RIGHT LEG (Dorsal View)

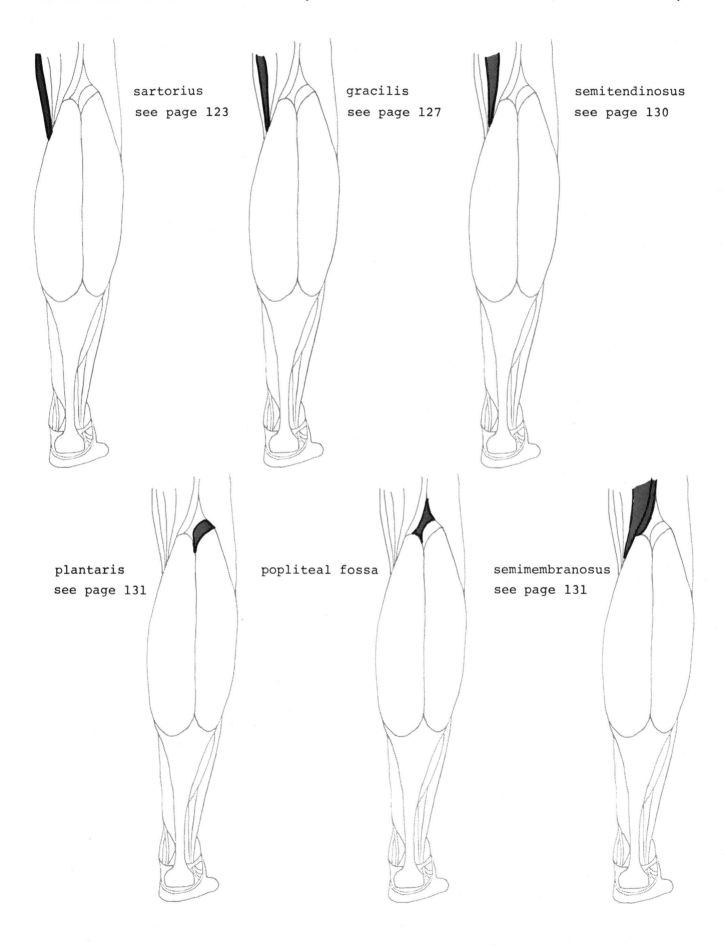

sartorius
see page 123

gracilis
see page 127

semitendinosus
see page 130

plantaris
see page 131

popliteal fossa

semimembranosus
see page 131

SUPERFICIAL MUSCLES, LOWER RIGHT LEG (Dorsal View)

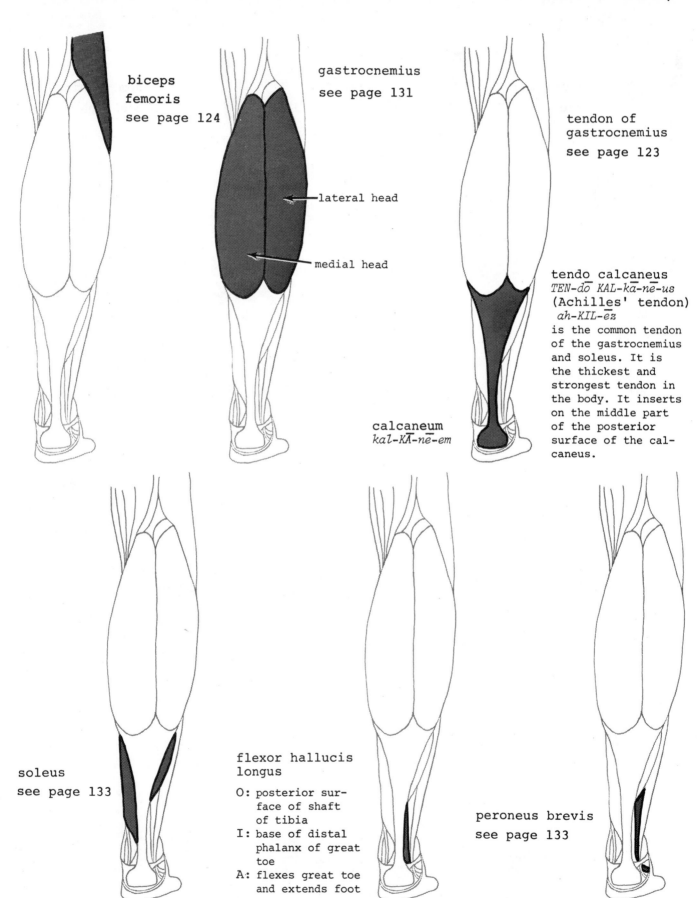

biceps
femoris
see page 124

gastrocnemius
see page 131

tendon of
gastrocnemius
see page 123

lateral head

medial head

tendo calcaneus
TEN-dō KAL-ka-nē-us
(Achilles' tendon)
ah-KIL-ēz
is the common tendon
of the gastrocnemius
and soleus. It is
the thickest and
strongest tendon in
the body. It inserts
on the middle part
of the posterior
surface of the cal-
caneus.

calcaneum
kal-KĀ-nē-em

soleus
see page 133

flexor hallucis
longus

O: posterior sur-
   face of shaft
   of tibia
I: base of distal
   phalanx of great
   toe
A: flexes great toe
   and extends foot

peroneus brevis
see page 133

SUPERFICIAL MUSCLES, LOWER RIGHT LEG (Dorsal View)

peroneus longus
see page 133

tibialis posterior

O: posterior surface
of interosseous
membrane between
tibia and fibula
I: three cuneiforms;
cuboid; navicular;
second, third and
fourth metatarsals
A: extends foot at
ankle and turns
in foot

flexor digitorum
longus

O: posterior surface
of shaft of tibia
I: distal phalanges
of lesser toes
A: flexes toes, ex-
tends foot

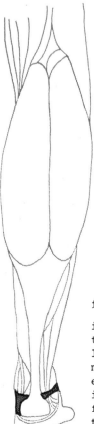

flexor retinaculum

is fascia. It attaches
to the tibial malleo-
lus proximally to the
margin of the calcan-
eus distally. Under
it pass tendons of
flexor muscles and
tibial vessels

calcaneum
(heel bone)

SECOND LAYER OF MUSCLES, LOWER RIGHT LEG
(Dorsal View)

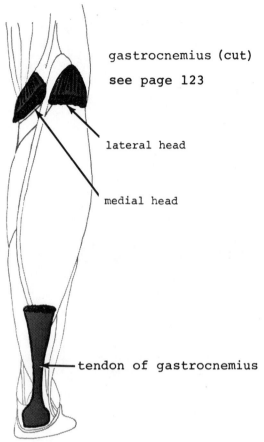

gastrocnemius (cut)

see page 123

lateral head

medial head

tendon of gastrocnemius

popliteus
pop-LIT-ē-us or
pop-li-TĒ-us

O: lateral condyle of
   femur
I: posterior surface
   of body of tibia
A: flexes leg; rotates
   flexed leg inward

plantaris
see page 131

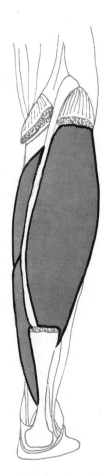

soleus
see page 133

INDEX OF ENGLISH TRANSLATION

abdominis
belly

ab/ductor
**away   to draw
from**

Achilles'
a Greek hero

acro/mion
tip   shoulder

ad/ductor
toward   to draw

alba
white

anconeus
elbow

auricul/aris
auricle of ear   adjective

bi/ceps
two   head

bi/cipit/al
two   head

brachi/alis
arm   adjective

brachii
arm

brachio/radi/alis
arm   radius

brevis
short

bursae
a purse

calcaneum
heel bone

caninus
dog

capitis
head

carp/al
wrist

carpi
wrist

cartilage
gristle

cervicis
neck

clavi/cle
key   little

clavicul/ar
clavicle

communis
common

coraco/brachi/alis
crow's beak   arm

corac/oid
crow's beak

cost/al
rib

cremaster
to suspend

cremaster/ic
to suspend

delt/oid
Greek "delta" Δ

di/gastr/ic
two   belly

digiti (pl.)
fingers, toes

digitorum
finger, toe

dors/alis
back

dorsi
back

epi/condyle
on, upon   knuckle

epi/cranius
on, upon   cranium

ex/tensor
out   to stretch

fascia
band

femoris
femur

fibula
buckle

flexor
to bend

flexores
to bend

fossa
ditch

front/alis
forehead

galea apo/neurotica
helmet   from   tendon

gastroc/nemius
belly   leg

gluteus
buttock

gracilis
slender, delicate

hallucis
great toe

humerus
shoulder

hy/oid
**letter
"U"**

ili/ac
hip bone

ilio/cost/alis
hip bone   rib

ilio/psoas
hip bone   muscles of loin

ilio/tibi/al
hip bone   tibia

ilium
hip bone

lumborum
loin

oculi
eye

indicis
one that points out

lumbricales
an earthworm

olecranon
elbow

inferioris
below, beneath

magnus
great

omo/hyoid
shoulder   hyoid bone
(Greek)

infra/orbit/al
beneath   orbit

malleolus
hammer

opponens
opposing

infra/spin/atus
beneath spine, ridge

mandible
a jaw

orbicul/aris
a small disk

inguin/al
groin

masseter
masticator

oris
mouth

inter/cost/als
between   rib

mast/oid
breast

palm/aris
palm of hand

inter/medius
between   middle

maximus
great

palpebrae
eyelid

inter/ossei
between   bone

medi/alis
middle

patella
pan (kneepan)

labii (pl.)
lip

medius
middle

pectineus
comb

lata
broad

ment/alis
chin

pector/alis
breast bone

later/alis
side

minimi
small

peroneus
brooch, fibula

latissimus
broad

mylo/hyoid
mill   hyoid bone
Pertains to the molar teeth and hyoid bone.

phalanx
line of soldiers

levator
lifter

nas/alis
nose

plant/aris
sole of foot

ligament
to bind

nuchae
back of neck

platysma
flat plate

ligamentum
to bind

oblique
slanting

pollicis
thumb

linea
line

occipit/al
back of head

poplite/al
back of knee

longissimus
long

occipit/alis
back of head

popliteus
back of knee

longus
long

occipito/front/alis
back of head   forehead

procerus
long

profundus
deep

pronator
to bend forward

proprius
one's own

quadratus
square

quinti
five

radi/alis
radius

radius
ray

rectus
straight

retinaculum
halter (Latin)

rhomboideus
a lozenge shaped figure

risorius
to laugh

sartorius
tailor

scalenus
uneven

scapula (sing.)
shoulder blade

scapulae (pl.)
shoulder blade

semi/membranosus
half   membrane

semi/spin/alis
half   spine

semi/tendinosus
half   tendon

serratus
a saw

soleus
sole of foot

spin/alis
spine

splenius
bandage
Wraps around the other muscles like a bandage.

stern/al
sternum

sterno/cleido/mastoid
sternum   clavicle   mastoid process

sterno/hyoid
sternum   hyoid bone

sternum
breast bone

stylo/hyoid
stake, pole hyoid bone

sub/cutane/ous
beneath   skin

sublimis
high, superficial

sub/scapul/aris
beneath   scapula

super/fici/alis
above   face

superioris
above

supinator
to bend backward

supra/spinatus
above   spine, ridge

tempor/al
time, temple

tendo
tendon

tendon
to stretch

tensor
to stretch

teres
long and round

thoracis
chest

thoraco/lumb/ar
chest   loin

thyro/hyoid
thyroid   hyoid bone

tibia
shin bone

tibi/alis
shin bone

trans/vers/alis
across   to turn

trans/versus
across   to turn

trapezius
table, counter

tri/angul/aris
three   angle

tri/ceps
three   head

ulna
elbow

uln/aris
elbow

umbilicus
the navel

vastus
huge

zygomat/ic
yoke together

zygomaticus
yoke together

CHAPTER IV
NERVOUS SYSTEM

The complex organization of structures of which the nervous system is composed, is divided into (1) the central nervous system and (2) the peripheral nervous system.

The central nervous system is composed of the brain and the spinal cord. It supplies all areas of the body except the viscera (thoracic and abdominal organs).

The nerves, ganglia and end organs which connect the central nervous system with all other parts of the body comprise the peripheral nervous system. Therefore, the peripheral nervous system consists of the cranial nerves, the spinal nerves and the sympathetic nervous system. The viscera are controlled partly by the sympathetic system and partly by portions of certain cranial and sacral nerves. This latter group is called the parasympathetic or craniosacral system. The sympathetic and parasympathetic systems make up the autonomic nervous system which is also called the visceral nervous system.

The peripheral nerves (like the arteries and veins) tend to take on the names of the areas which they serve. For example, the radial nerve and ulnar nerve of the forearm, the femoral nerve of the thigh and the peroneal nerve of the leg. Many of the nerves have "cutaneous" added to designate that these nerves are distributed to the skin.

Because the nervous system is so complex the material in this chapter is concerned only with the general anatomy of each of the divisions.

LOBES AND FISSURES OF CEREBRUM

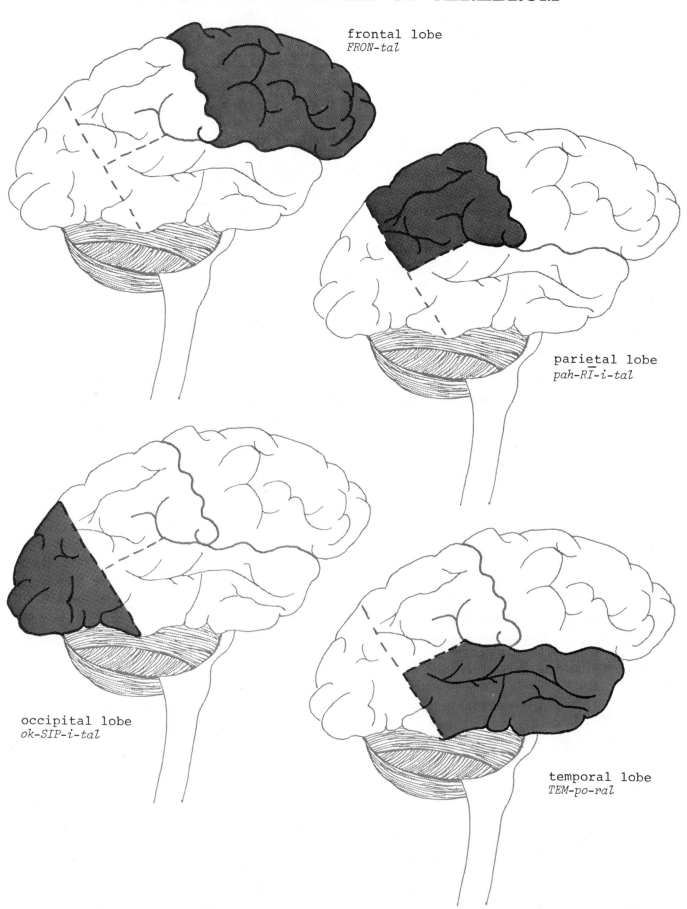

frontal lobe
FRON-tal

parietal lobe
pah-RĪ-i-tal

occipital lobe
ok-SIP-i-tal

temporal lobe
TEM-po-ral

LOBES AND FISSURES OF CEREBRUM

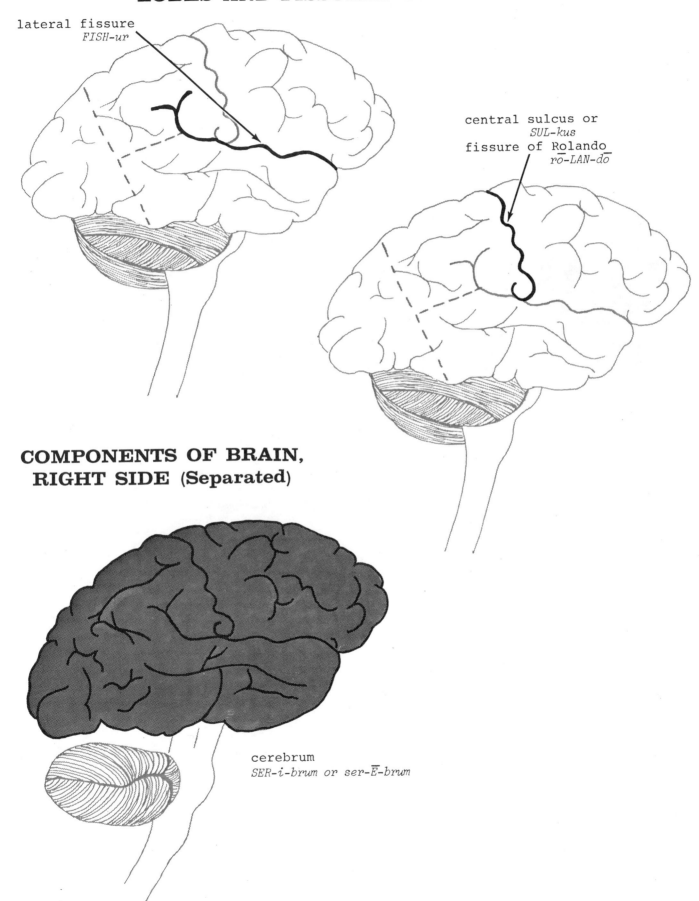

lateral fissure
FISH-ur

central sulcus or
SUL-kus
fissure of Rolando
rō-LAN-dō

COMPONENTS OF BRAIN,
RIGHT SIDE (Separated)

cerebrum
SER-i-brum or ser-Ē-brum

COMPONENTS OF BRAIN, RIGHT SIDE (Separated)

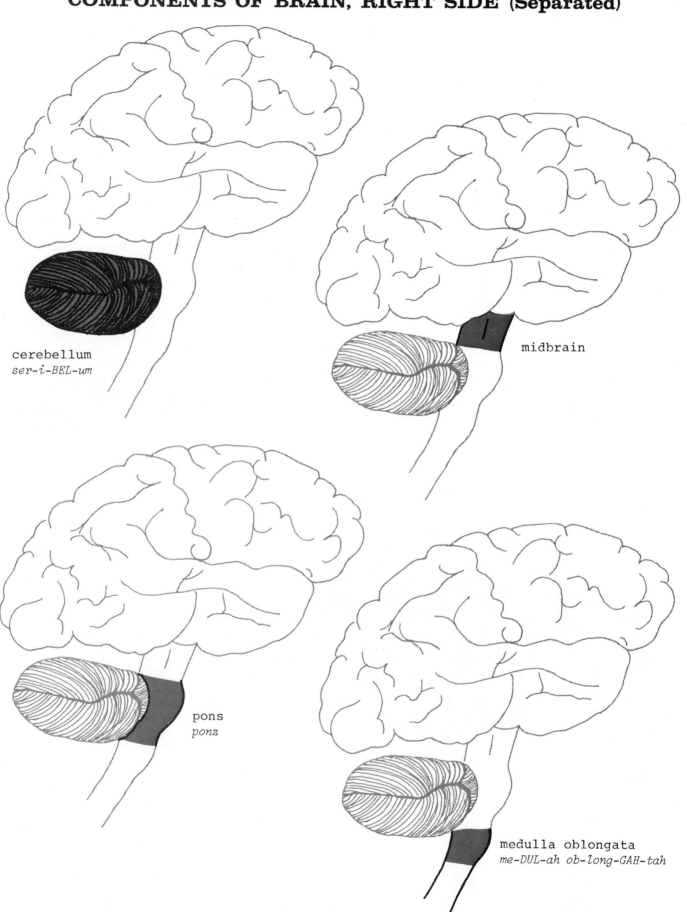

cerebellum
ser-i-BEL-um

midbrain

pons
ponz

medulla oblongata
me-DUL-ah ob-long-GAH-tah

VENTRICLES OF THE CEREBRUM

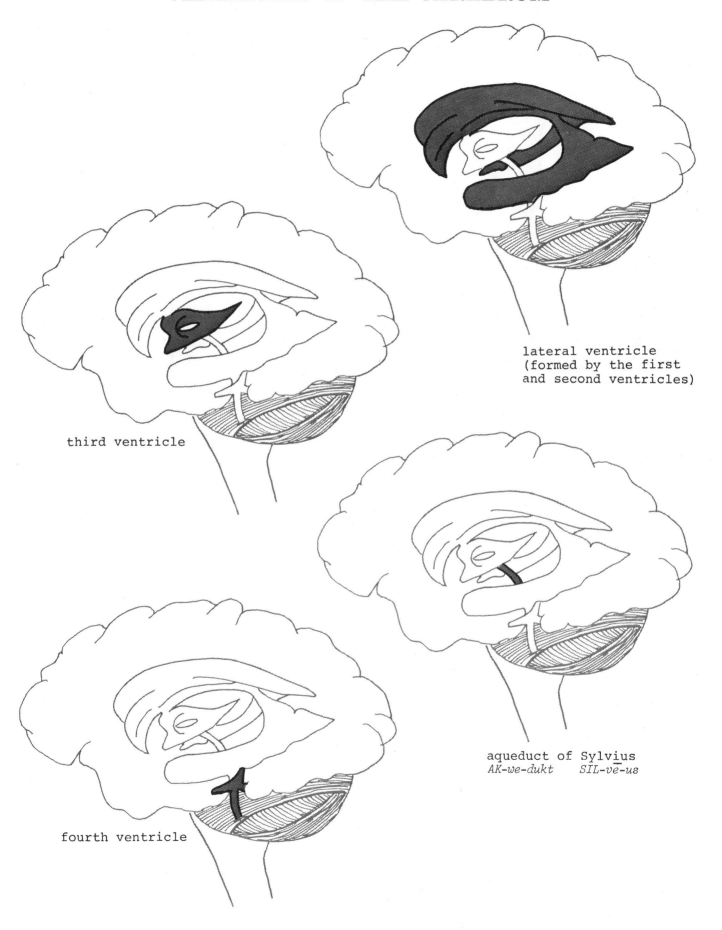

lateral ventricle
(formed by the first
and second ventricles)

third ventricle

aqueduct of Sylvius
AK-we-dukt   SIL-vē-us

fourth ventricle

THE CEREBRUM

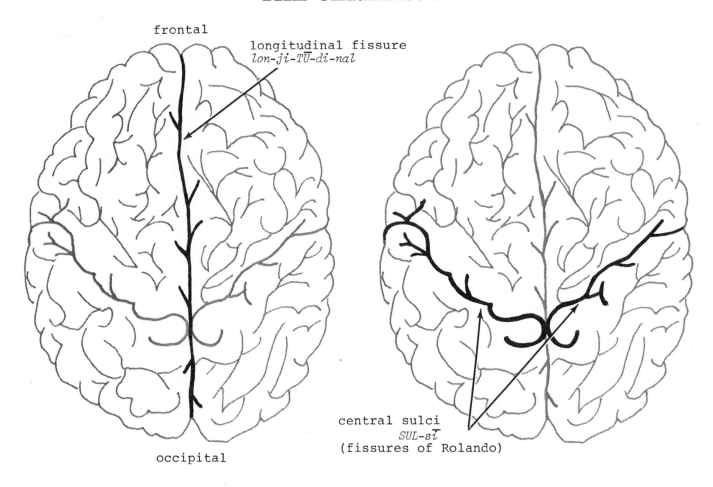

frontal

longitudinal fissure
lon-ji-TŪ-di-nal

occipital

central sulci
SUL-sī
(fissures of Rolando)

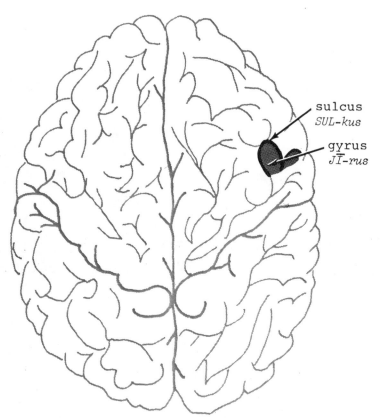

sulcus
SUL-kus

gyrus
JĪ-rus

LEFT HALF OF BRAIN (Sagittal Section)

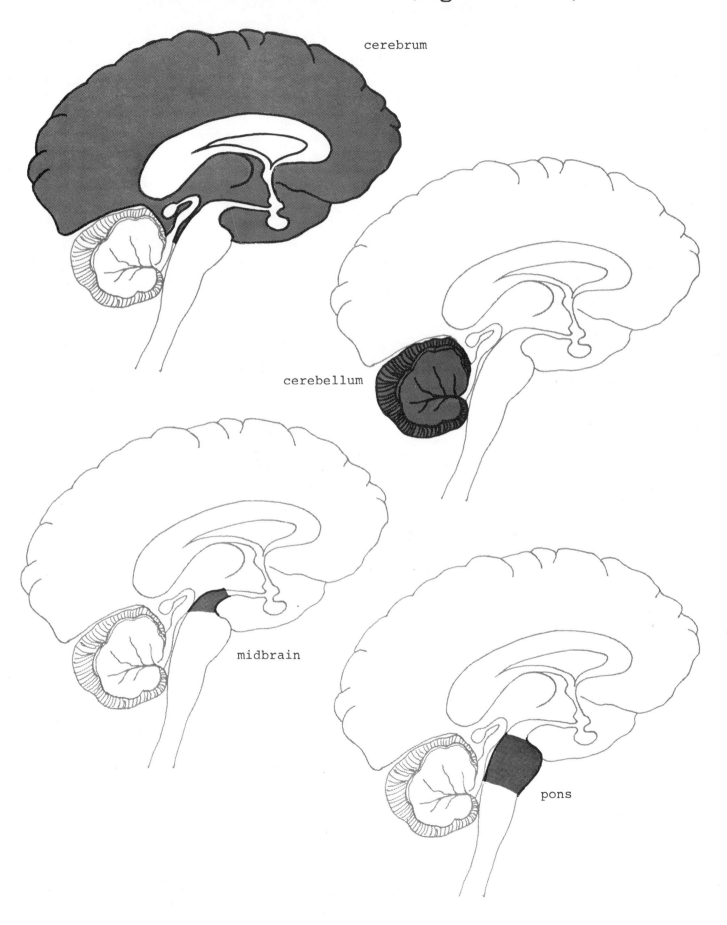

cerebrum

cerebellum

midbrain

pons

LEFT HALF OF BRAIN (Sagittal Section)

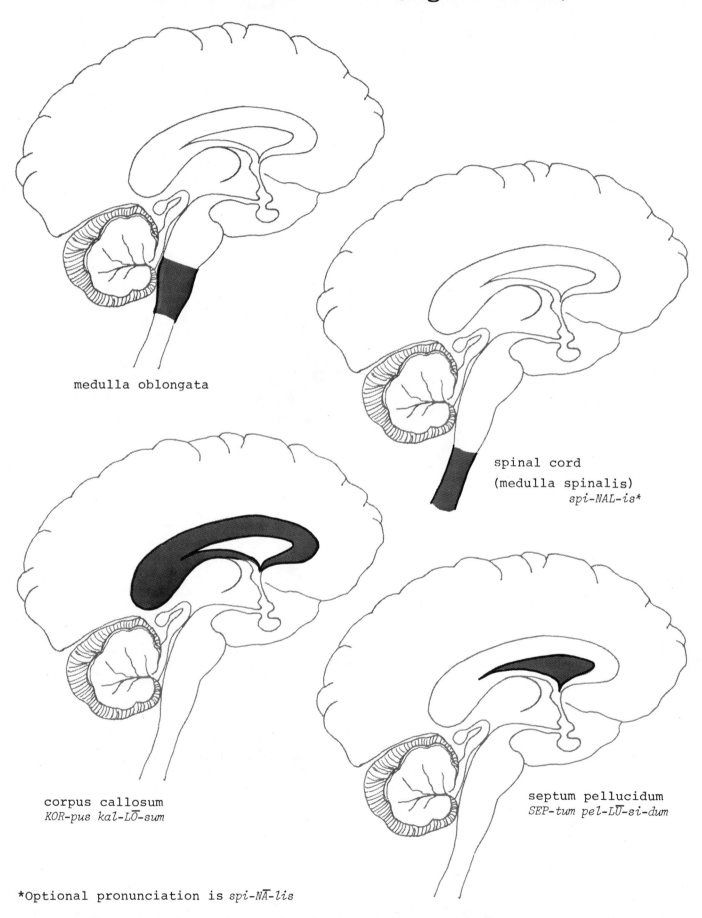

medulla oblongata

spinal cord
(medulla spinalis)
*spi-NAL-is**

corpus callosum
KOR-pus kal-LŌ-sum

septum pellucidum
SEP-tum pel-LŪ-si-dum

*Optional pronunciation is *spi-NĀ-lis*

LEFT HALF OF BRAIN (Sagittal Section)

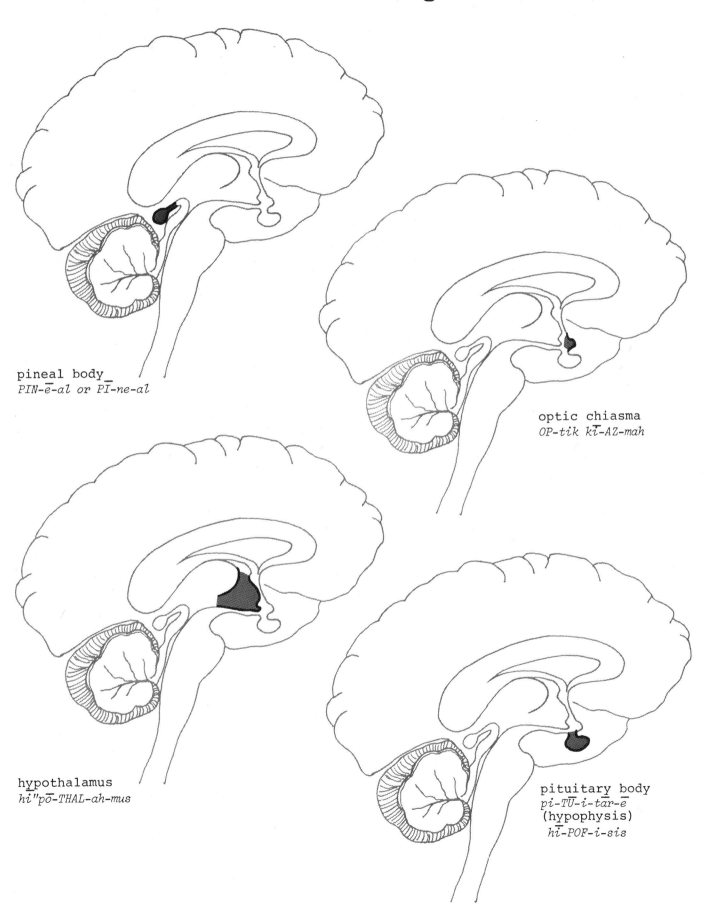

pineal body
PIN-ē-al or PĪ-ne-al

optic chiasma
OP-tik kī-AZ-mah

hypothalamus
hi"pō-THAL-ah-mus

pituitary body
pi-TŪ-i-tār-ē
(hypophysis)
hī-POF-i-sis

LEFT HALF OF BRAIN (Sagittal Section)

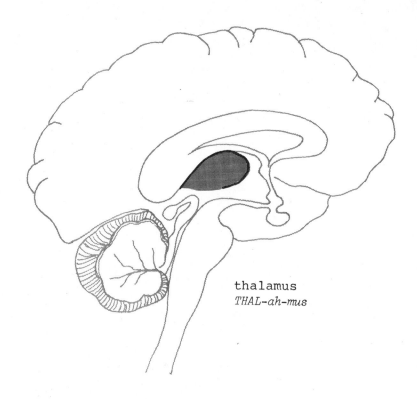

thalamus
THAL-ah-mus

SPINAL CORD, VERTEBRAE AND MENINGES

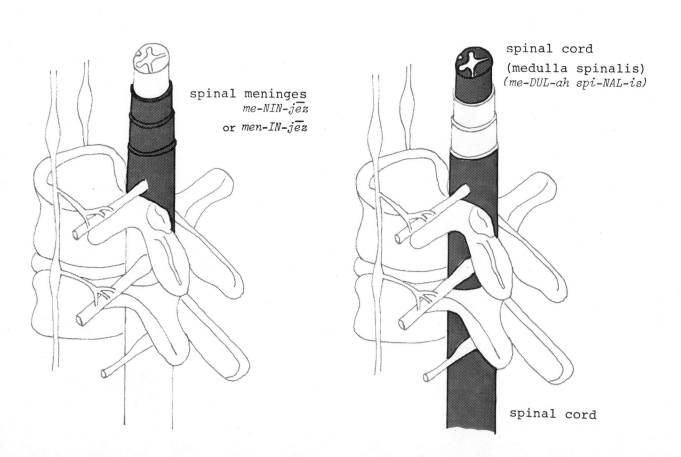

spinal meninges
me-NIN-jēz
or *men-IN-jēz*

spinal cord
(medulla spinalis)
(me-DUL-ah spi-NAL-is)

spinal cord

SPINAL CORD, VERTEBRAE AND MENINGES

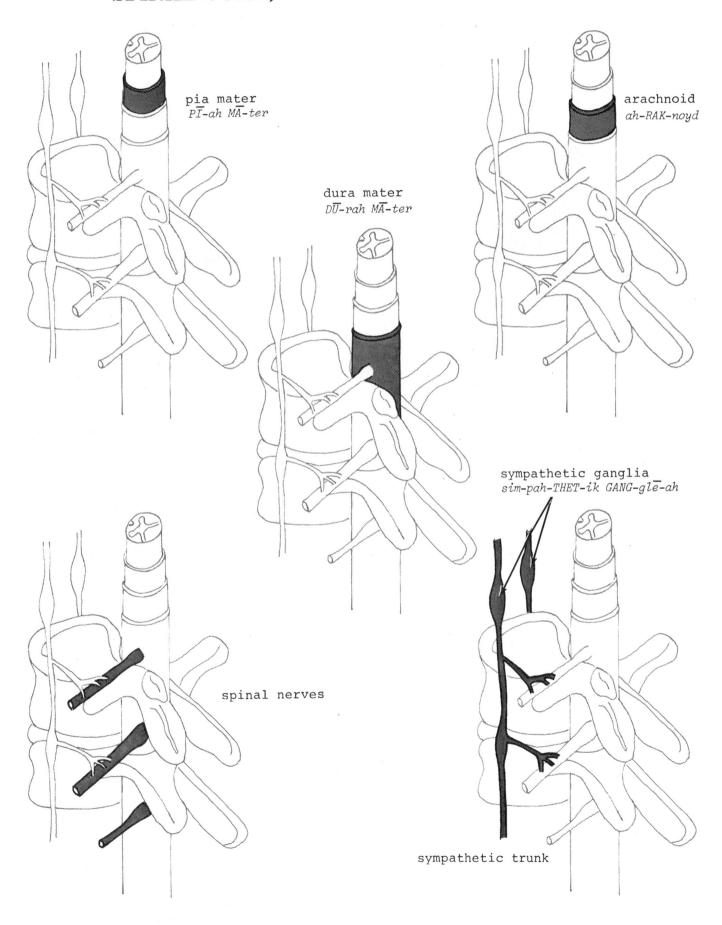

pia mater
PĪ-ah MĀ-ter

arachnoid
ah-RAK-noyd

dura mater
DŪ-rah MĀ-ter

sympathetic ganglia
sim-pah-THET-ik GANG-glē-ah

spinal nerves

sympathetic trunk

SPINAL CORD, VERTEBRAE AND MENINGES

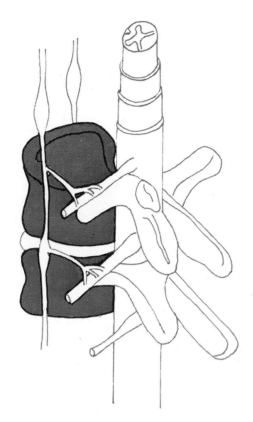

body of vertebrae

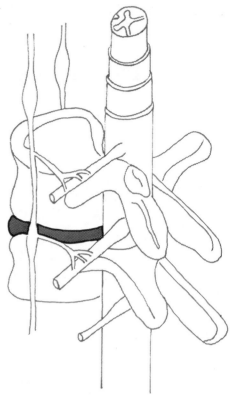

intervertebral disk (disc)
in-ter-VER-te-bral

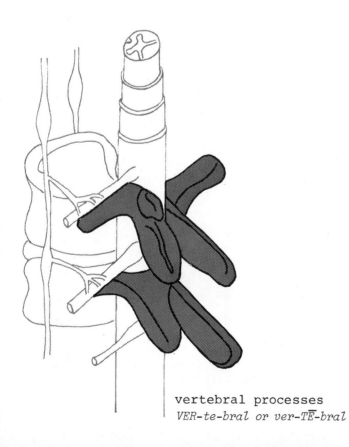

vertebral processes
VER-te-bral or ver-TĒ-bral

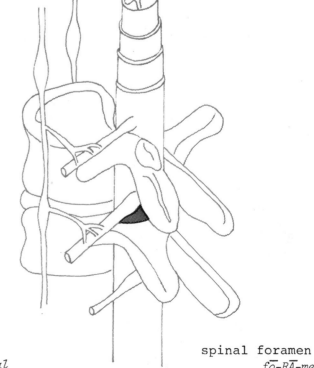

spinal foramen
fō-RĀ-men

CRANIAL NERVES

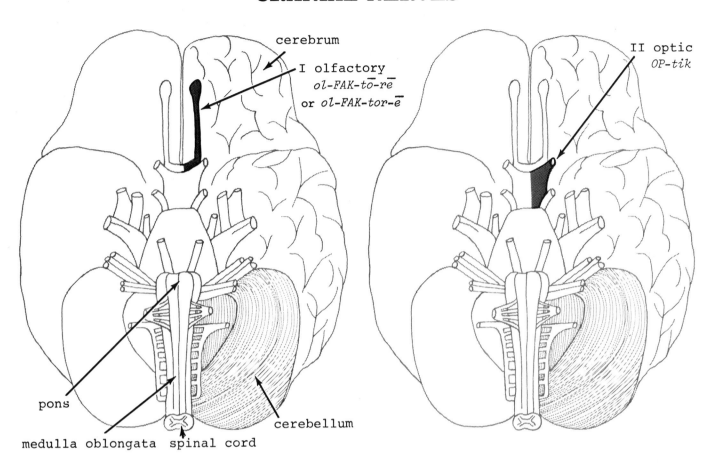

cerebrum

I olfactory
ol-FAK-tō-rē
or *ol-FAK-tor-ē*

pons

medulla oblongata   spinal cord

cerebellum

II optic
OP-tik

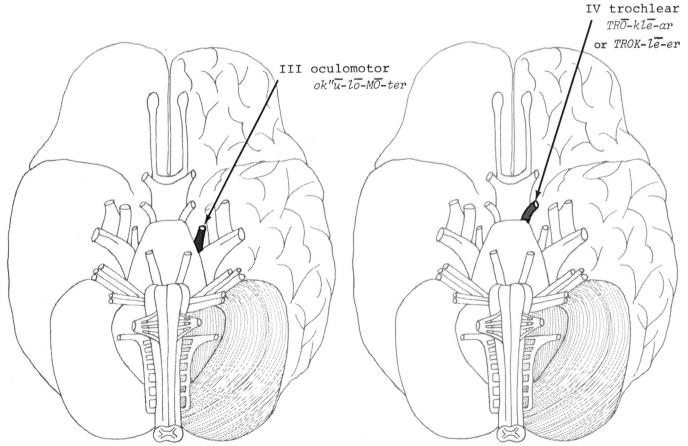

III oculomotor
ok″ū-lō-MŌ-ter

IV trochlear
TRŌ-klē-ar
or *TROK-lē-er*

CRANIAL NERVES

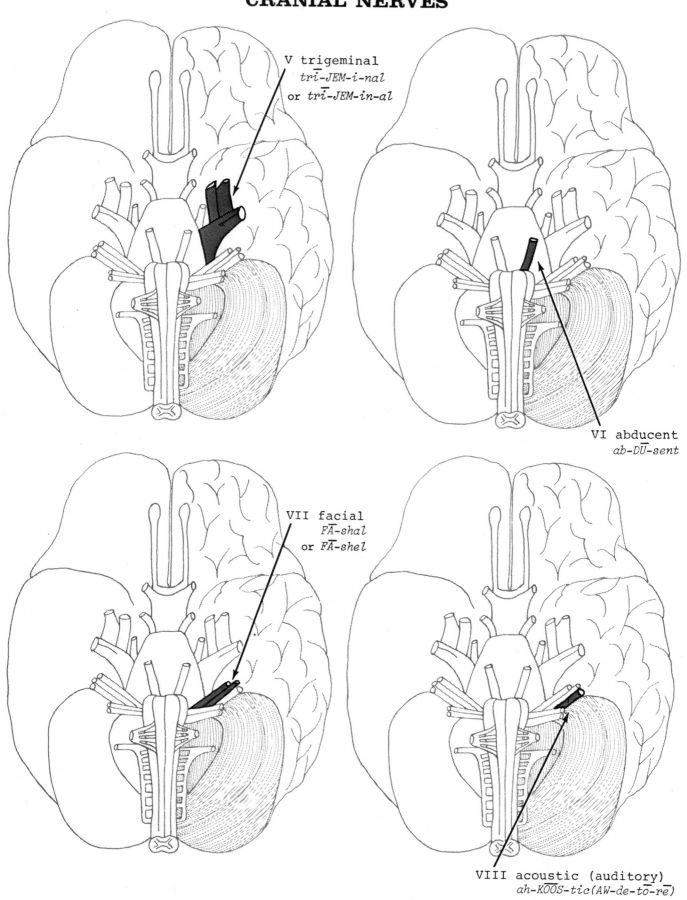

V trigeminal
trī-JEM-i-nal
or *trī-JEM-in-al*

VI abducent
ab-DŪ-sent

VII facial
FĀ-shal
or *FĀ-shel*

VIII acoustic (auditory)
ah-KOOS-tic(AW-de-tō-rē)

CRANIAL NERVES

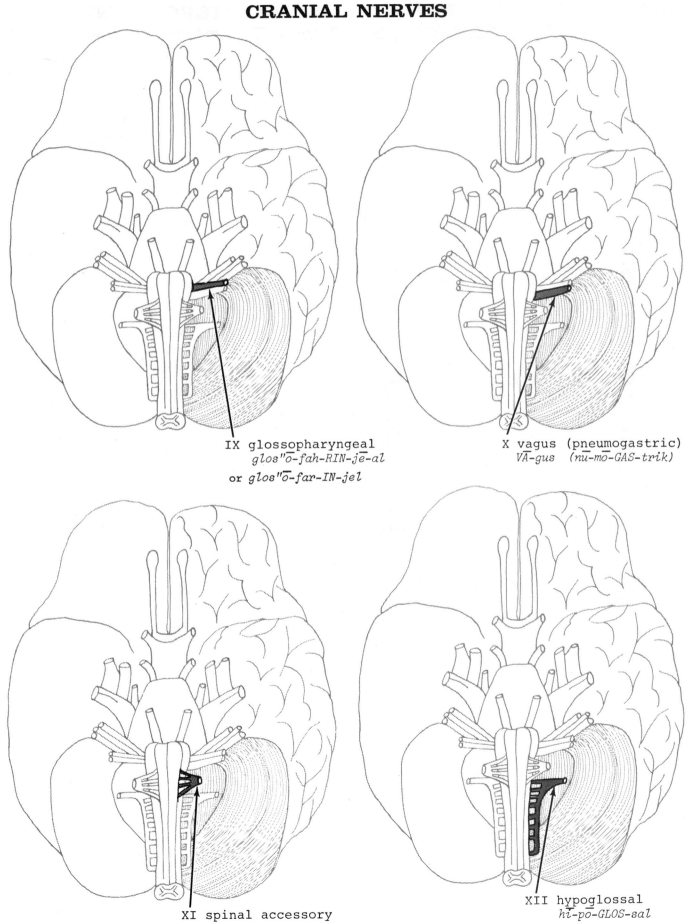

IX glossopharyngeal
glos"ō-fah-RIN-jē-al

or *glos"ō-far-IN-jel*

X vagus (pneumogastric)
VĀ-gus (nū-mō-GAS-trik)

XI spinal accessory

XII hypoglossal
hī-pō-GLOS-sal

DORSAL VIEW OF SPINAL CORD AND EMERGING NERVES
CENTRAL NERVOUS SYSTEM – (Voluntary Nerves)

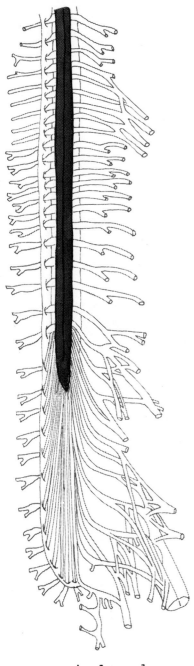

spinal cord

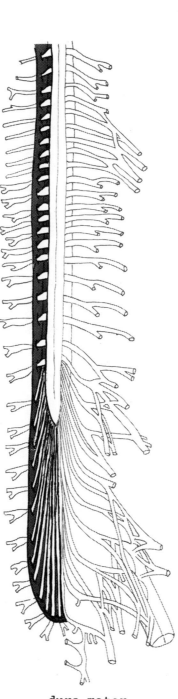

dura mater

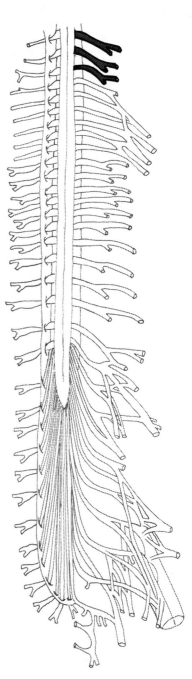

cervical plexus *PLEK-ses*
(while there are seven
cervical vertebrae there
are eight cervical nerves
--three are shown here)

DORSAL VIEW OF SPINAL CORD AND EMERGING NERVES
CENTRAL NERVOUS SYSTEM – (Voluntary Nerves)

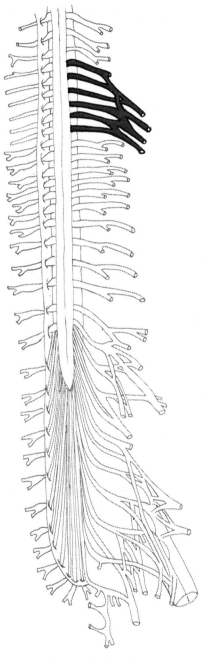

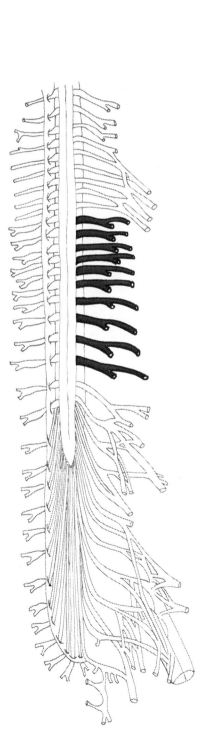

brachial plexus
BRĀ-kē-al

lumbar plexus
LUM-bar

intercostal (thoracic) nerves
in-ter-KOS-tal (thō-RAS-ik)

DORSAL VIEW OF SPINAL CORD AND EMERGING NERVES
CENTRAL NERVOUS SYSTEM – (Voluntary Nerves)

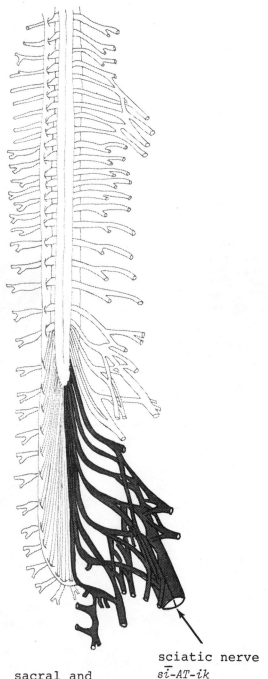

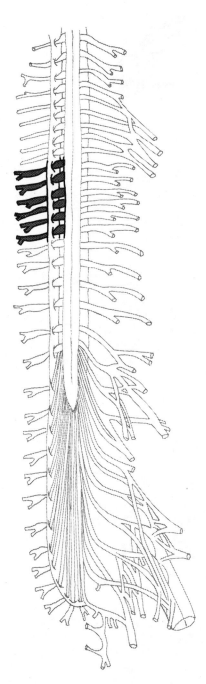

sciatic nerve
sī-AT-ik

sacral and
SĀ-kral
coccygeal plexus
kok-SIJ-ē-al

thoracic spinal nerves
emerging through opened
dura

DORSAL VIEW OF SPINAL CORD AND EMERGING NERVES
CENTRAL NERVOUS SYSTEM – (Voluntary Nerves)

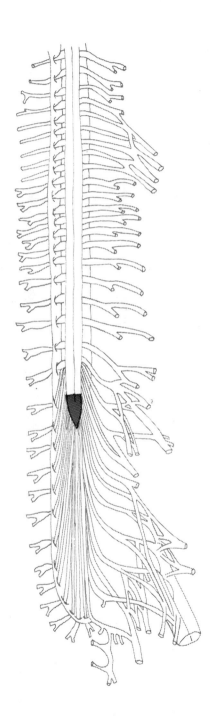

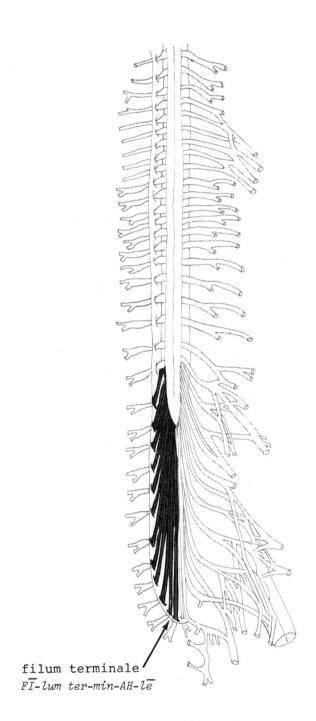

filum terminale
FĪ-lum ter-min-AH-lē

conus medullaris
KŌ-nus med-ū-LAR-is

cauda equina
KAW-dah ē-KWĪ-nah

SYMPATHETIC NERVOUS SYSTEM
(Involuntary Nerves)

superior cervical
ganglion
GANG-glē-on

middle cervical
ganglion

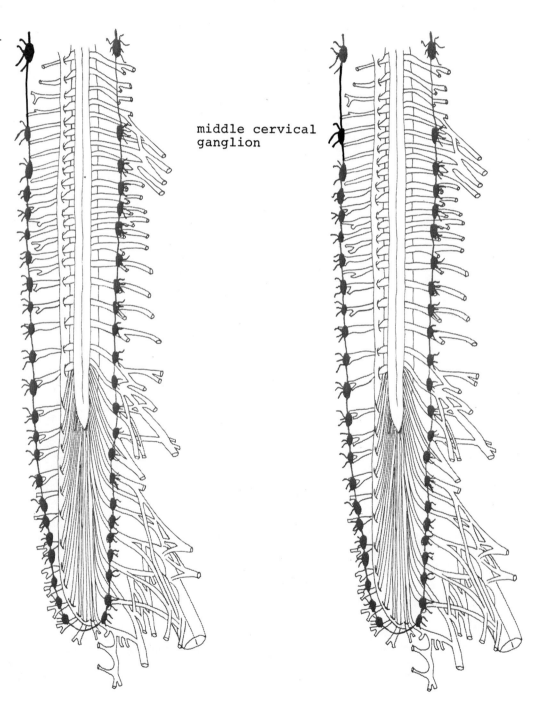

SYMPATHETIC NERVOUS SYSTEM
(Involuntary Nerves)

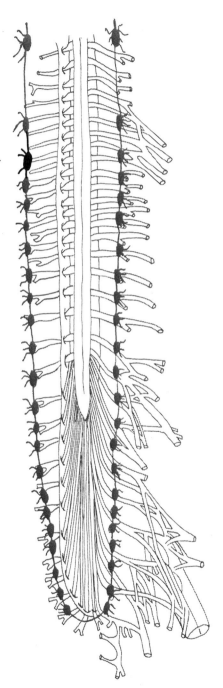

inferior cervical
ganglion

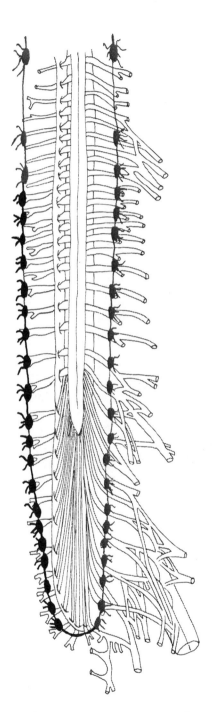

chain of sympathetic
ganglia
GANG-glē-ah

AUTONOMIC NERVOUS SYSTEM (Involuntary Nerves)

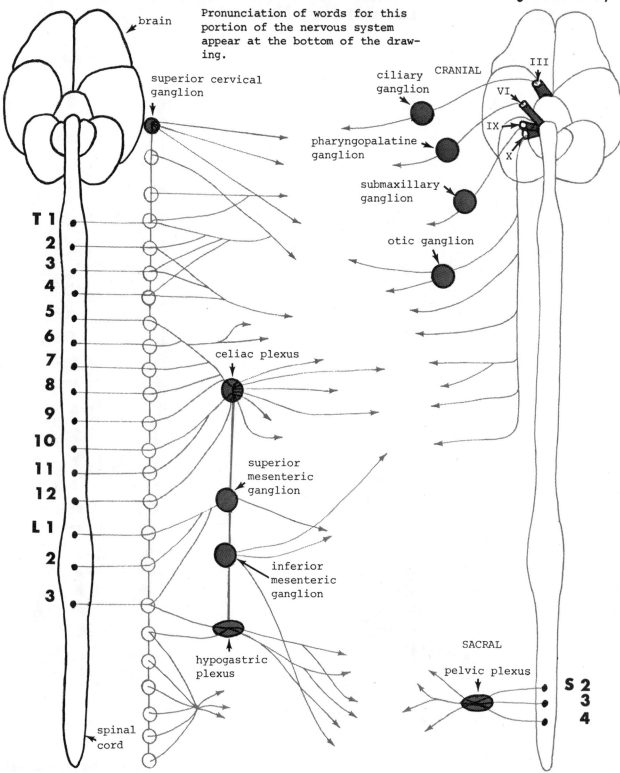

Pronunciation of words for this portion of the nervous system appear at the bottom of the drawing.

brain

superior cervical ganglion

T 1
2
3
4
5
6
7
8
9
10
11
12

L 1
2
3

spinal cord

celiac plexus

superior mesenteric ganglion

inferior mesenteric ganglion

hypogastric plexus

CRANIAL

ciliary ganglion

III
VI
IX
X

pharyngopalatine ganglion

submaxillary ganglion

otic ganglion

SACRAL

pelvic plexus

S 2
3
4

Distribution of Sympathetic or Thoracolumbar Division

Distribution of Parasympathetic or Craniosacral Division

The functions of these two systems are opposite and antagonistic in effect. For example, the parasympathetic (cranial portion) slows the heart, the sympathetic accelerates the heart's action.

celiac
SĒ-lē-ak

ciliary
SIL-ē-ār-ē

hypogastric
hī-po-GAS-trik

mesenteric
mes-en-TER-ik

otic
Ō-tik

pharyngopalatine
fah-ring"gō-PAL-ah-tin

submaxillary
sub-MAK-si-lār-ē

AUTONOMIC NERVOUS SYSTEM (Involuntary Nerves)

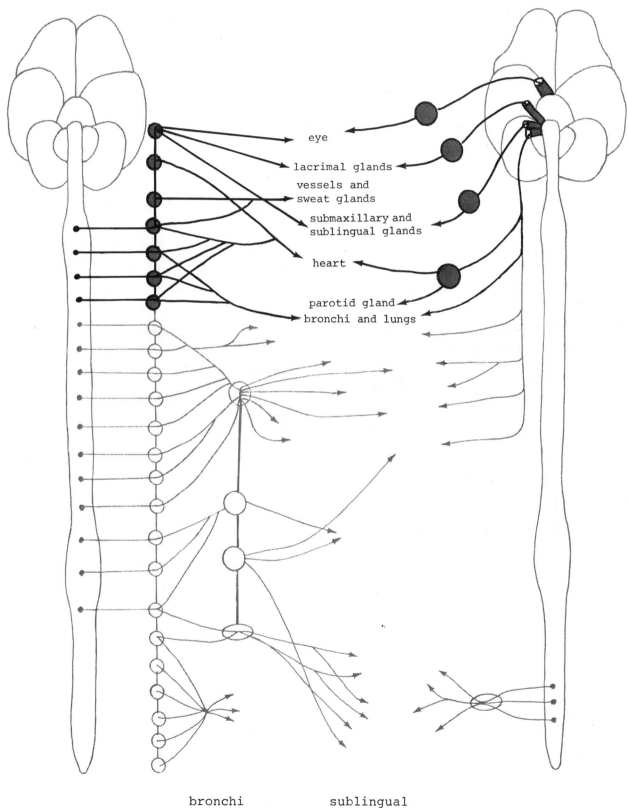

eye

lacrimal glands

vessels and
sweat glands

submaxillary and
sublingual glands

heart

parotid gland

bronchi and lungs

bronchi
BRONG-kī

lacrimal
LAK-re-mal

parotid
pah-ROT-id

sublingual
sub-LING-gwal

submaxillary
sub-MAK-si-lār-ē

AUTONOMIC NERVOUS SYSTEM (Involuntary Nerves)

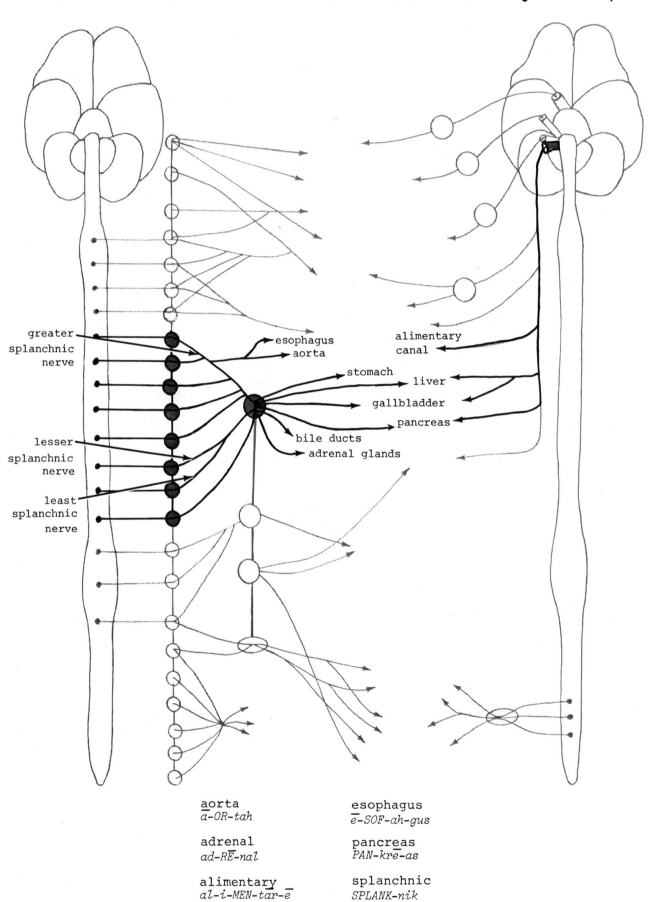

greater
splanchnic
nerve

lesser
splanchnic
nerve

least
splanchnic
nerve

esophagus
aorta

stomach
liver
gallbladder
pancreas
bile ducts
adrenal glands

alimentary
canal

aorta
$\overline{a}$-OR-tah

adrenal
ad-R$\overline{E}$-nal

alimentary
al-i-MEN-tar-$\overline{e}$

esophagus
$\overline{e}$-SOF-ah-gus

pancreas
PAN-kr$\overline{e}$-as

splanchnic
SPLANK-nik

AUTONOMIC NERVOUS SYSTEM (Involuntary Nerves)

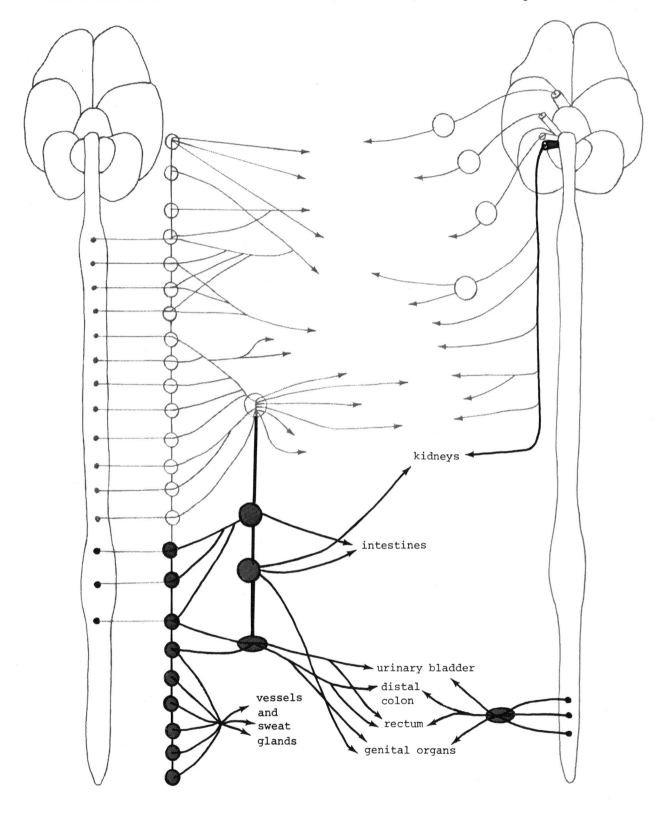

kidneys

intestines

urinary bladder

distal colon

rectum

vessels and sweat glands

genital organs

genital
JEN-i-tal

urinary
Ū-ri-nār-ē

INDEX OF ENGLISH TRANSLATION

ab/ducent
away   to draw
from

acoust/ic
hearing

ad/ren/al
near   kidney

aliment/ary
to nourish

aorta
to lift up

aque/duct
water   canal

arachn/oid
spider web

audit/ory
to hear

brachi/al
arm

bronchi
windpipe

callosum
hard

cauda
tail

celi/ac
belly

cerebellum
small brain

cerebrum
brain

cervic/al
neck

chiasma
crossing in form of an X

cili/ary
eyelash

coccyge/al
coccyx

conus
cone

corpus
body

dura mater
hard   mother

equina
horse

esophagus
gullet

faci/al
face

filum
thread

fissure
to split

foramen
opening

front/al
forehead, front

ganglia (pl.)
knot

ganglion (sing.)
knot

genit/al
organs of reproduction

glosso/pharynge/al
tongue   pharynx

gyrus
ring, circle

hypo/gastr/ic
below   stomach

hypo/gloss/al
below   tongue

hypo/physis
below   to grow

hypo/thalamus
below   chamber

inter/cost/al
between   rib

inter/vertebr/al
between   vertebra

lacrim/al
tear

later/al
side

lumb/ar
loin

medulla
marrow

medull/aris
marrow

meninges
membrane

mes/enter/ic
middle intestine (mesentery)

oblongata
rather long

occipit/al
back of head

oculo/motor
eye   mover

olfact/ory
to smell

opt/ic
eye

ot/ic
ear

pan/creas
all   flesh

pariet/al
wall

par/otid
beside   ear

pel/lucidum
through   to shine

pharyngo/palatine
pharynx   palate

pia mater
tender   mother

pine/al
pine cone

pituit/ary
phlegm
It was believed that this gland caused the
formation of mucus (phlegm).

plexus
network

pneumo/gastr/ic
lungs   stomach

pons
bridge

sacr/al
sacrum

sciat/ic
hip joint

septum
wall off

spin/al
spine

spin/alis
spine

splanchn/ic
viscera

sub/lingu/al
below   tongue

sub/mandibul/ar
below   mandible

sub/maxill/ary
below   maxilla

sulci (pl.)
trench

sulcus (sing.)
trench

sym/pathet/ic
together   suffer

tempor/al
time, temple

terminale
end

thalamus
chamber

thorac/ic
chest

tri/gemin/al
three   twin

trochle/ar
pulley

urin/ary
urine

vagus
wandering

ventr/icle
belly   small

vertebrae (pl.)
to turn

vertebr/al
vertebra

CHAPTER V

CIRCULATORY SYSTEM

This system can be considered as a very efficient pumping station (the heart) and distributing conduits or pipes (the arteries, arterioles, venules and veins). The fluid (the blood) is circulated throughout this system to give up oxygen to the cells and take on the carbon dioxide which the cells give up, as one of its functions. Other functions consist of the transportation of food from the alimentary canal to the cells, carrying waste products from the cells to the organs of excretion, regulation of body temperature, carrying hormones from the endocrine glands, maintaining a proper acid-base balance and protecting the body from invading organisms.

The veins of the Circulatory System are supplied with valves to prevent regurgitation of blood into the capillaries.

VALVES OF VEINS
EXTERNAL VIEW   INSIDE VIEWS

DILATATION AT SITE OF VALVE | VALVES OPENED | VALVES CLOSED

In addition to the blood circulation there is a lymphatic circulation which is composed of lymph vessels and nodes. The nodes are small, oval and bean-shaped. Situated in the course of the lymphatic vessels they filter the lymph as it passes through them removing bacteria, dead cells and clots to be consumed by the phagocytes. Antibodies are produced to return the lymph in a "purified" state.

There are three organs which are closely related to the lymphatic system. They are the spleen, tonsils and thymus. Although once classed as an endocrine gland the thymus is now included in the lymphatic system as it is known to produce small lymphocytes. No evidence of endocrine function has been discovered to date.

A LYMPH NODE WITH ITS AFFERENT AND EFFERENT VESSELS

The lymph vascular system helps return proteins to the blood, furnishes drainage to the lymph nodes to filter out toxic or malignant products and helps in the absorption of digested fats.

PRINCIPAL ARTERIES OF HEAD, NECK AND BODY

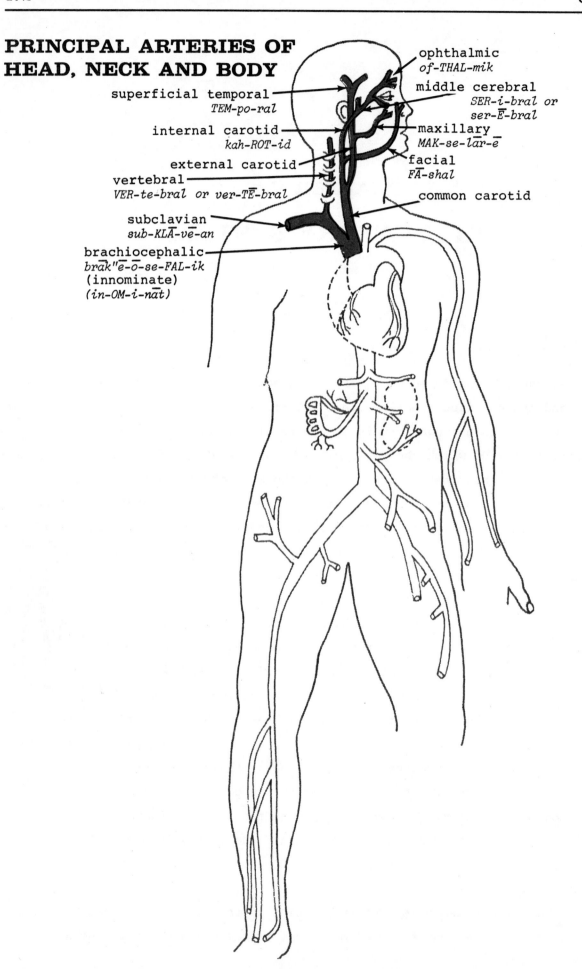

ophthalmic
of-THAL-mik

middle cerebral
SER-i-bral or
ser-Ē-bral

superficial temporal
TEM-po-ral

internal carotid
kah-ROT-id

maxillary
MAK-se-lār-ē

external carotid

facial
FĀ-shal

vertebral
VER-te-bral or ver-TĒ-bral

common carotid

subclavian
sub-KLĀ-vē-an

brachiocephalic
brāk"ē-ō-se-FAL-ik
(innominate)
(in-OM-i-nāt)

PRINCIPAL ARTERIES OF HEAD, NECK AND BODY

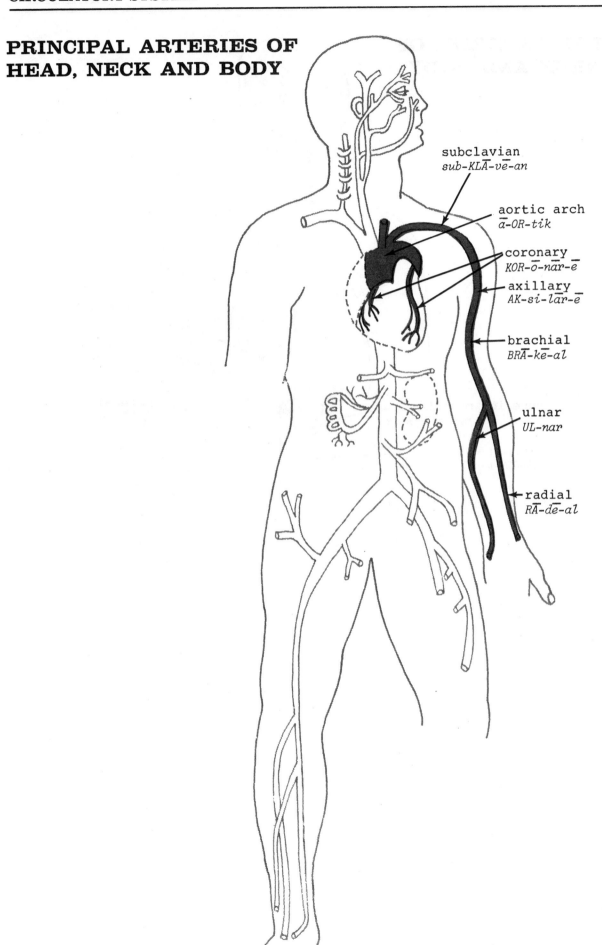

subclavian
sub-KLĀ-vē-an

aortic arch
ā-OR-tik

coronary
KOR-ō-nār-ē

axillary
AK-si-lār-ē

brachial
BRĀ-kē-al

ulnar
UL-nar

radial
RĀ-dē-al

PRINCIPAL ARTERIES OF
HEAD, NECK AND BODY

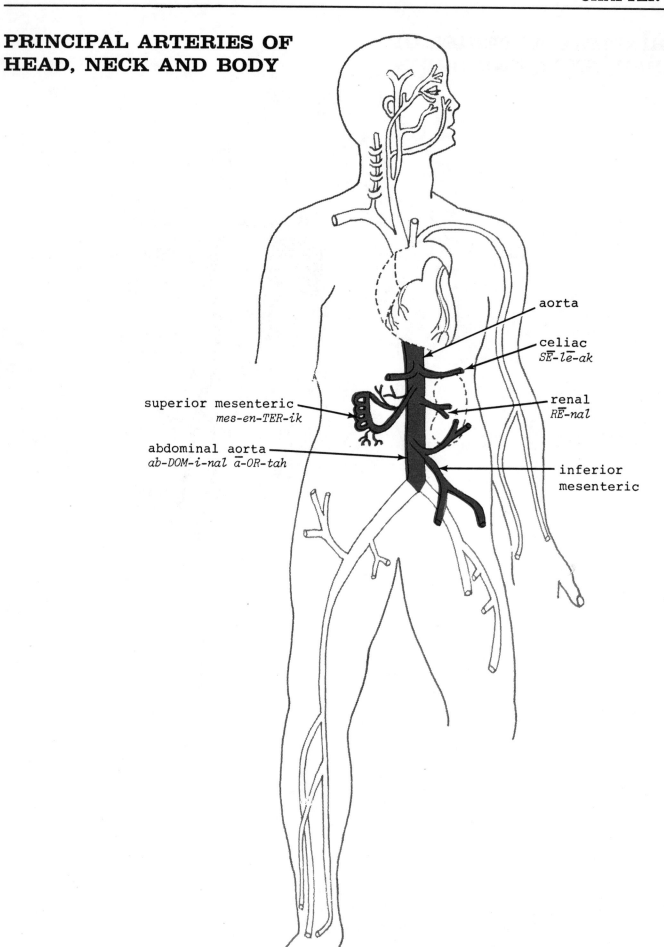

aorta

celiac
SĒ-lē-ak

renal
RĒ-nal

superior mesenteric
mes-en-TER-ik

abdominal aorta
ab-DOM-i-nal ā-OR-tah

inferior
mesenteric

PRINCIPAL ARTERIES OF
HEAD, NECK AND BODY

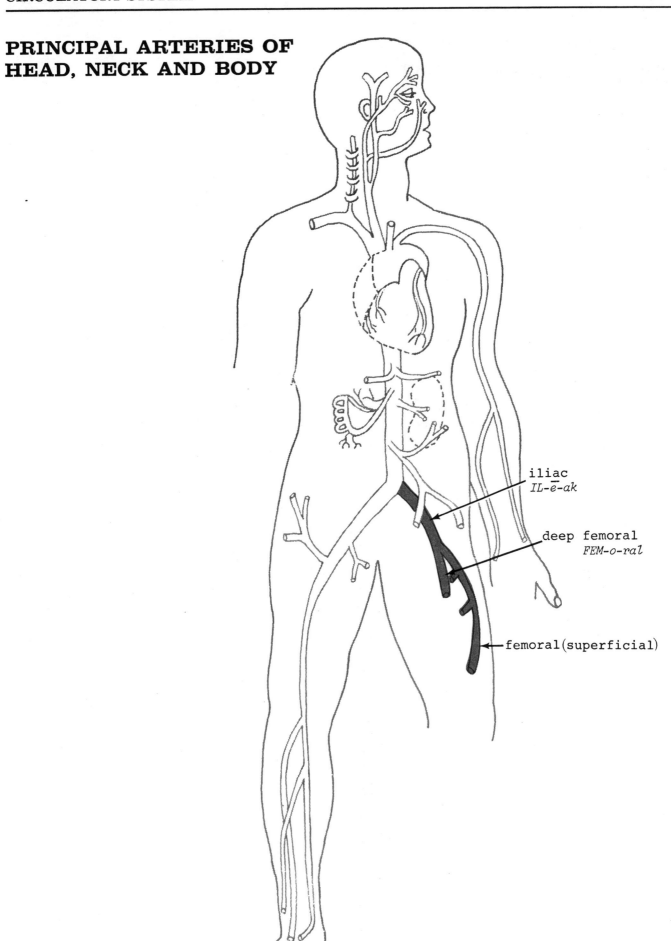

iliac
IL-ē-ak

deep femoral
FEM-o-ral

femoral(superficial)

PRINCIPAL ARTERIES OF
HEAD, NECK AND BODY

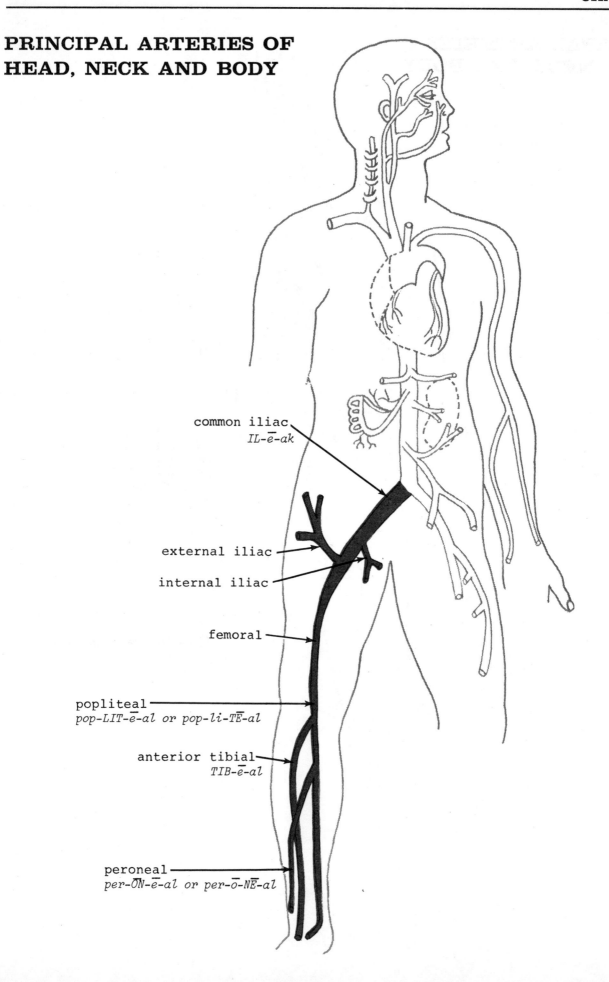

common iliac
IL-ē-ak

external iliac

internal iliac

femoral

popliteal
pop-LIT-ē-al or pop-li-TĒ-al

anterior tibial
TIB-ē-al

peroneal
per-ŌN-ē-al or per-ō-NĒ-al

PRINCIPAL VEINS OF HEAD, NECK AND BODY

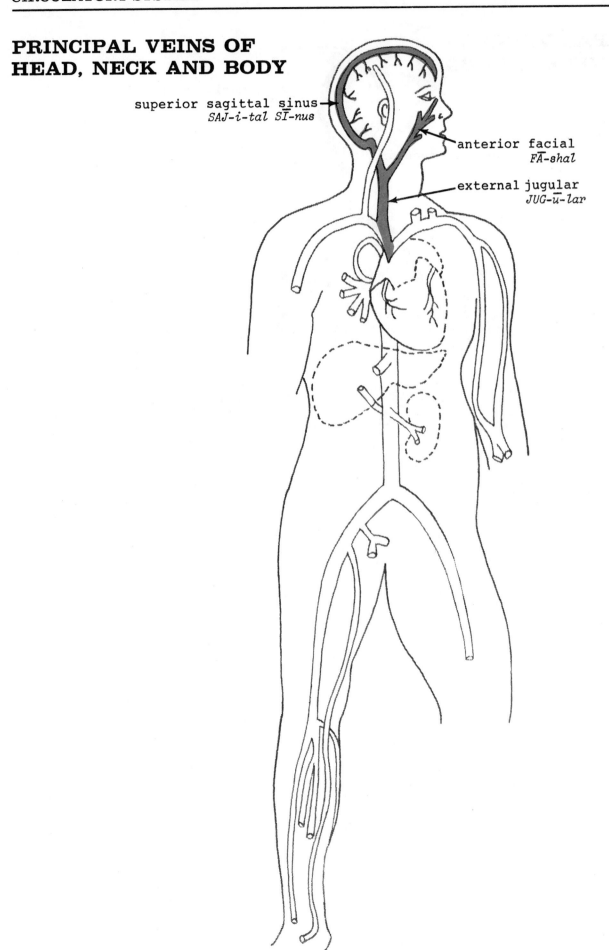

superior sagittal sinus
SAJ-i-tal SĪ-nus

anterior facial
FĀ-shal

external jugular
JUG-ū-lar

PRINCIPAL VEINS OF
HEAD, NECK AND BODY

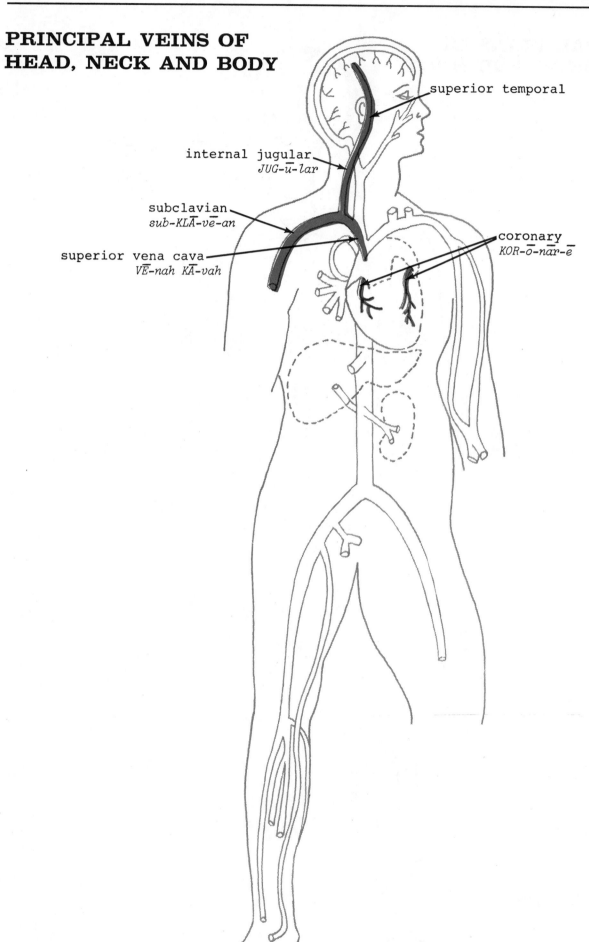

superior temporal

internal jugular
JUG-u̅-lar

subclavian
sub-KLA̅-ve̅-an

superior vena cava
VE̅-nah KA̅-vah

coronary
KOR-o̅-nar-e̅

PRINCIPAL VEINS OF
HEAD, NECK AND BODY

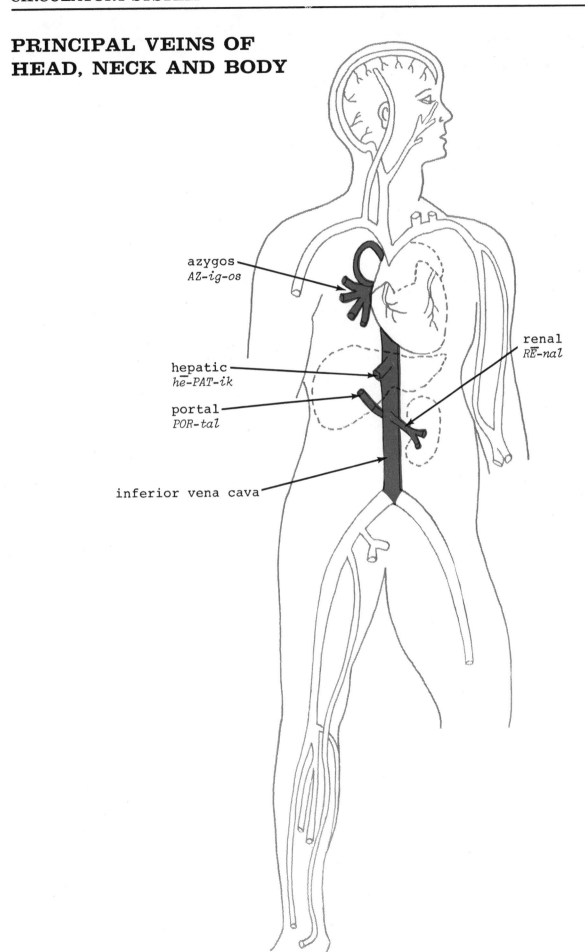

azygos
AZ-ig-os

hepatic
hē-PAT-ik

portal
POR-tal

inferior vena cava

renal
RĒ-nal

PRINCIPAL VEINS OF
HEAD, NECK AND BODY

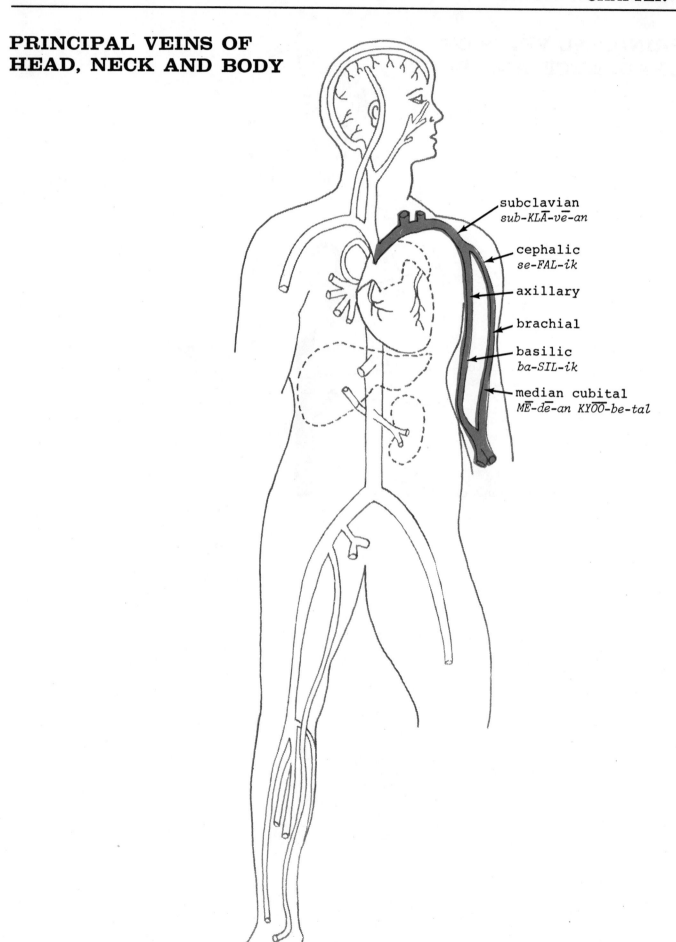

subclavian
sub-KLĀ-vē-an

cephalic
se-FAL-ik

axillary

brachial

basilic
ba-SIL-ik

median cubital
MĒ-dē-an KYOO-be-tal

PRINCIPAL VEINS OF HEAD, NECK AND BODY

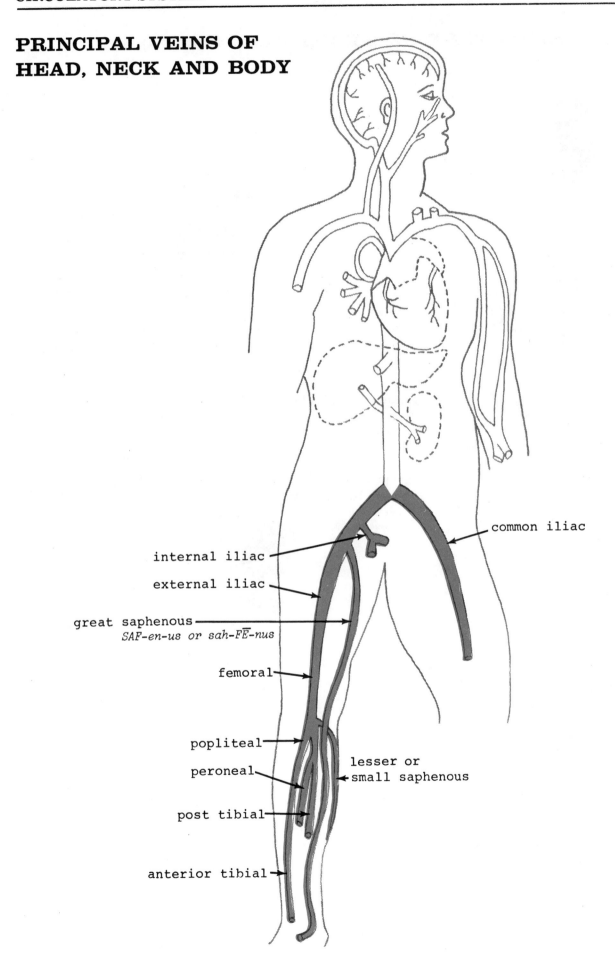

common iliac

internal iliac

external iliac

great saphenous
SAF-en-us or sah-FĒ-nus

femoral

popliteal

peroneal

post tibial

anterior tibial

lesser or small saphenous

STRUCTURE OF HEART (Cross Section)

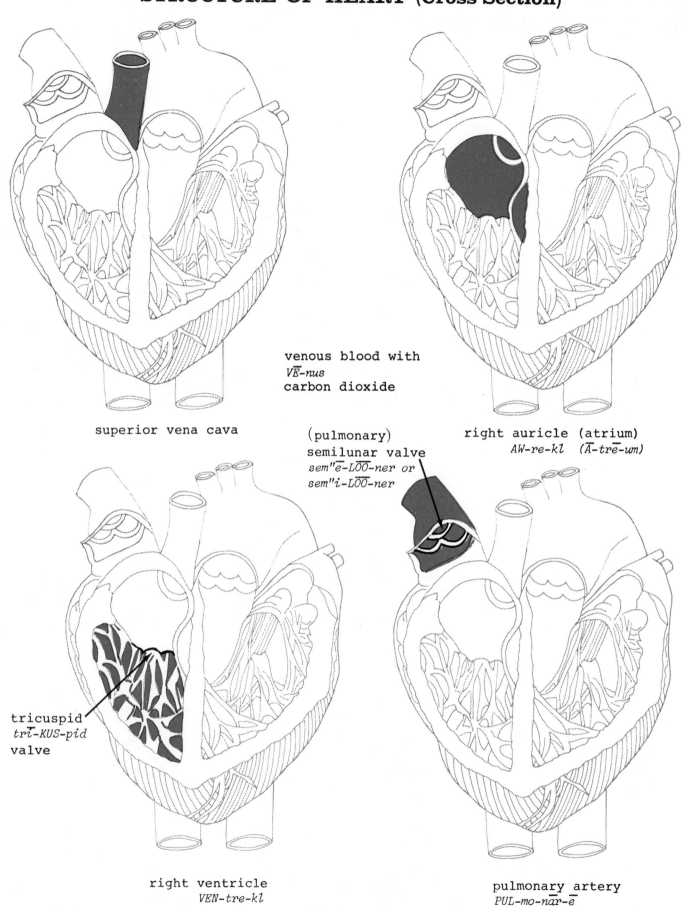

venous blood with
VĒ-nus
carbon dioxide

superior vena cava

(pulmonary)
semilunar valve
sem"ē-LOO-ner or
sem"i-LOO-ner

right auricle (atrium)
AW-re-kl (Ā-trē-um)

tricuspid
trī-KUS-pid
valve

right ventricle
VEN-tre-kl

pulmonary artery
PUL-mo-nār-ē

STRUCTURE OF HEART (Cross Section)

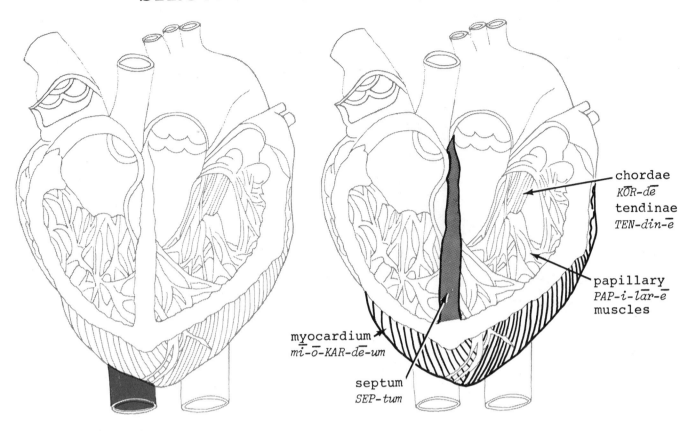

chordae
KOR-de
tendinae
TEN-din-e

papillary
PAP-i-lar-e
muscles

myocardium
mi-o-KAR-de-um

septum
SEP-tum

inferior vena cava

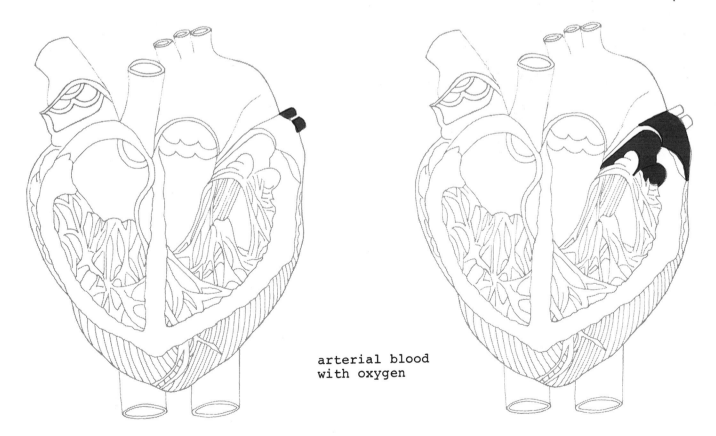

arterial blood
with oxygen

left pulmonary veins   left auricle

STRUCTURE OF HEART (Cross Section)

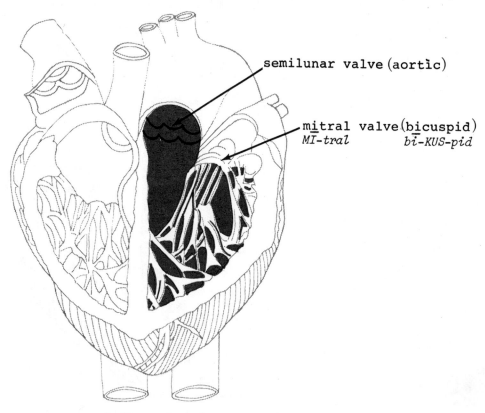

semilunar valve (aortic)

mitral valve (bicuspid)
MĪ-tral   *bī-KUS-pid*

left ventricle

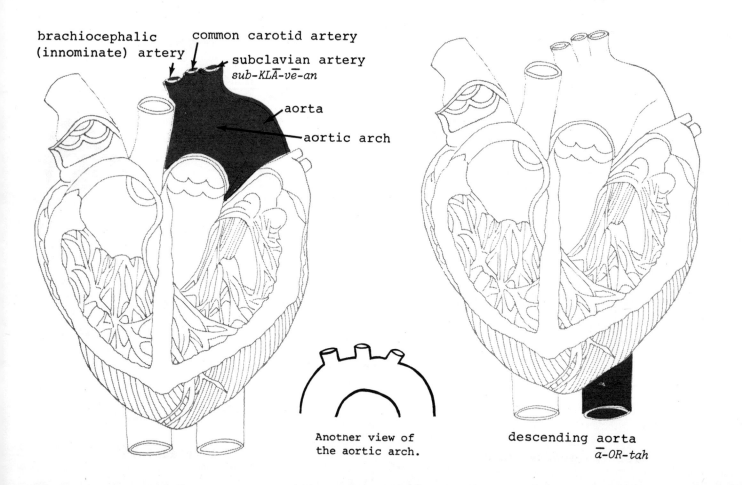

brachiocephalic
(innominate) artery

common carotid artery

subclavian artery
sub-KLĀ-vē-an

aorta

aortic arch

Another view of
the aortic arch.

descending aorta
ā-OR-tah

CIRCULATION OF BLOOD THROUGH THE HEART

The blood enters the right auricle from the superior and inferior venae cavae. Contraction of the right auricle expresses blood into the right ventricle from where it is forced by contraction of the ventricle through the semilunar valve into the pulmonary artery. Upon its return from the lungs where it gave up its carbon dioxide the blood is now oxygenated and enters the left auricle by way of the pulmonary veins. It is then forced into the left ventricle and from there into the aorta.

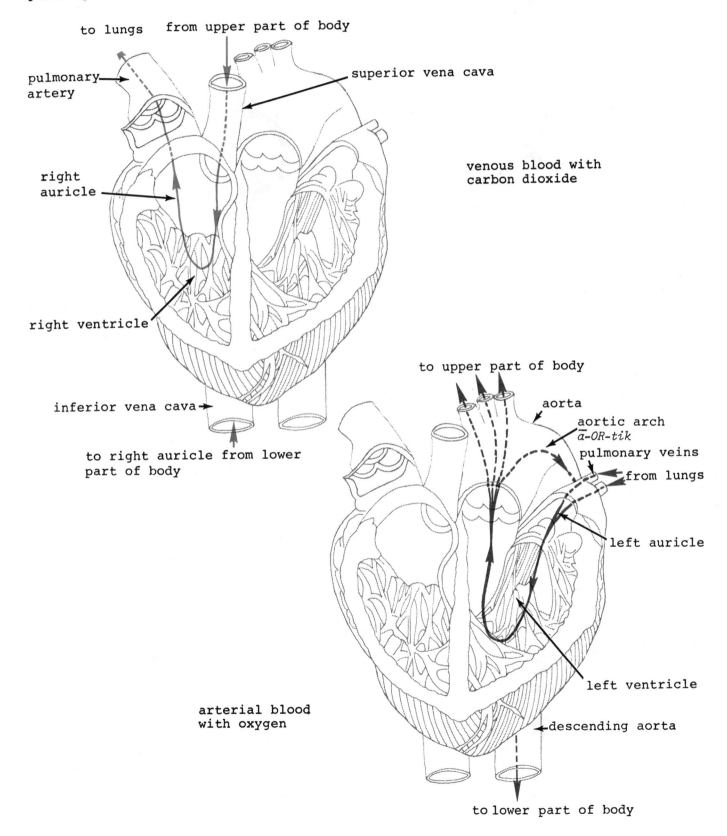

to lungs

from upper part of body

pulmonary artery

superior vena cava

right auricle

venous blood with carbon dioxide

right ventricle

inferior vena cava

to right auricle from lower part of body

to upper part of body

aorta

aortic arch
ā-OR-tik

pulmonary veins

from lungs

left auricle

left ventricle

arterial blood with oxygen

descending aorta

to lower part of body

SIMPLIFIED VERSION OF CIRCULATION
(arrow indicates flow of blood)

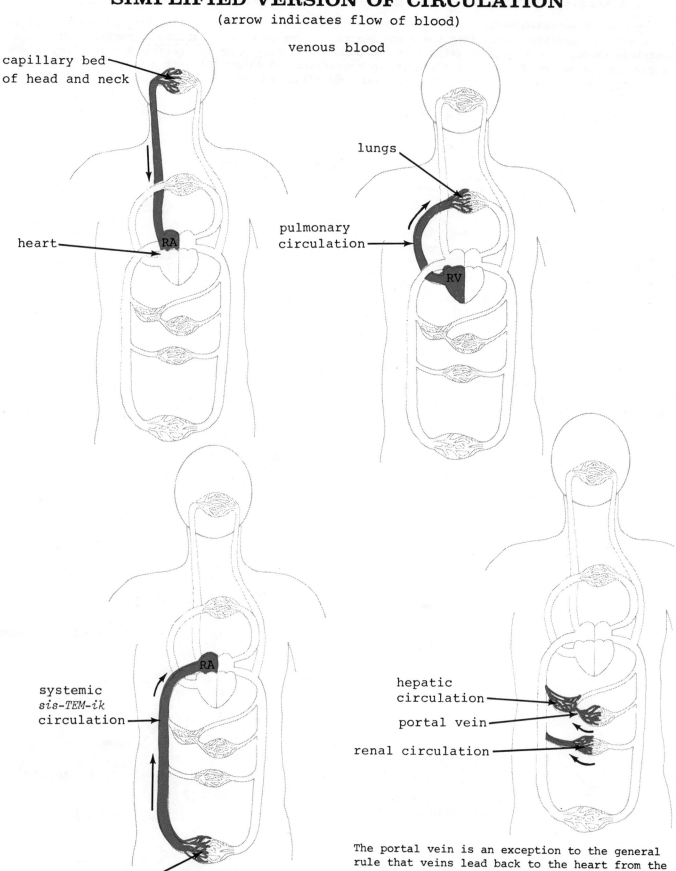

The portal vein is an exception to the general
rule that veins lead back to the heart from the
capillaries. This vein leads from the digestive
organs to the liver where it breaks down into
capillaries again.

SIMPLIFIED VERSION OF CIRCULATION
(arrow indicates flow of blood)

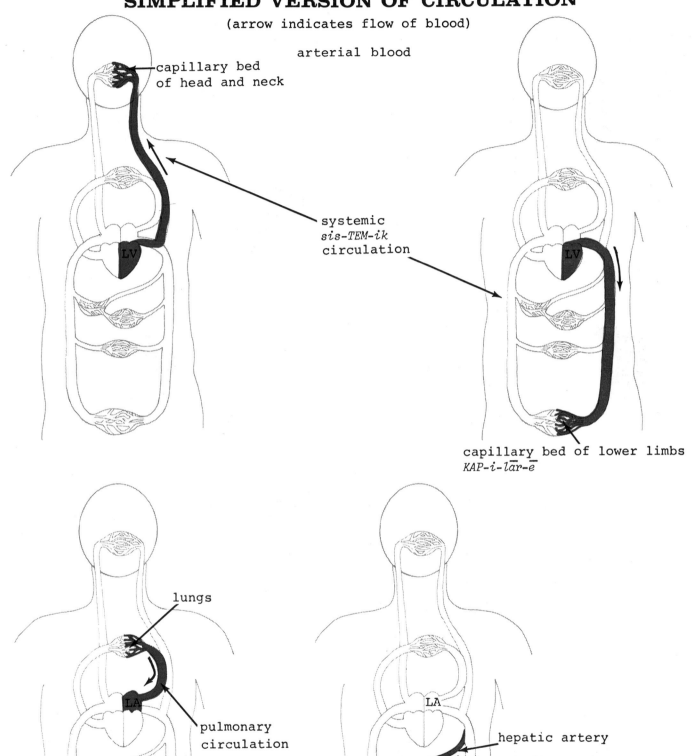

arterial blood

capillary bed
of head and neck

systemic
sis-TEM-ik
circulation

capillary bed of lower limbs
KAP-i-lār-ē

lungs

pulmonary
circulation

hepatic artery

arteries to
gastro-intestinal
tract

renal circulation

CORONARY CIRCULATION – ARTERIAL

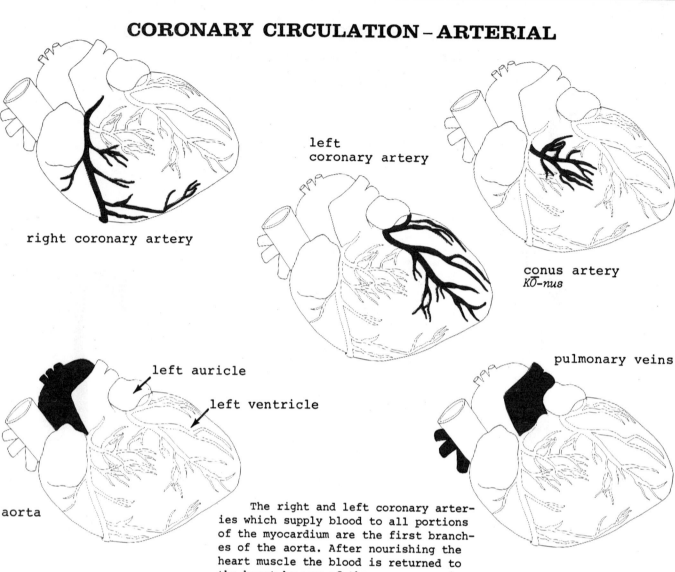

right coronary artery

left
coronary artery

conus artery
KŌ-nus

left auricle

left ventricle

pulmonary veins

aorta

The right and left coronary arteries which supply blood to all portions of the myocardium are the first branches of the aorta. After nourishing the heart muscle the blood is returned to the heart by way of the coronary veins which merge to form a coronary blood sinus from which the venous blood is delivered to the right auricle.

CORONARY CIRCULATION
VENOUS

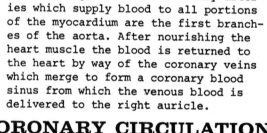

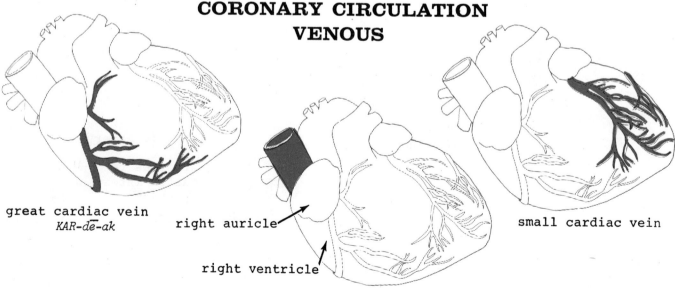

great cardiac vein
KAR-dē-ak

right auricle

right ventricle

small cardiac vein

superior vena cava

NORMAL BLOOD CELLS

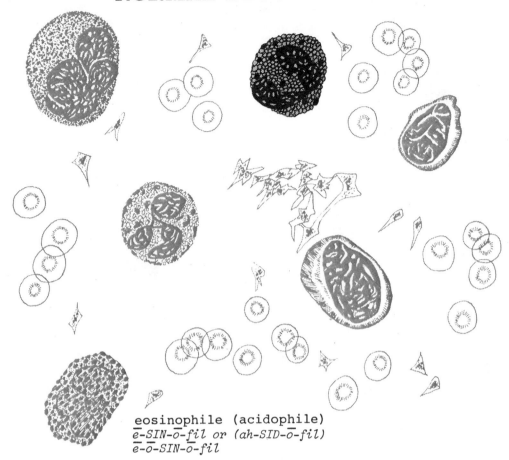

eosinophile (acidophile)
ē-SIN-ō-fil or (ah-SID-ō-fil)
ē-ō-SIN-ō-fil

MICROSCOPIC COMPOSITION OF BLOOD

Blood is made up of cells or corpuscles and plasma, a liquid.

One of the blood cell groups is the erythrocytes or red blood cells (RBC) which contain hemoglobin. It is the hemoglobin which has the power to combine with oxygen. The oxygen is given up in the tissues and replaced by carbon dioxide given off by the tissues. The carbon dioxide is then released by the blood in the lungs and replaced by oxygen.

There are approximately 20 trillion erythrocytes in the human blood. This number varies with altitude, nutrition, the constitution of the individual, temperature, the age of the individual and even before and after meals.

Another group of blood cells are the leukocytes or white blood cells (WBC). They are classified into different types of cells: lymphocytes, monocytes, and granulocytes. The granulocytes are further broken down into neutrophiles*, eosinophiles* and baso-philes*.

The most important function of the white cells is their ability to help protect the body against infection.

The third group of cells is the blood platelets or thrombocytes. Their most important function is to aid in the clotting of the blood.

The plasma of the blood is a clear amber color and is about nine-tenths water. The plasma serves as a source of nutrition and as a means of removing waste products from the tissues.

*These terms may also be spelled neutrophils, eosinophils and basophils.

NORMAL BLOOD CELLS

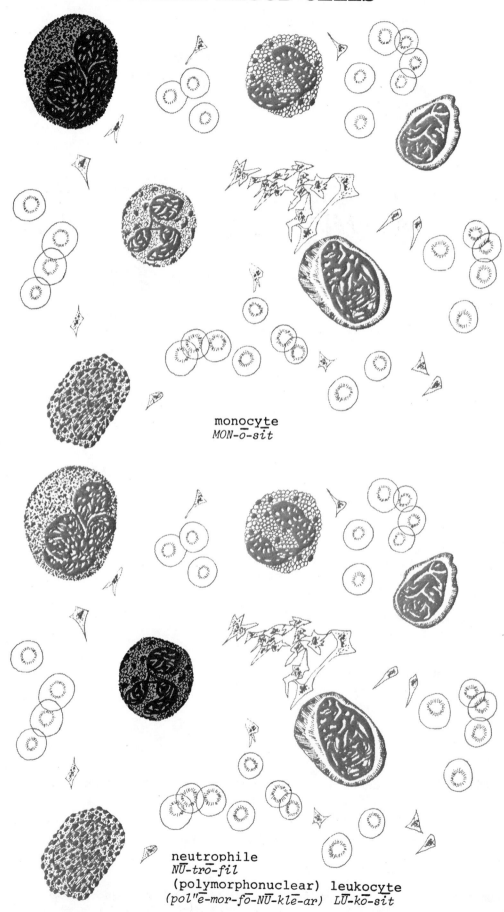

monocyte
MON-ō-sĭt

neutrophile
NŪ-trō-fĭl
(polymorphonuclear) leukocyte
(pol"ē-mor-fō-NŪ-klē-ar) LŪ-kō-sĭt

NORMAL BLOOD CELLS

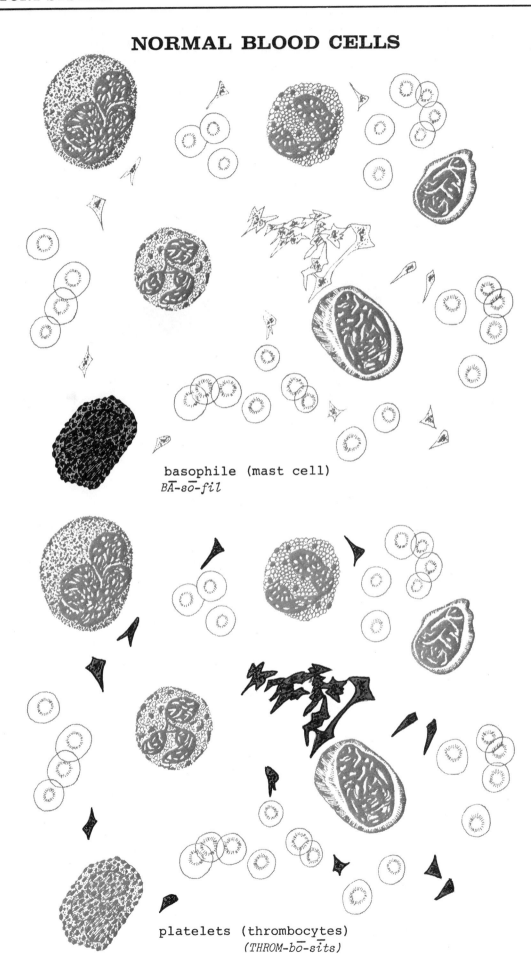

basophile (mast cell)
BĀ-sō-fil

platelets (thrombocytes)
(THROM-bō-sĭts)

NORMAL BLOOD CELLS

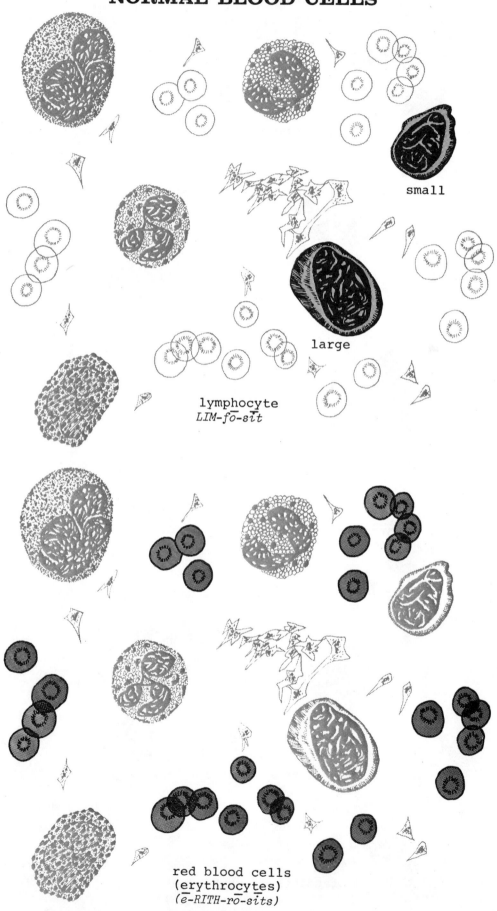

small

large

lymphocyte
LIM-fō-sĭt

red blood cells
(erythrocytes)
(*ē-RITH-rō-sĭts*)

LYMPHATIC SYSTEM
lim-FAT-ik

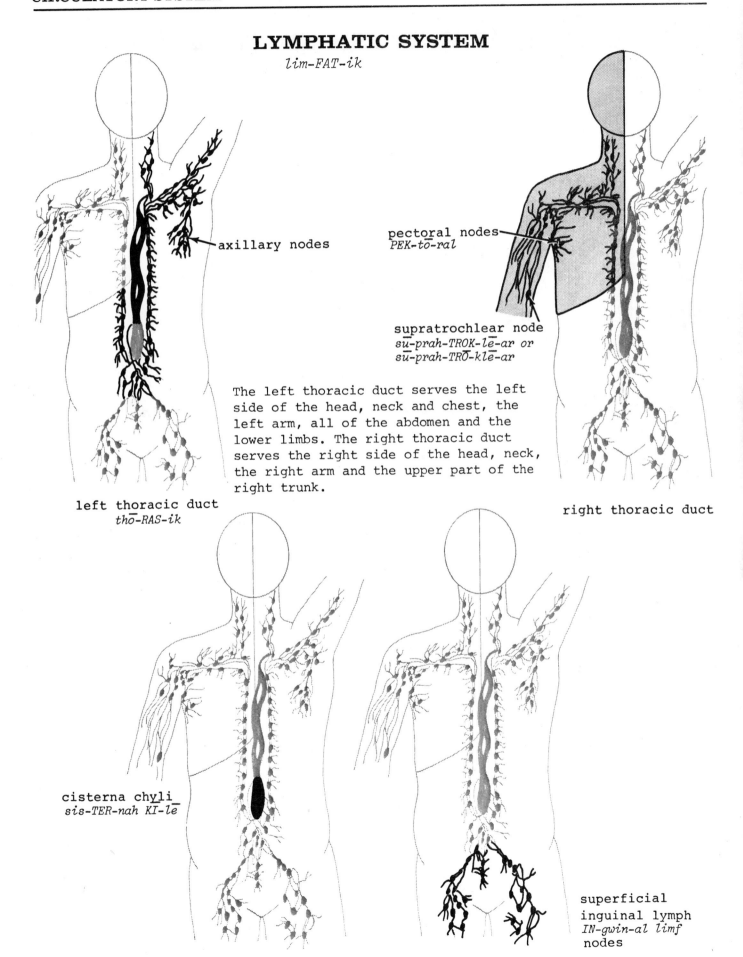

axillary nodes

pectoral nodes
PEK-tō-ral

supratrochlear node
sū-prah-TROK-lē-ar or
sū-prah-TRŌ-klē-ar

The left thoracic duct serves the left
side of the head, neck and chest, the
left arm, all of the abdomen and the
lower limbs. The right thoracic duct
serves the right side of the head, neck,
the right arm and the upper part of the
right trunk.

left thoracic duct
thō-RAS-ik

right thoracic duct

cisterna chyli
sis-TER-nah KI-lē

superficial
inguinal lymph
IN-gwin-al limf
nodes

INDEX OF ENGLISH TRANSLATION

abdomin/al
belly

acido/phile
sour, acid   to love

aorta
to lift up

aort/ic
aorta

arteri/al
artery

atrium
chamber

aur/icle
ear   small

axill/ary
armpit

a/zygos
not   a yoke

basil/ic
royal

baso/phile
base   to love

brachi/al
arm

brachio/cephal/ic
arm   head

capillary
hairlike

cardi/ac
heart

carotid
to put to sleep

celi/ac
belly

cephal/ic
head

cerebr/al
brain

cisterna
cistern

chordae
cords

chyli
juice

conus
cone

coron/ary
crown

cubit/al
ulna

eosino/phile
dawn   to love

erythro/cytes
red   cell

faci/al
face

femor/al
femur (thigh)

gastro/-intestin/al
stomach   intestines

hepat/ic
liver

ili/ac
hip bone

inguin/al
groin

in/nominate
not   name
(nameless)

jugul/ar
neck

leuko/cyte
white   cell

lymph
water

lympho/cyte
water   cell

maxill/ary
jawbone

medi/an
middle

mesenter/ic
mesentery

mitr/al
miter
(a tall cap resembling a pointed arch)

mono/cyte
single   cell

myo/cardium
muscle   heart

neutro/phile
neither   to love

ophthalm/ic
eye

papill/ary
nipple

pector/al
breast bone

perone/al
fibula

poly/morpho/nucle/ar
many   shape   nucleus

poplite/al
back of knee

port/al
gate

pulmon/ary
lung

radi/al
radius

ren/al
kidney

sagitt/al
straight, arrow

saphen/ous
manifest

semi/lun/ar
half   moon

septum
wall off

sinus
hollow

sub/clavi/an
below   clavicle adjective

supra/trochle/ar
above   trochlea

system/ic
affecting the whole body

tempor/al
time, temple

tendinae
tendons

thorac/ic
chest

thrombo/cytes
clot   cells

tibi/al
shin bone

tri/cuspid
three   point

uln/ar
ulna

vena cava
vein   hollow

ven/ous
vein

ventr/icle
belly   small

vertebr/al
vertebra

CHAPTER VI

ENDOCRINE SYSTEM

To the original seven endocrine glands (also called the ductless glands) — the thyroid, parathyroids, adrenals, pituitary, pineal*, thymus** and the gonads (ovaries and testes) has been added the islet (or island) cells of Langerhans of the pancreas. These cells secrete insulin.

All of these glands together weigh between four and seven ounces. Yet they are of vital importance in the regulation of all the other cells of our bodies.

While a great deal is now known about these glands and their secretions there is still much to be learned. Their over- or under-production can create some baffling and long-lasting conditions.

Of great importance to the well being of the body are the hormones produced by most of the glands. The major hormones are as follows:

pituitary body (hypophysis):

    anterior lobe: somatotrophin (STH), adrenocorticotrophin (ACTH), follicle stimulating hormone (FSH), luteinizing hormone (LH) or interstitial cell-stimulating hormone (ICSH), prolactin

    posterior lobe: pituitrin, vasopressin or pitressin

thyroid: thyroxin

parathyroids: parathyrin or parathormone

island cells of Langerhans: insulin

adrenal glands (suprarenal glands):

    cortex: aldosterone, cortisone

    medulla: epinephrin (adrenalin)

ovary: estrogen, progesterone

testis: testosterone

* see page 198

** see page 201

ENDOCRINE SYSTEM SHOWING LOCATION OF GLANDS

EN-dō-crin

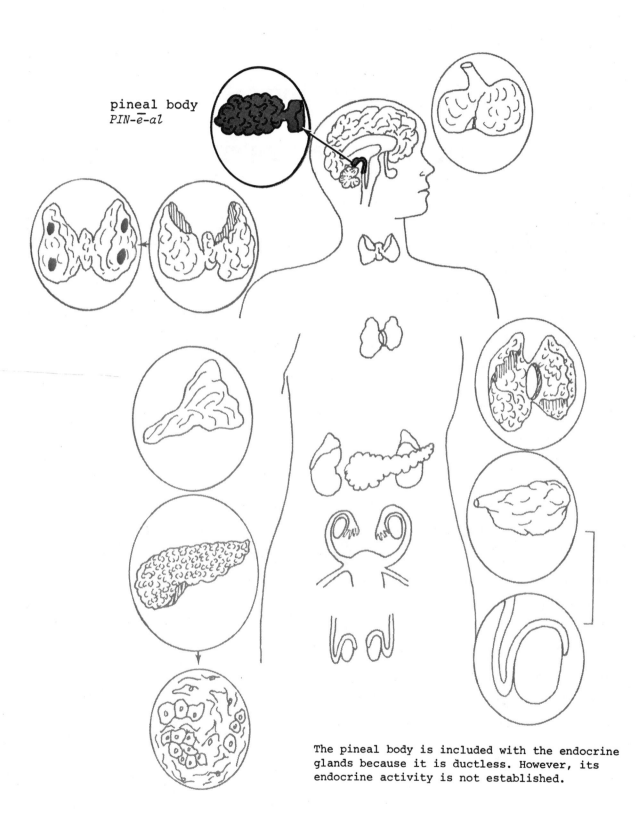

pineal body
PIN-ē-al

The pineal body is included with the endocrine
glands because it is ductless. However, its
endocrine activity is not established.

ENDOCRINE SYSTEM SHOWING LOCATION OF GLANDS

pituitary body (hypophysis)
pi-TŪ-i-tār-ē   *(hī-POF-i-sis)*

posterior lobe   anterior lobe

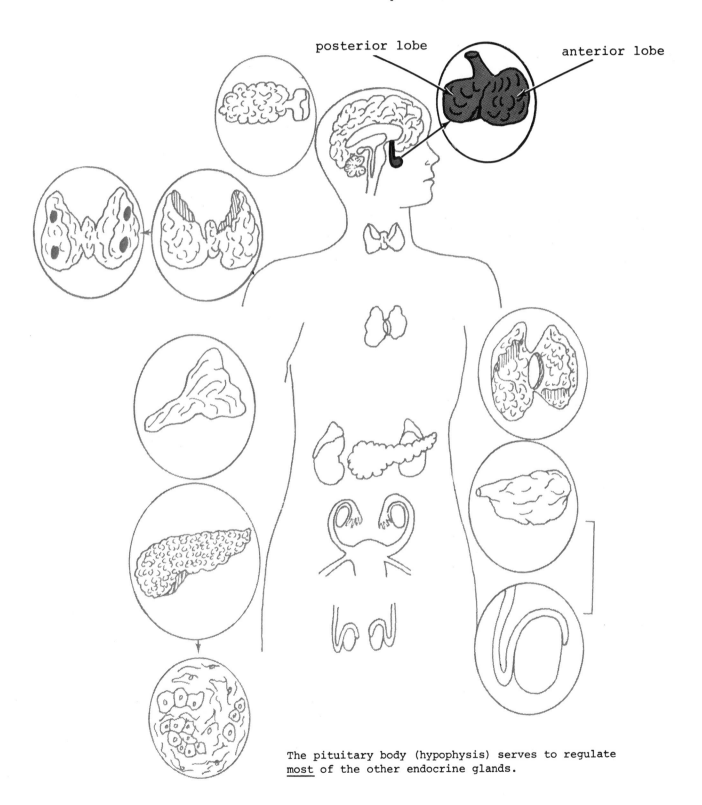

The pituitary body (hypophysis) serves to regulate <u>most</u> of the other endocrine glands.

ENDOCRINE SYSTEM SHOWING LOCATION OF GLANDS

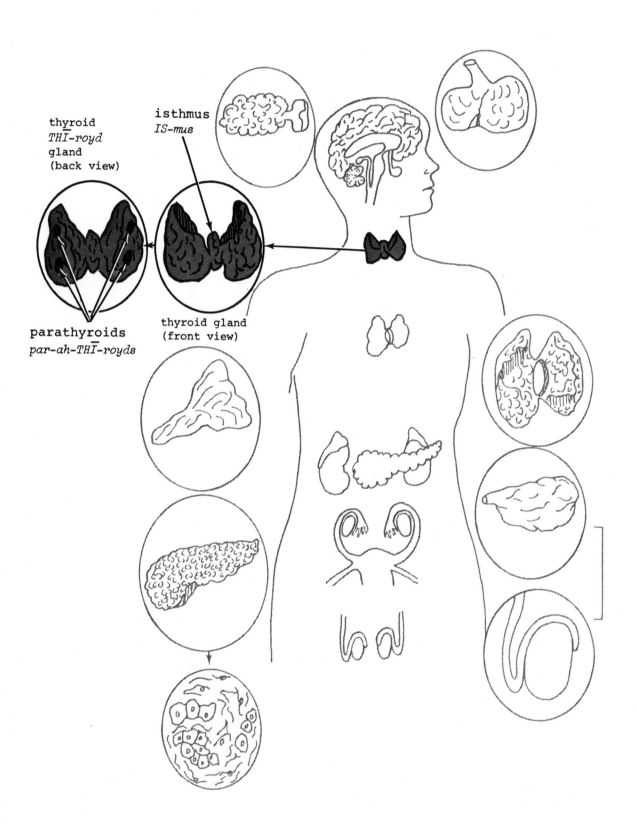

thyroid
THĪ-royd
gland
(back view)

isthmus
IS-mus

parathyroids
par-ah-THĪ-royds

thyroid gland
(front view)

ENDOCRINE SYSTEM SHOWING LOCATION OF GLANDS

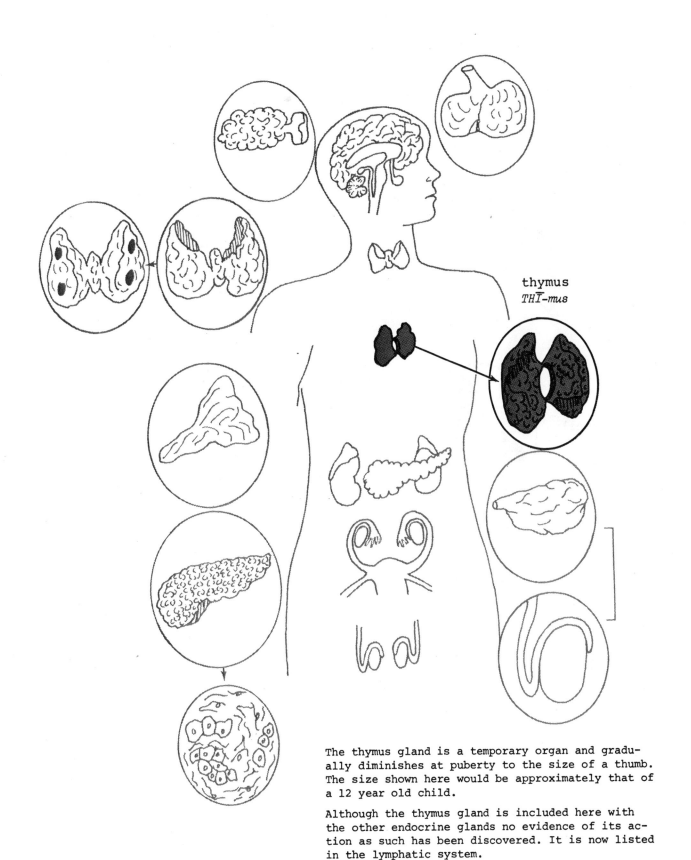

thymus
THĪ-mus

The thymus gland is a temporary organ and gradu-
ally diminishes at puberty to the size of a thumb.
The size shown here would be approximately that of
a 12 year old child.

Although the thymus gland is included here with
the other endocrine glands no evidence of its action as such has been discovered. It is now listed
in the lymphatic system.

ENDOCRINE SYSTEM SHOWING LOCATION OF GLANDS

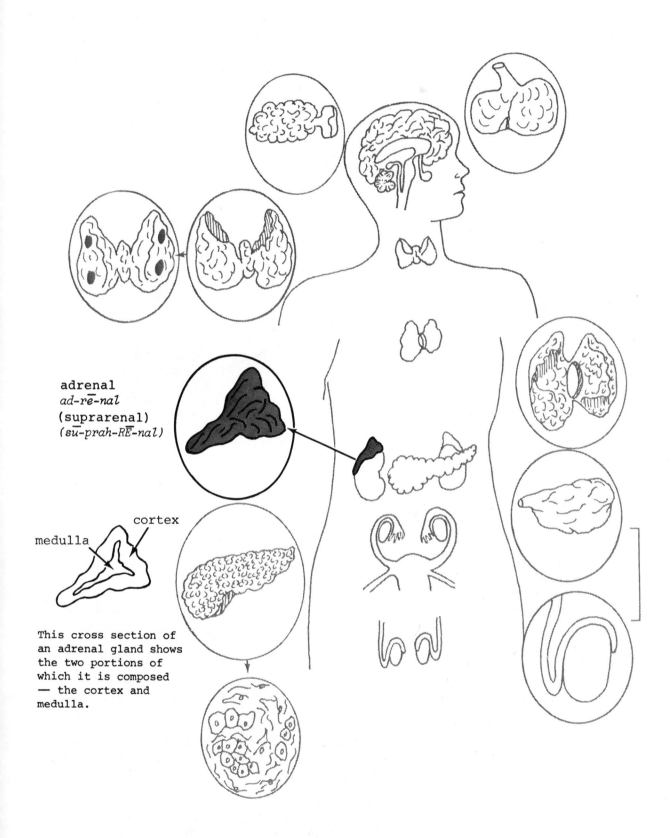

adrenal
ad-rē-nal
(suprarenal)
(sū-prah-RĒ-nal)

medulla   cortex

This cross section of
an adrenal gland shows
the two portions of
which it is composed
— the cortex and
medulla.

ENDOCRINE SYSTEM SHOWING LOCATION OF GLANDS

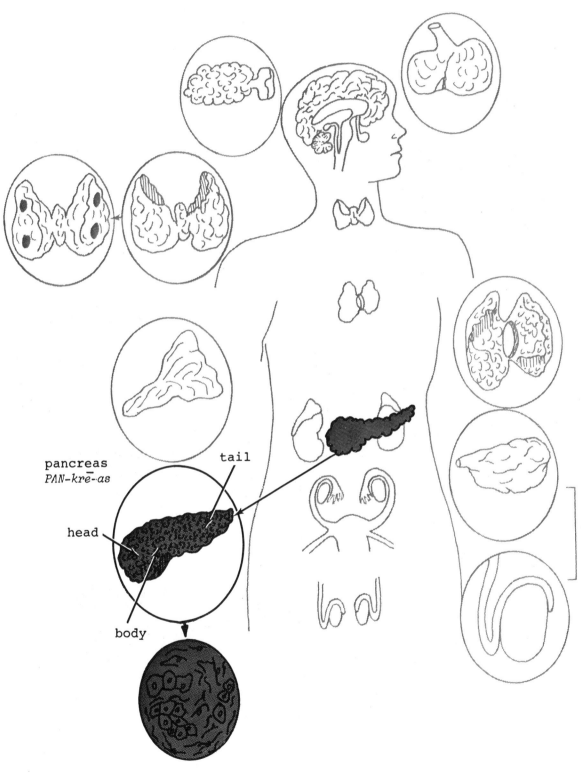

pancreas
PAN-kre̅-as

tail

head

body

islands of Langerhans
LAHNG-er-hanz

ENDOCRINE SYSTEM SHOWING LOCATION OF GLANDS

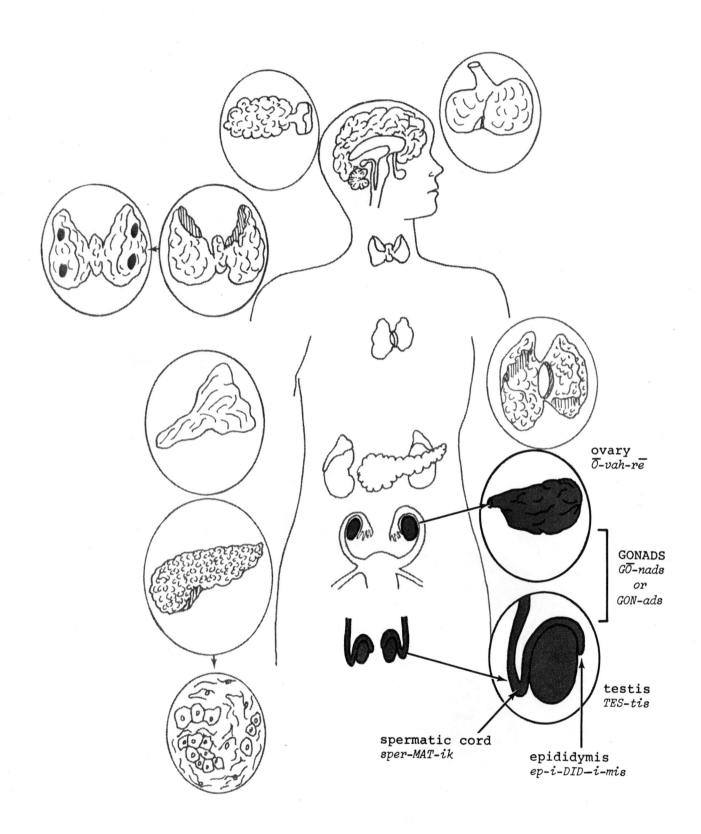

ovary
Ō-vah-rē

GONADS
GŌ-nads
or
GON-ads

testis
TES-tis

spermatic cord
sper-MAT-ik

epididymis
ep-i-DID-i-mis

INDEX OF ENGLISH TRANSLATION

ad/ren/al
near kidney

epi/didymis
on, upon   twin

gonads
from seed

hypo/physis
below   to grow

isthmus
adjoining strip

Langerhans
after a 19th Century German pathologist

ovary
egg bearer

pan/creas
all   flesh

para/thyr/oids
beside   shield

pine/al
pine cone

pituit/ary
phlegm

spermat/ic
seed

supra/ren/al
above   kidney

testis
shell

thymus
excrescence

thyr/oid
shield

CHAPTER VII

RESPIRATORY SYSTEM

In this age of air pollution everyone is considerably more aware of this particular system than they were decades ago. Compounded with air pollution is the complication of smoking with its attendant danger to this system.

Starting in the skull with the sinuses, the nose and then down through the larynx, trachea, bronchi and lungs any bacterial or fungus invasion can create temporary and, sometimes, permanent disability.

Without the lungs the blood would not be able to give up the tissue waste of carbon dioxide and take on the fresh oxygen needed to replenish the cells. To make this as effective as possible the lungs are placed in communication with the nose and mouth by means of the bronchi, trachea and larynx.

Our ability to speak is possible through the use of the vocal cords in the larynx, the nose helps in phonation, and the tongue aids in articulation.

RESPIRATORY ORGANS

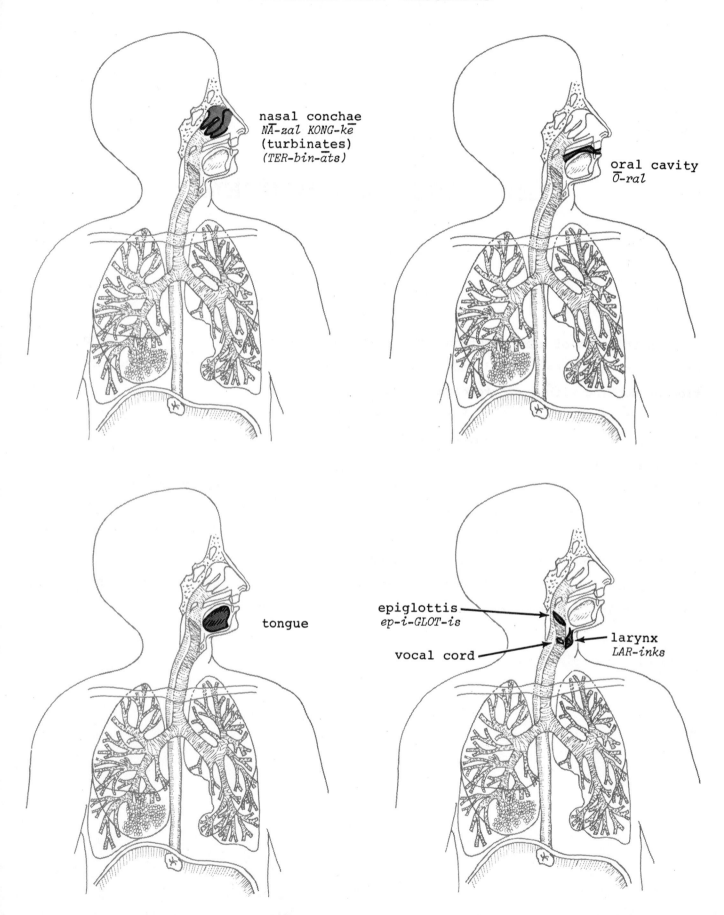

nasal conchae
NĀ-zal KONG-kē
(turbinates)
(TER-bin-āts)

oral cavity
Ō-ral

tongue

epiglottis
ep-i-GLOT-is

vocal cord

larynx
LAR-inks

RESPIRATORY ORGANS

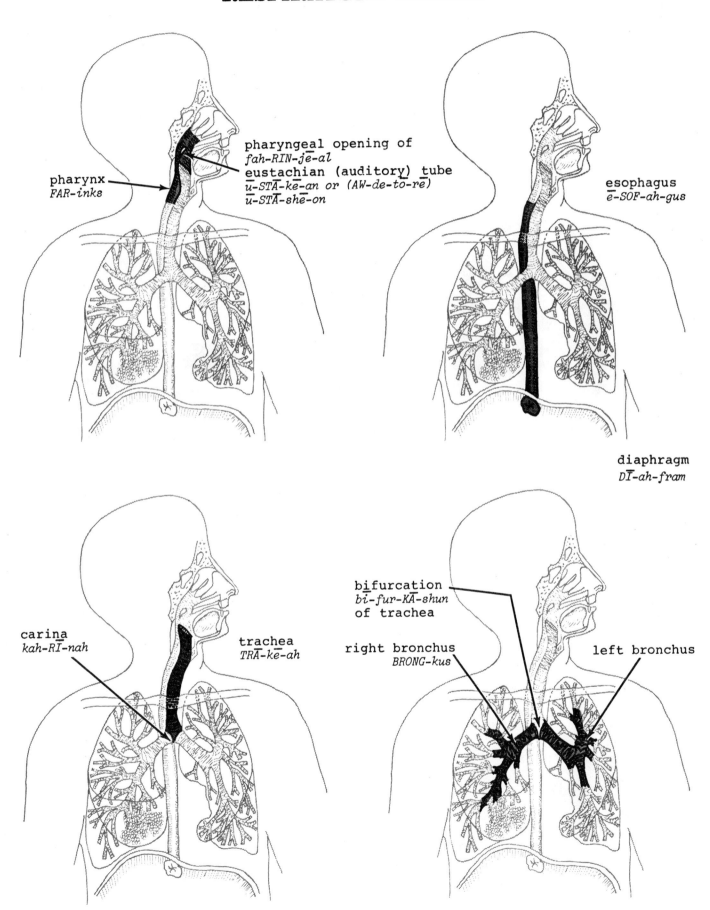

pharynx
FAR-inks

pharyngeal opening of
fah-RIN-jē-al
eustachian (auditory) tube
u-STĀ-kē-an or (AW-de-to-rē)
u-STĀ-shē-on

esophagus
ē-SOF-ah-gus

diaphragm
DĪ-ah-fram

carina
kah-RĪ-nah

trachea
TRĀ-kē-ah

bifurcation
bi-fur-KĀ-shun
of trachea

right bronchus
BRONG-kus

left bronchus

RESPIRATORY ORGANS

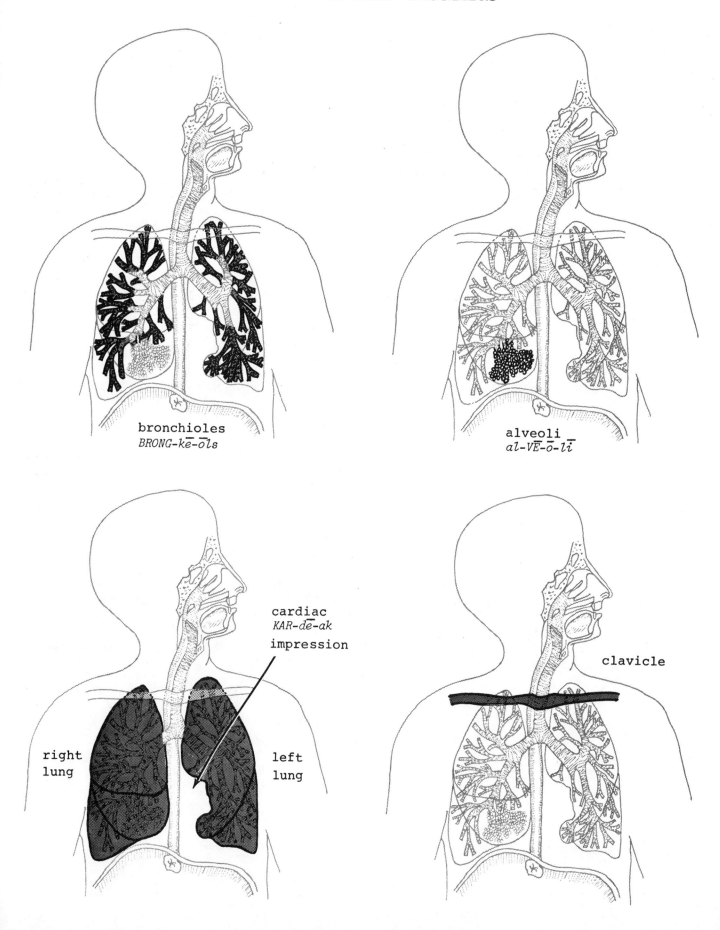

bronchioles
BRONG-kē-ōls

alveoli
al-VĒ-ō-lī

cardiac
KAR-dē-ak
impression

right
lung

left
lung

clavicle

UPPER RESPIRATORY ORGANS

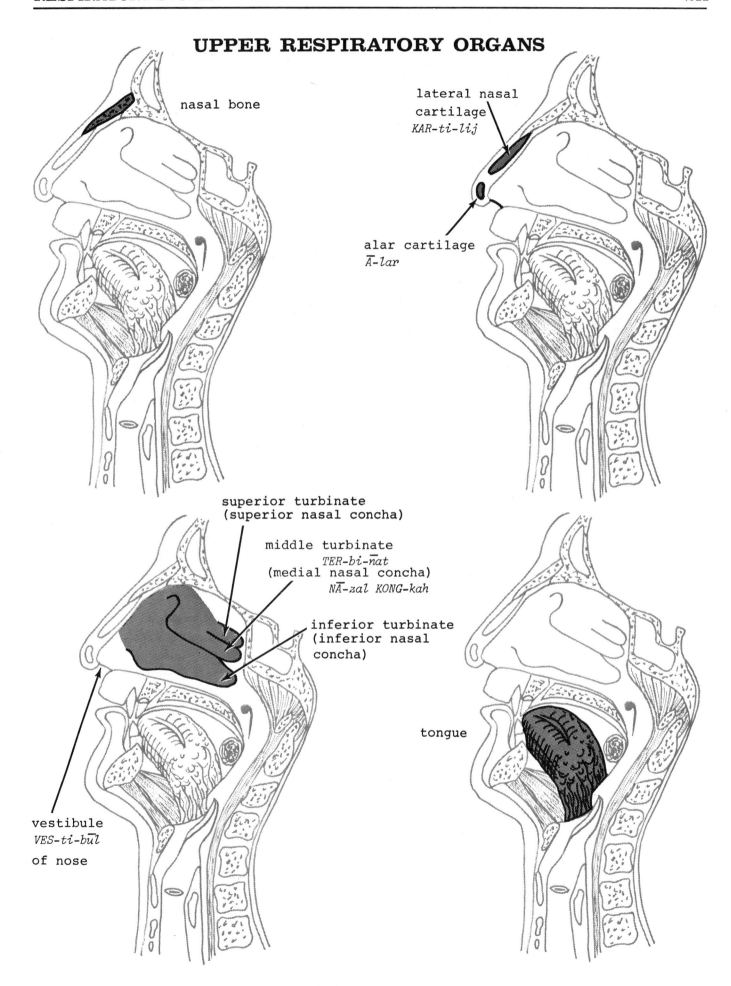

nasal bone

lateral nasal
cartilage
KAR-ti-lij

alar cartilage
Ā-lar

superior turbinate
(superior nasal concha)

middle turbinate
TER-bi-nāt
(medial nasal concha)
NĀ-zal KONG-kah

inferior turbinate
(inferior nasal
concha)

vestibule
VES-ti-bŭl
of nose

tongue

UPPER RESPIRATORY ORGANS

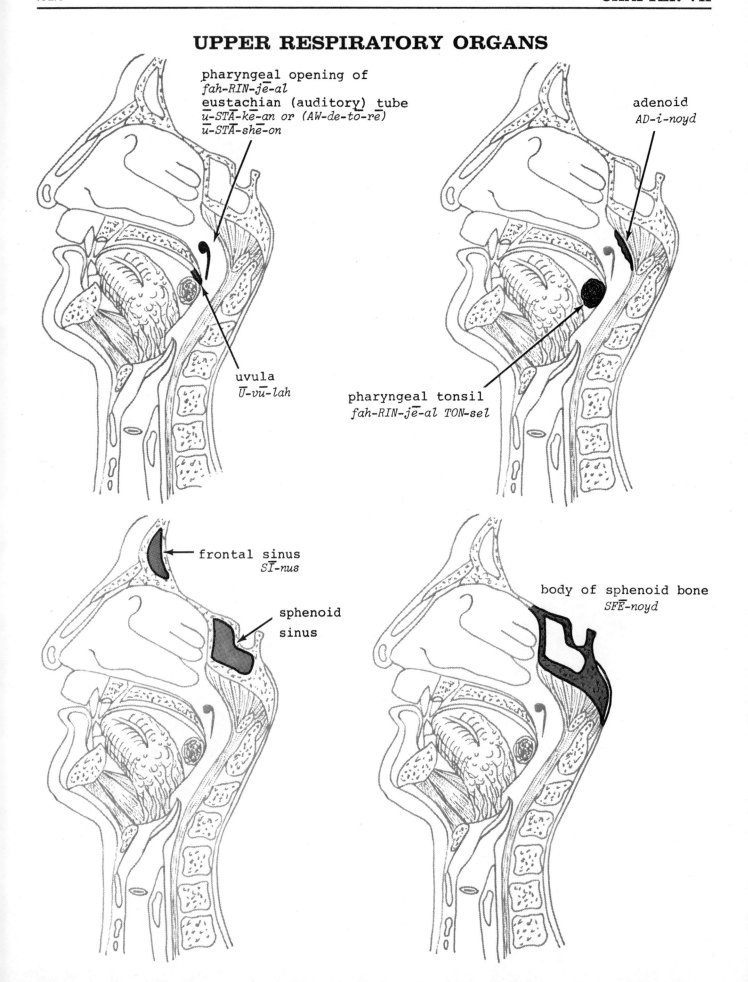

pharyngeal opening of
fah-RIN-jē-al
eustachian (auditory) tube
ū-STĀ-kē-an or (AW-de-tō-rē)
ū-STĀ-shē-on

adenoid
AD-i-noyd

uvula
Ū-vū-lah

pharyngeal tonsil
fah-RIN-jē-al TON-sel

frontal sinus
SĪ-nus

sphenoid
sinus

body of sphenoid bone
SFĒ-noyd

UPPER RESPIRATORY ORGANS

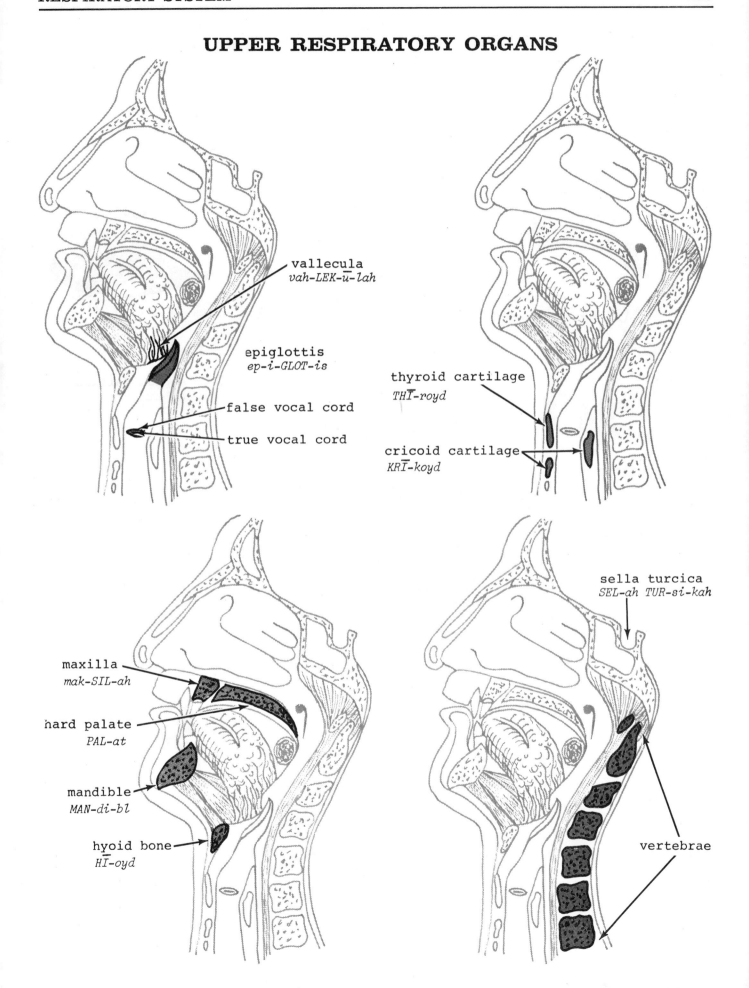

vallecula
vah-LEK-ū-lah

epiglottis
ep-i-GLOT-is

false vocal cord

true vocal cord

thyroid cartilage
THĪ-royd

cricoid cartilage
KRĪ-koyd

maxilla
mak-SIL-ah

hard palate
PAL-at

mandible
MAN-di-bl

hyoid bone
HĪ-oyd

sella turcica
SEL-ah TUR-si-kah

vertebrae

SEPTUM OF NOSE (Right Side)

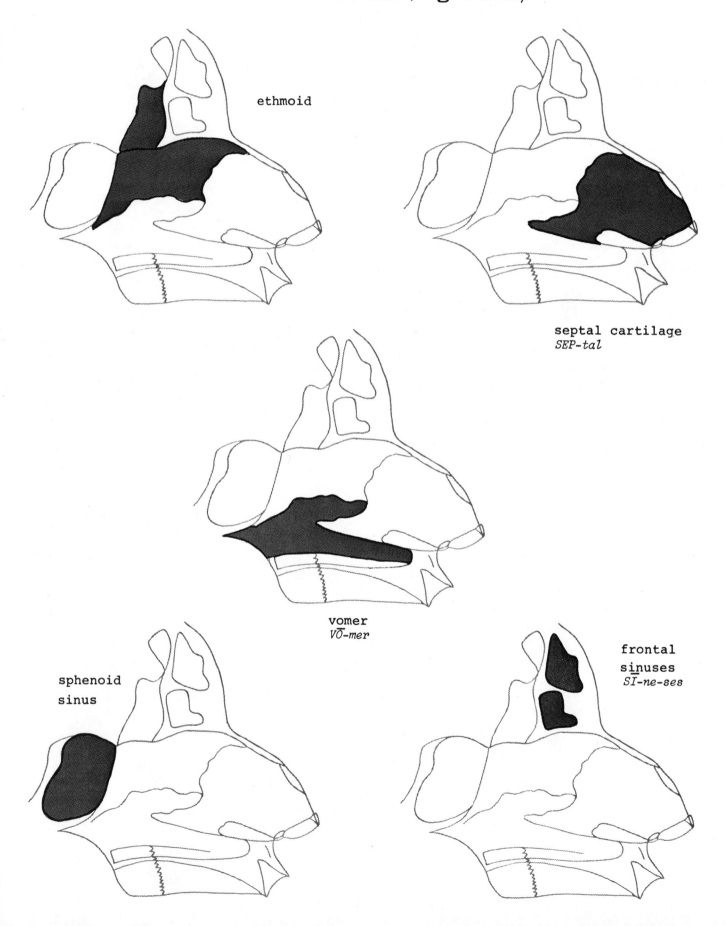

ethmoid

septal cartilage
SEP-tal

vomer
VŌ-mer

sphenoid
sinus

frontal
sinuses
SĪ-ne-ses

SEPTUM OF NOSE (Right Side)

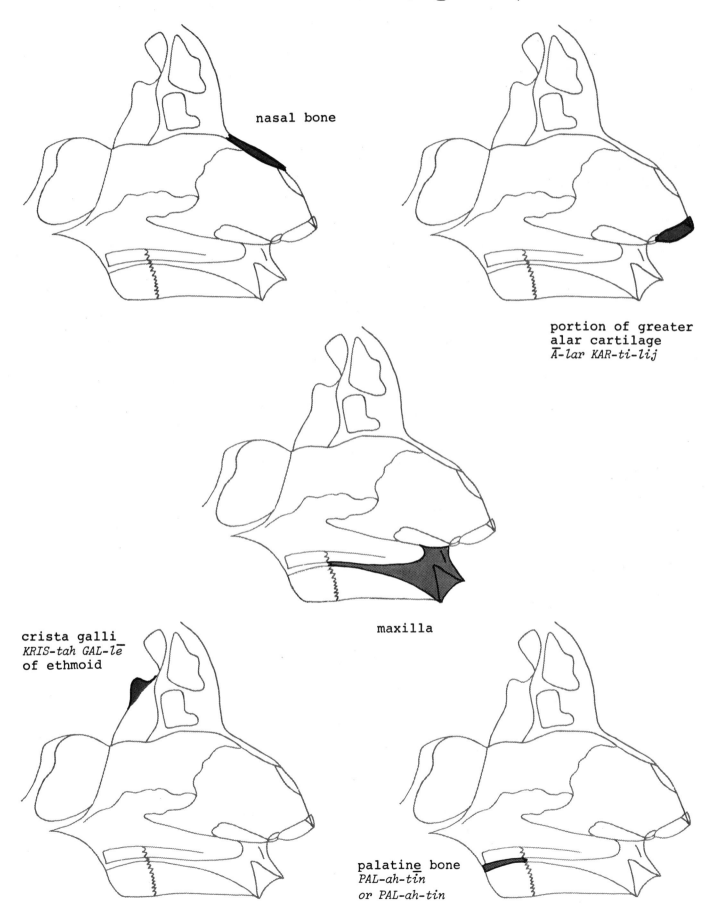

nasal bone

portion of greater
alar cartilage
Ā-lar KAR-ti-lij

maxilla

crista galli
KRIS-tah GAL-lē
of ethmoid

palatine bone
PAL-ah-tĭn
or *PAL-ah-tin*

CARTILAGE OF NOSE (Inferior View)

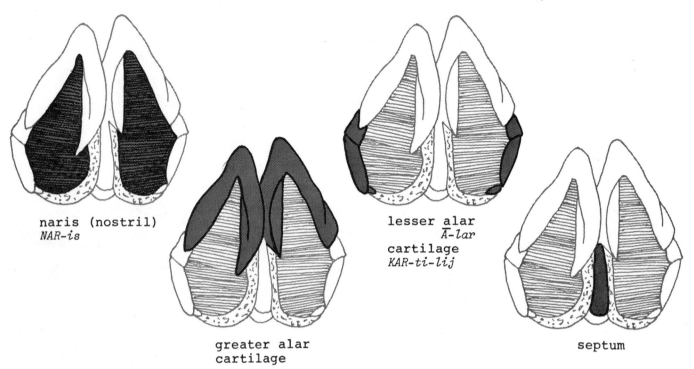

naris (nostril)
NAR-is

greater alar
cartilage

lesser alar
Ā-lar
cartilage
KAR-ti-lij

septum

NASAL CAVITIES (Coronal Section)

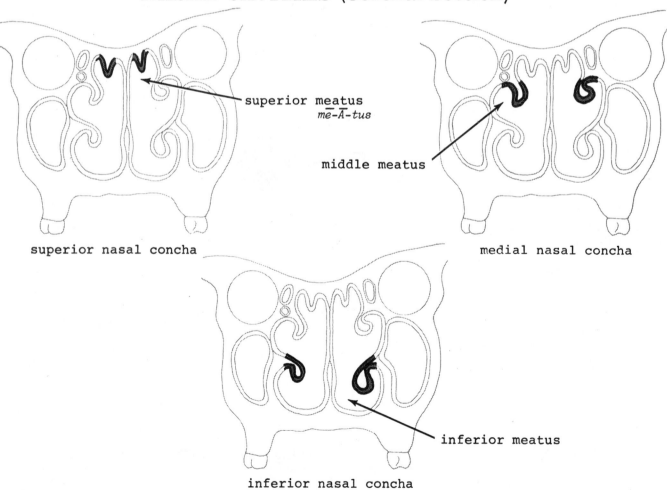

superior meatus
mē-Ā-tus

middle meatus

inferior meatus

superior nasal concha

medial nasal concha

inferior nasal concha

NASAL CAVITIES (Coronal Section)

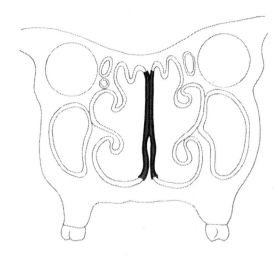

nasal septum

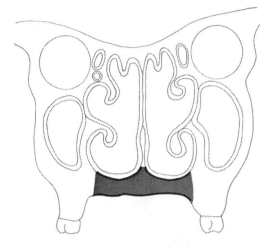

hard palate

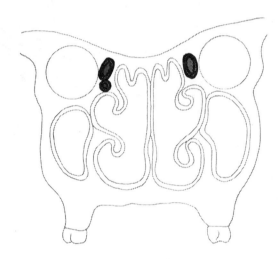

ethmoidal air cell
eth-MOYD-al

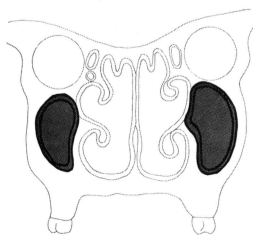

maxillary sinus
MAK-se-lar-e

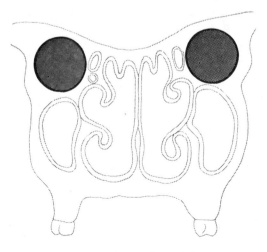

contents of orbit

NERVE SUPPLY OF NOSE (Right Side)

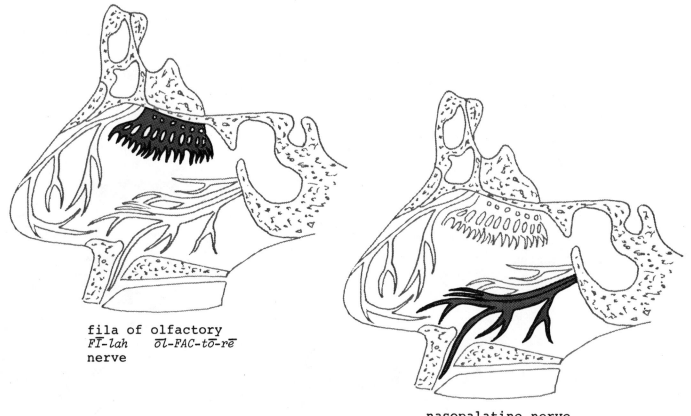

fila of olfactory
$F\overline{I}$-lah $\overline{ol}$-FAC-$t\overline{o}$-$r\overline{e}$
nerve

nasopalatine nerve
$n\overline{a}''z\overline{o}$-$PAL$-$ah$-$t\overline{i}n$ or
$n\overline{a}''z\overline{o}$-$PAL$-$ah$-$tin$

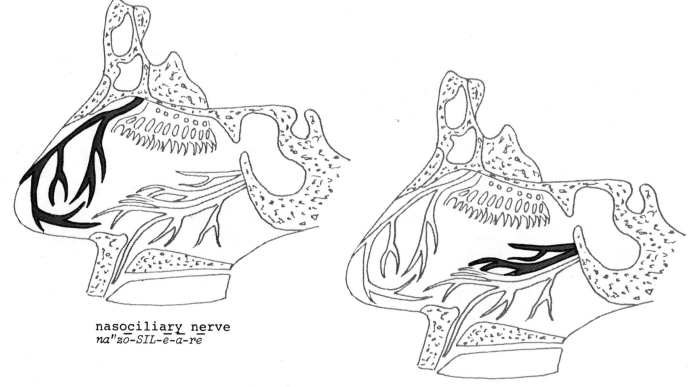

nasociliary nerve
$na''z\overline{o}$-SIL-$\overline{e}$-$\overline{a}$-$r\overline{e}$

posterior superior
nasal branches
nasopalatine nerve

THE LUNGS (Anterior View)

trachea
TRĀ-kē-ah

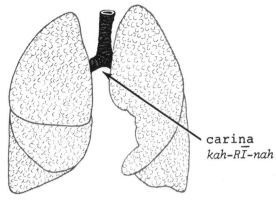

carina
kah-RĪ-nah

right lung   left lung

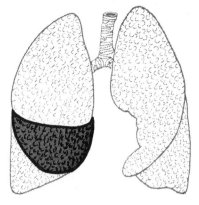

middle lobe

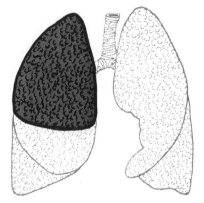

upper lobe
(superior)

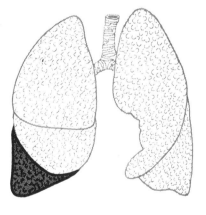

lower lobe
(inferior)

cardiac impression
KAR-dē-ak

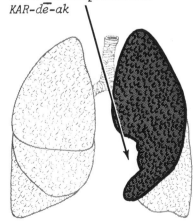

upper lobe
(superior)

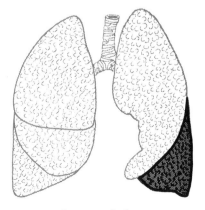

lower lobe
(inferior)

BRONCHIAL TREE AND TRACHEA
SHOWING
CARTILAGINOUS STRUCTURE

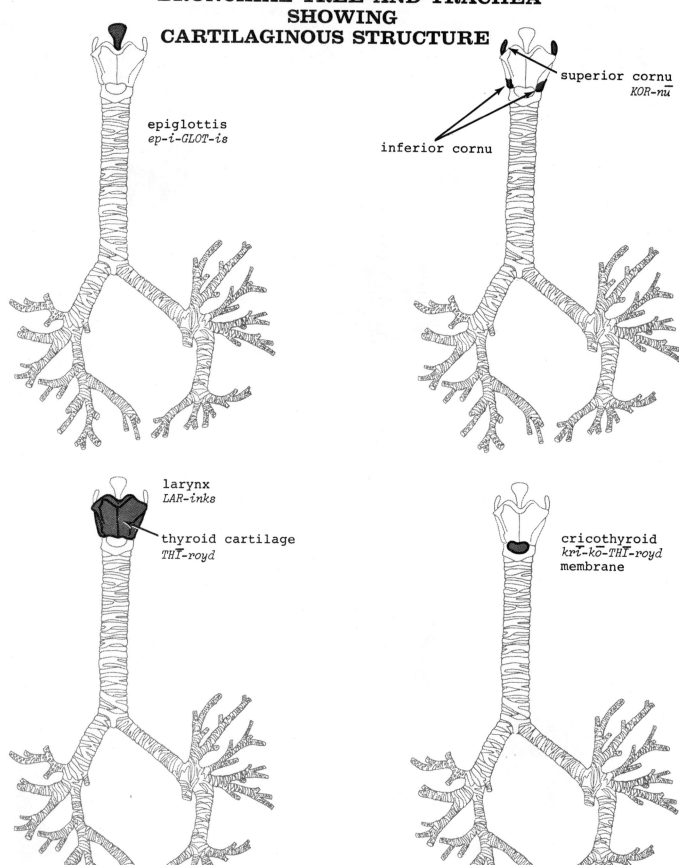

epiglottis
ep-i-GLOT-is

superior cornu
KOR-nū

inferior cornu

larynx
LAR-inks

thyroid cartilage
THĪ-royd

cricothyroid
krī-kō-THĪ-royd
membrane

BRONCHIAL TREE AND TRACHEA
SHOWING CARTILAGINOUS STRUCTURE

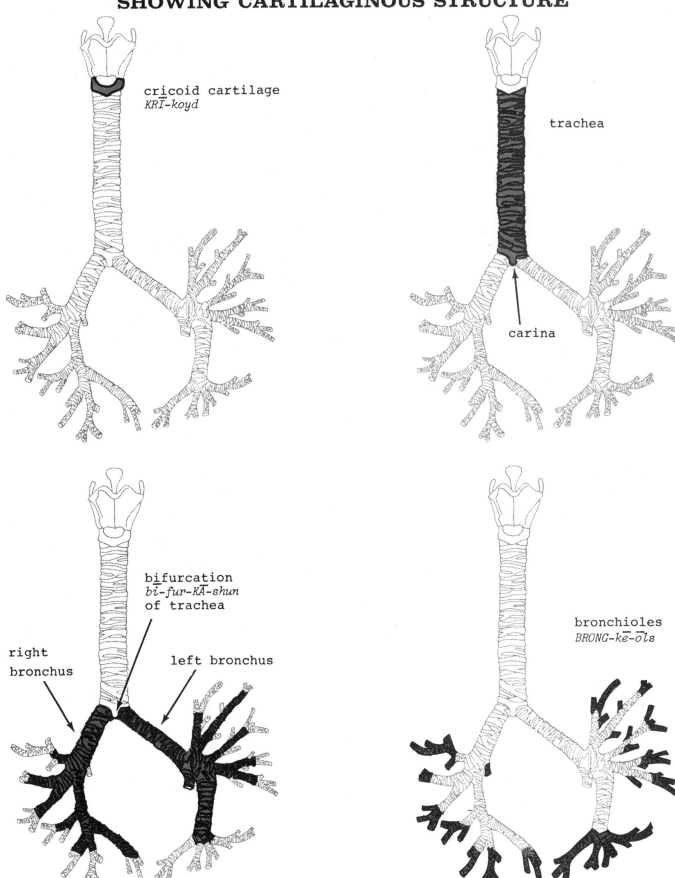

cricoid cartilage
KRĪ-koyd

trachea

carina

bifurcation
bī-fur-KĀ-shun
of trachea

right
bronchus

left bronchus

bronchioles
BRONG-kē-ōls

LOBULE OF THE LUNG

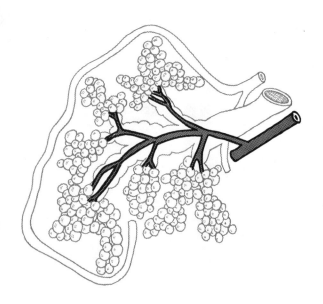

pulmonary artery
PUL-mo-nar̄-ē

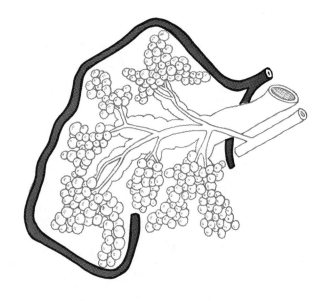

pulmonary vein

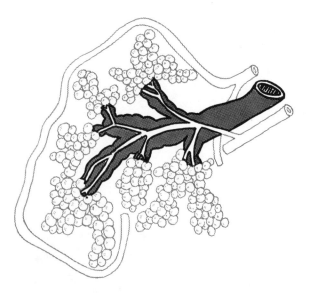

bronchiole
BRONG-kē-ōl

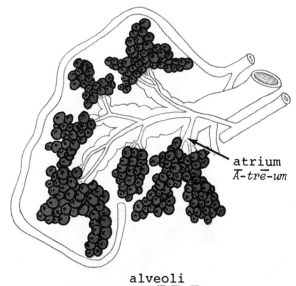

atrium
Ā-trē-um

alveoli
al-VĒ-ō-lī
(air sacs)

PLEURA AND MEDIASTINUM (Cross Section from Above)
*mē"dē-AS-ti-num**

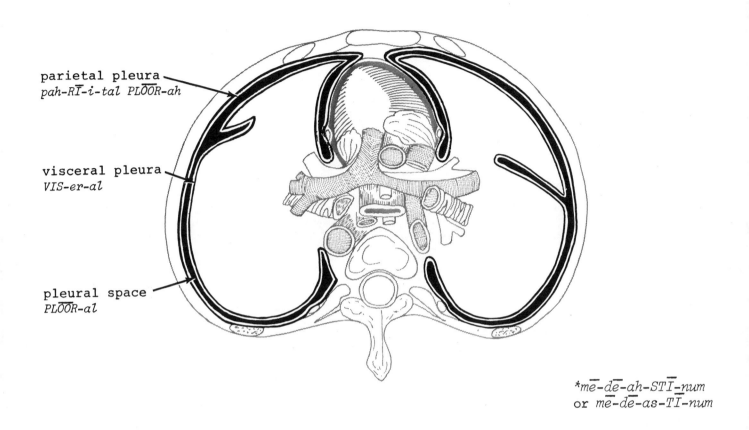

parietal pleura
pah-RĪ-i-tal PLŌŌR-ah

visceral pleura
VIS-er-al

pleural space
PLŌŌR-al

*mē-dē-ah-STĪ-num
or *mē-dē-as-TĪ-num*

pericardial cavity
per-ē-KAR-dē-al

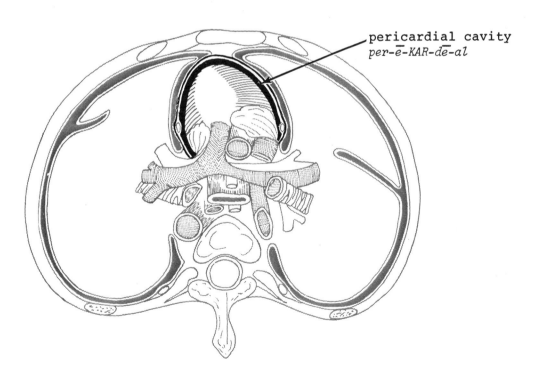

The pleural and pericardial cavities have been exaggerated to make them easier to visualize. Actually, there is little if any space between the pleura. A small amount of fluid serves as a lubricant between the pleura. As only the mediastinum and the pleura are concerned in this illustration the lungs and the extension of the blood vessels and bronchi are not included.

PLEURA AND MEDIASTINUM (Cross Section from Above)

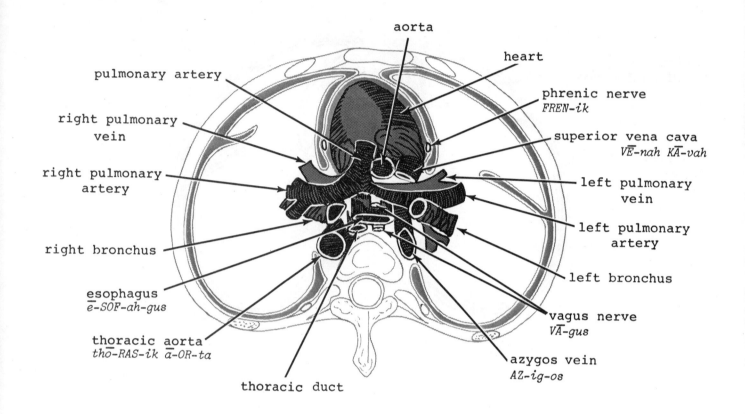

aorta

heart

phrenic nerve
FREN-ik

superior vena cava
VĒ-nah KĀ-vah

left pulmonary
vein

left pulmonary
artery

left bronchus

vagus nerve
VĀ-gus

azygos vein
AZ-ig-os

pulmonary artery

right pulmonary
vein

right pulmonary
artery

right bronchus

esophagus
ē-SOF-ah-gus

thoracic aorta
thō-RAS-ik ā-OR-ta

thoracic duct

Contents of Mediastinum

The contents of the mediastinum include the heart, the pulmonary trunk and arteries, the superior and inferior venae cavae, the thoracic aorta and its branches, the trachea and part of the bronchi, the esophagus, the vagus nerves, the phrenic nerves, the thoracic duct, many lymph nodes and vessels, the azygos vein and the thymus gland or its fibrous remainder. The view shown here is of the middle and posterior portion of the mediastinum. Other divisions are the superior mediastinum and the anterior mediastinum.

PLEURA AND MEDIASTINUM (Cross Section from Above)

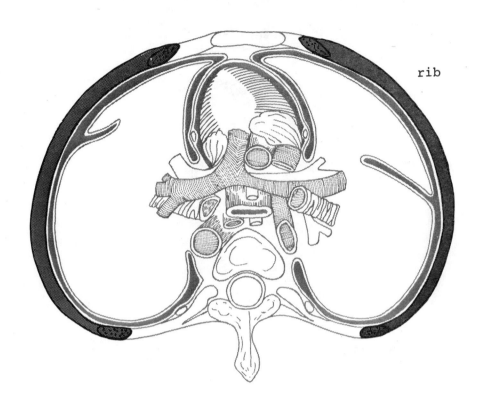

rib

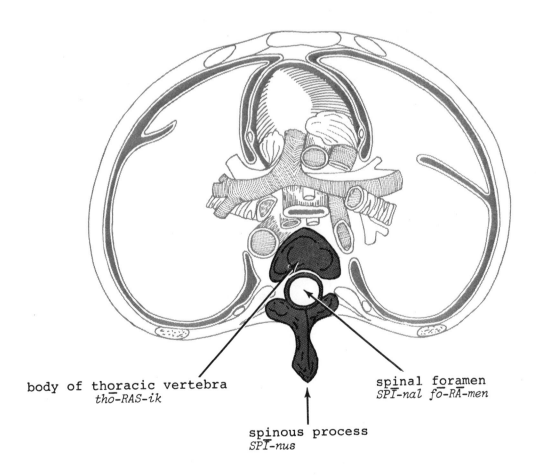

body of thoracic vertebra
thō-RAS-ik

spinal foramen
SPĪ-nal fō-RĀ-men

spinous process
SPĪ-nus

PLEURA AND MEDIASTINUM (Cross Section from Above)

sternum
STER-num

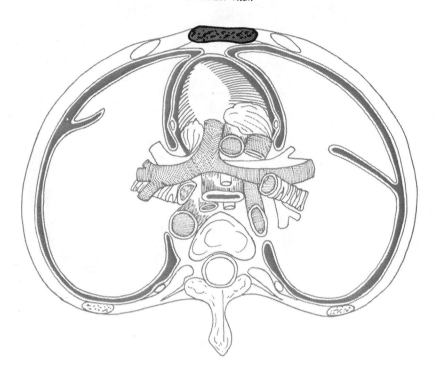

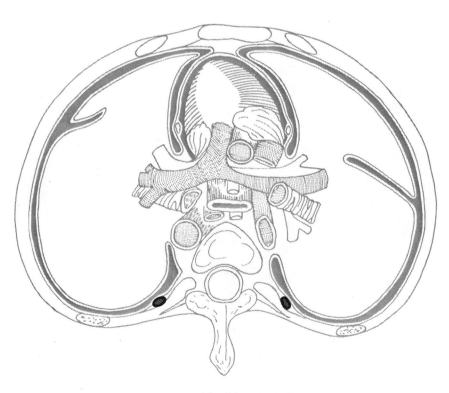

sympathetic trunk
sim-pah-THET-ik

INDEX OF ENGLISH TRANSLATION

aden/oid
gland

alar
wing

alveol/ar
cavity

alveoli
cavity

aorta
to lift up

atrium
chamber

audit/ory
to hear

a/zygos
not   a yoke

bi/furc/ation
two   forked

bronchi/ole
bronchus   small

bronchus
windpipe

cardi/ac
heart

carina
keel

cartilage
gristle

clav/icle
key   little

concha (sing.)
shell

conchae (pl.)
shell

cornu
horn

cric/oid
a ring

crico/thyr/oid
a ring   a shield

crista galli
crest   a cock's comb

epi/glottis
on, upon   mouth of windpipe

esophagus
gullet

ethm/oid
sieve

ethm/oid/al
sieve

eustachian
after a 16th Century Italian anatomist

fila
threadlike structure

foramen
opening

hy/oid
letter
"U"

larynx
windpipe

mandible
a jaw

maxilla
jawbone

maxill/ary
jawbone

meatus
passage

media/stinum
middle   septum

naris
nose

nas/al
nose

naso/cili/ary
nose   eyebrow

naso/palatine
nose   palate

olfact/ory
to smell

or/al
mouth

palate
roof of mouth

palatine
palate

pariet/al
wall

peri/cardi/al
around   heart

pharynge/al
pharynx

pharynx
throat

phren/ic
diaphragm

pleura
rib

pleur/al
pleura

pulmon/ary
lung

sella turcica
saddle   Turkish

sept/al
septum

septum
wall off

sinus (sing.)
hollow

sinuses (pl.)
hollow

sphen/oid
wedge

spin/al
spine

spin/ous
spine

sternum
the chest

sym/pathet/ic
together   to suffer

thorac/ic
chest

thyr/oid
shield

trachea
rough

turbinates
shaped like a top

uvula
little grape

vagus
wandering

vallecula
a depression

vena cava
vein   hollow

vertebrae
to turn

vestibule
an entrance court

viscer/al
organ

vomer
ploughshare

CHAPTER VIII

GASTRO-INTESTINAL SYSTEM

This system extends from the mouth in a continuous tube approximately 30 feet long to terminate in the anus. The 30 feet of alimentary canal undergo many twists and turns particularly in the small intestine which fills a large portion of the abdominal cavity.

The three portions of the small intestine have interesting translations. The *duodenum* was so-called because it means 12 and is, actually, about 12 inches in length. The next portion, the *jejunum*, translates as "empty" and received its name because early anatomists always found it empty. The last portion of the small intestine, the *ileum*, means "twisted intestine" and was so named because of its peristaltic action.

In the large intestine the *sigmoid* colon is shaped somewhat like an "S" and hence received its name as "sigm" means "S" and "-oid" means "to resemble." The *rectum* is the straightest part of the large intestine and the term actually means "straight."

The term *alimentary* means "to nourish, to feed" and this is the function of the Gastro-Intestinal System.

ALIMENTARY ORGANS

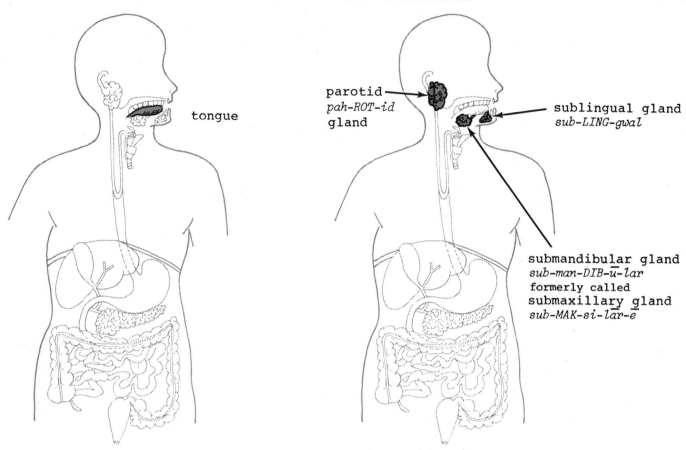

tongue

parotid
pah-ROT-id
gland

sublingual gland
sub-LING-gwal

submandibular gland
sub-man-DIB-ū-lar
formerly called
submaxillary gland
sub-MAK-si-lar-e

The organs are spread out slightly for easier viewing.

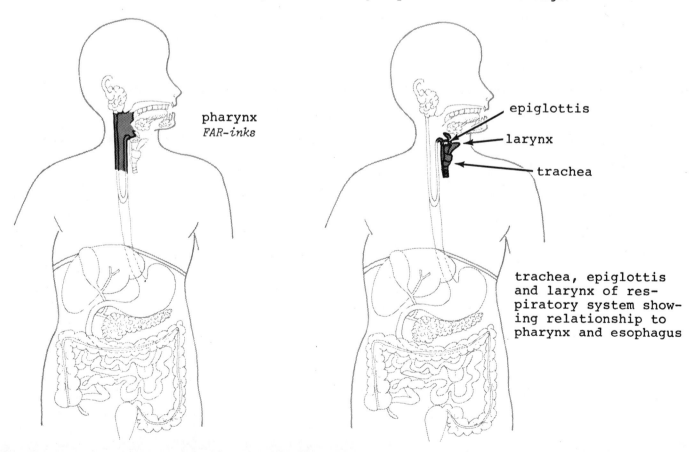

pharynx
FAR-inks

epiglottis

larynx

trachea

trachea, epiglottis
and larynx of res-
piratory system show-
ing relationship to
pharynx and esophagus

ALIMENTARY ORGANS

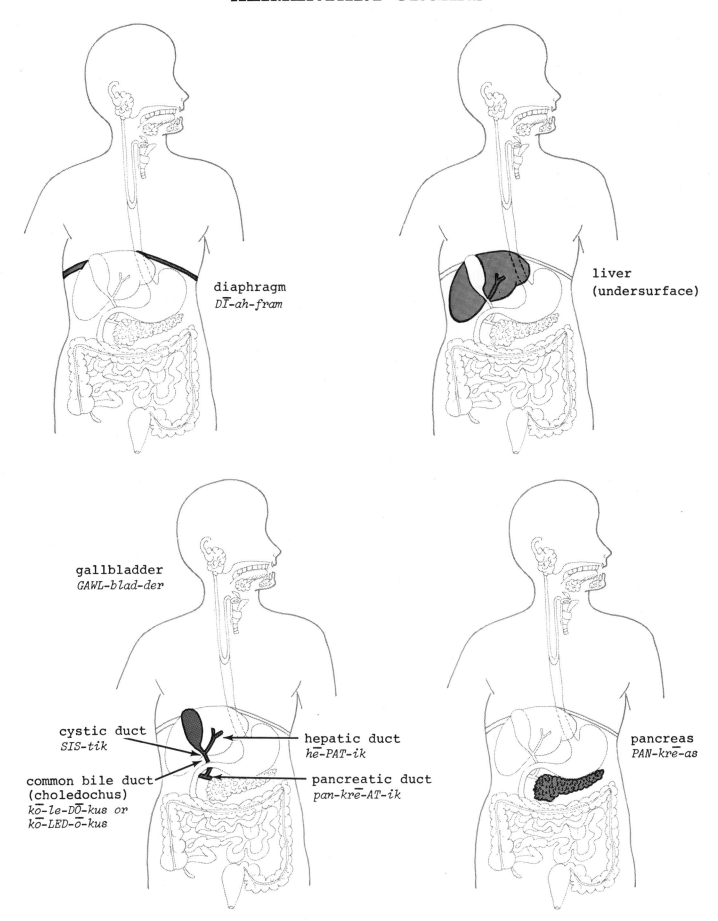

diaphragm
DĪ-ah-fram

liver
(undersurface)

gallbladder
GAWL-blad-der

cystic duct
SIS-tik

hepatic duct
hē-PAT-ik

common bile duct
(choledochus)
kō-le-DŌ-kus or
kō-LED-ō-kus

pancreatic duct
pan-krē-AT-ik

pancreas
PAN-krē-as

ALIMENTARY ORGANS

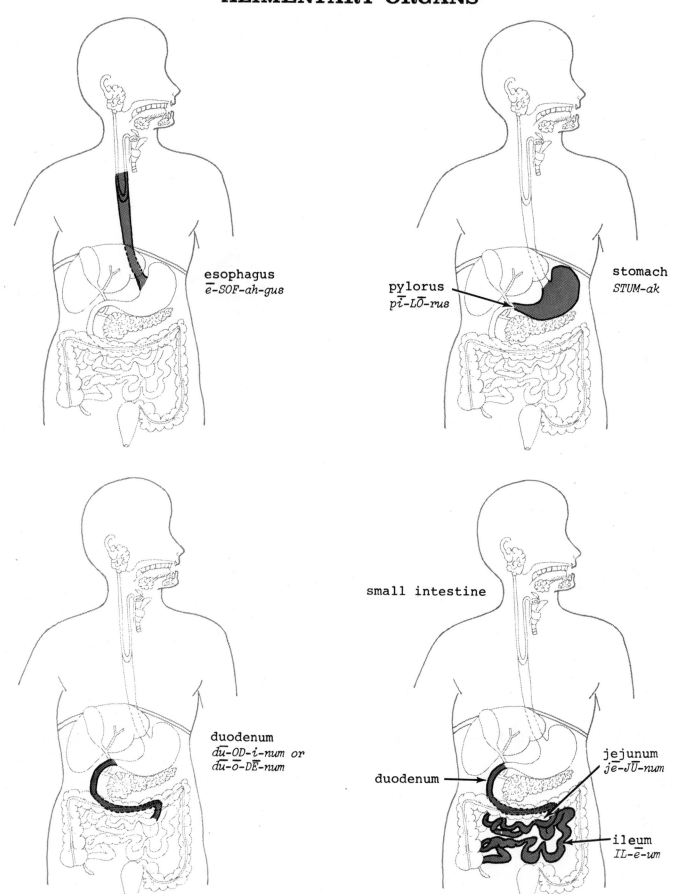

esophagus
ē-SOF-ah-gus

pylorus
pī-LŌ-rus

stomach
STUM-ak

duodenum
dū-OD-i-num or
dū-ō-DĒ-num

small intestine

duodenum

jejunum
jē-JŪ-num

ileum
IL-ē-um

ALIMENTARY ORGANS

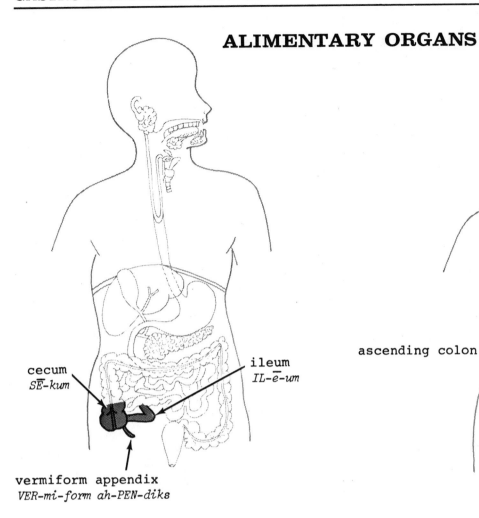

cecum
$S\overline{E}$-kum

ileum
IL-$\overline{e}$-um

vermiform appendix
VER-mi-form ah-PEN-diks

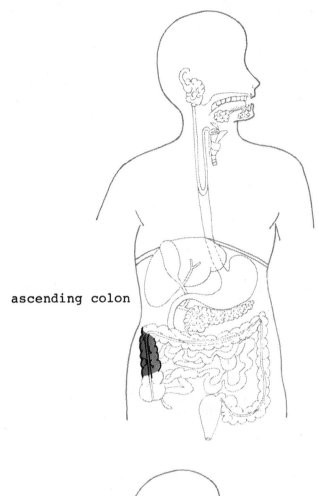

ascending colon

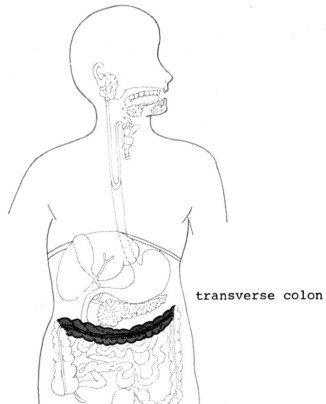

transverse colon

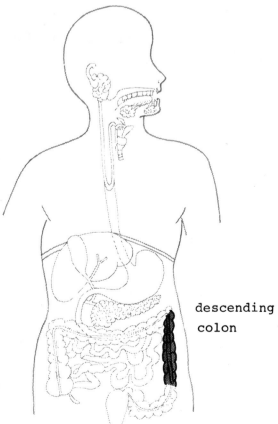

descending
colon

ALIMENTARY ORGANS

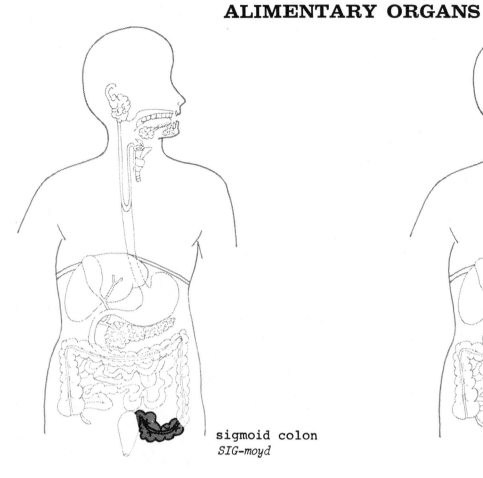

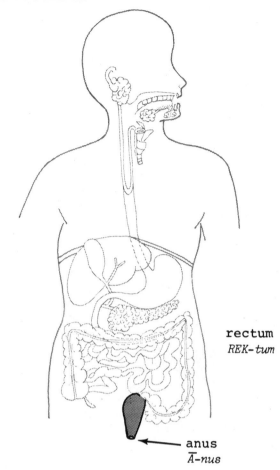

sigmoid colon
SIG-moyd

rectum
REK-tum

anus
Ā-nus

SALIVARY GLANDS AND RELATED STRUCTURES

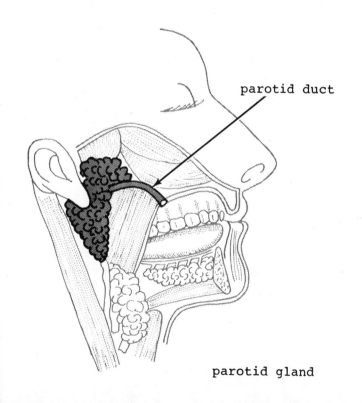

parotid duct

parotid gland

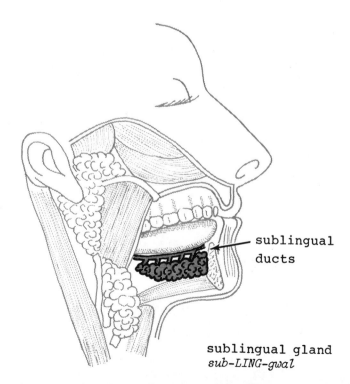

sublingual
ducts

sublingual gland
sub-LING-gwal

SALIVARY GLANDS AND RELATED STRUCTURES

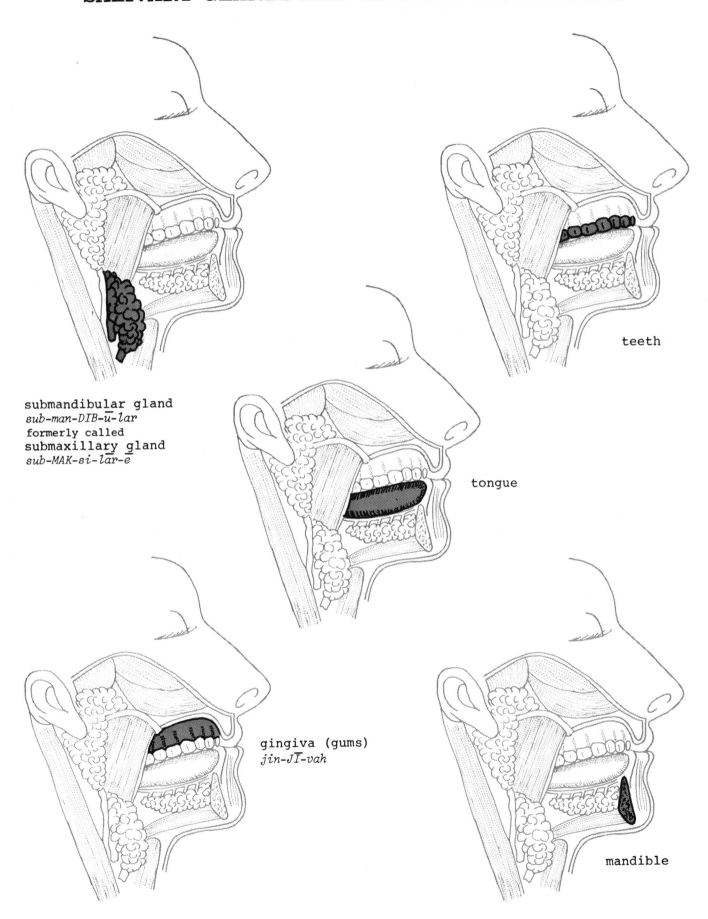

submandibular gland
sub-man-DIB-ū-lar
formerly called
submaxillary gland
sub-MAK-si-lar-ē

teeth

tongue

gingiva (gums)
jin-JĪ-vah

mandible

SALIVARY GLANDS AND RELATED STRUCTURES

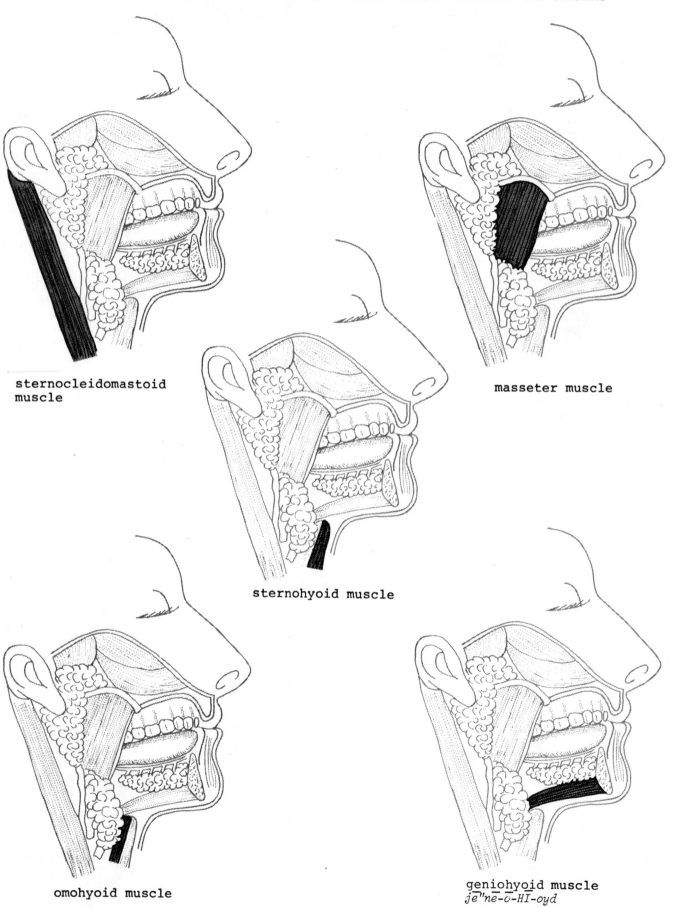

sternocleidomastoid
muscle

masseter muscle

sternohyoid muscle

omohyoid muscle

geniohyoid muscle
je͞"ne͞-o͞-HI͞-oyd

SALIVARY GLANDS AND RELATED STRUCTURES

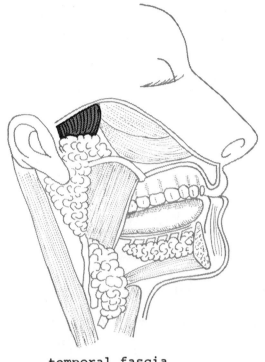

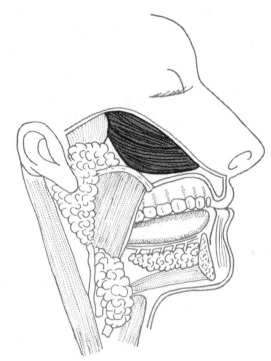

temporal fascia
TEM-po-ral FASH-ē-ah

orbicularis
oculi muscle

MOUTH CAVITY

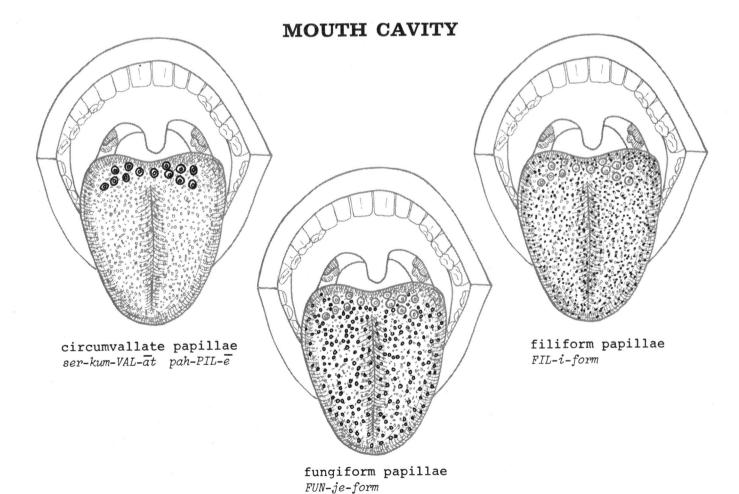

circumvallate papillae
ser-kum-VAL-āt pah-PIL-ē

filiform papillae
FIL-i-form

fungiform papillae
FUN-je-form

MOUTH CAVITY

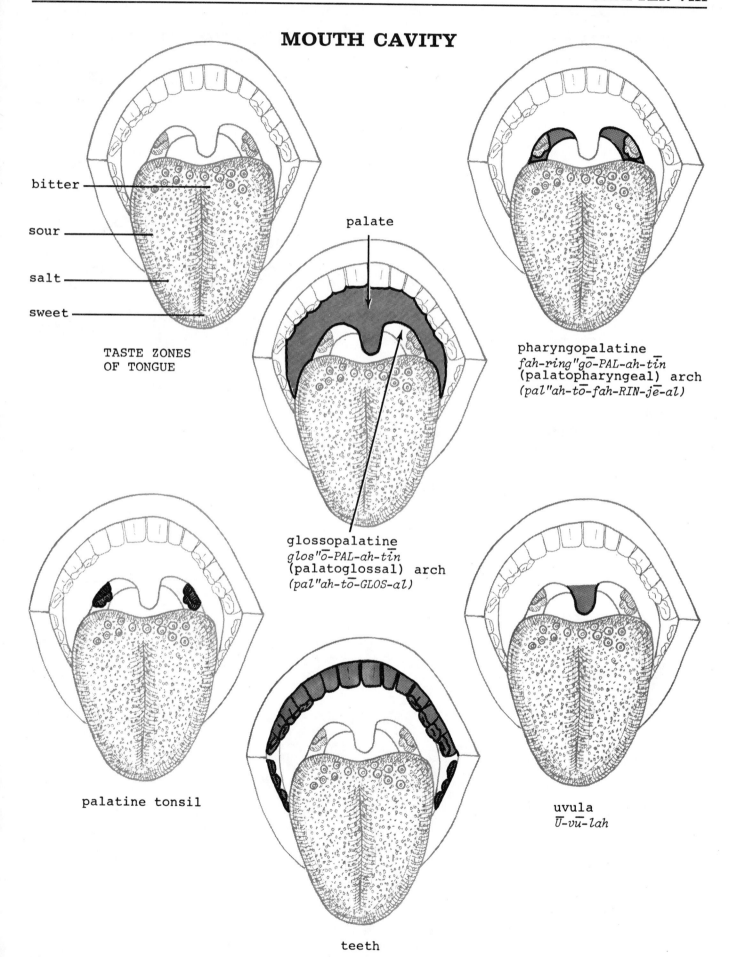

bitter

sour

salt

sweet

TASTE ZONES
OF TONGUE

palate

pharyngopalatine
fah-ring"gō-PAL-ah-tĭn
(palatopharyngeal) arch
(pal"ah-tō-fah-RIN-jē-al)

glossopalatine
glos"ō-PAL-ah-tĭn
(palatoglossal) arch
(pal"ah-tō-GLOS-al)

palatine tonsil

uvula
Ū-vū-lah

teeth

LONGITUDINAL SECTION OF TOOTH
AND SUPPORTING STRUCTURES

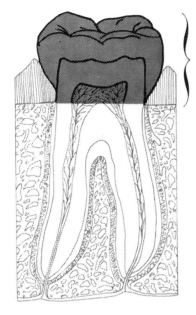

ANATOMICAL
CROWN

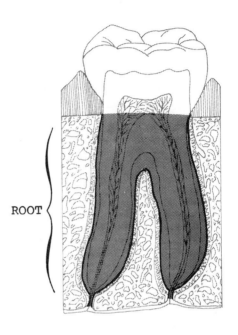

ROOT

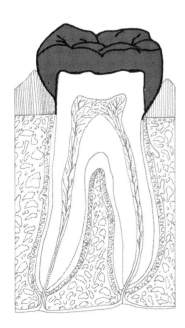

enamel

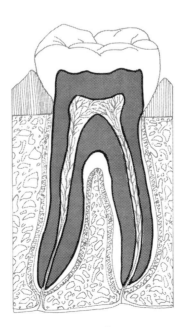

dentin
DEN-tin

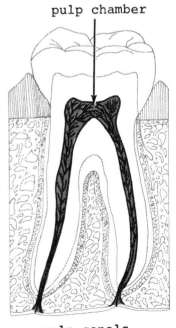

pulp chamber

pulp canals
contain blood vessels and nerves

LONGITUDINAL SECTION OF TOOTH
AND SUPPORTING STRUCTURES

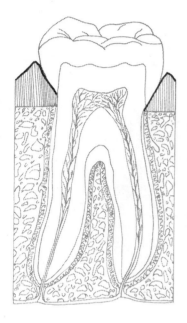

gingival line
JIN-je-val

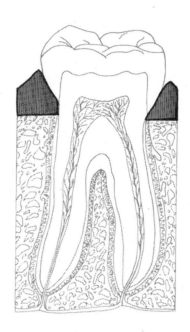

gingiva (gums)
jin-JĪ-vah

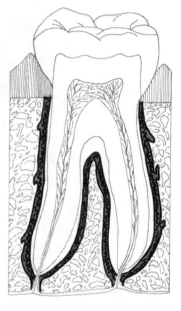

peridontal membrane
per-ē-DON-tal

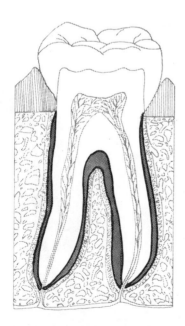

cementum
se-MEN-tum

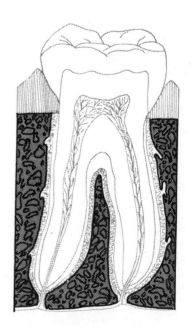

alveolar bone
al-VĒ-ō-lar

UPPER JAW OF ADULT (Right Side)

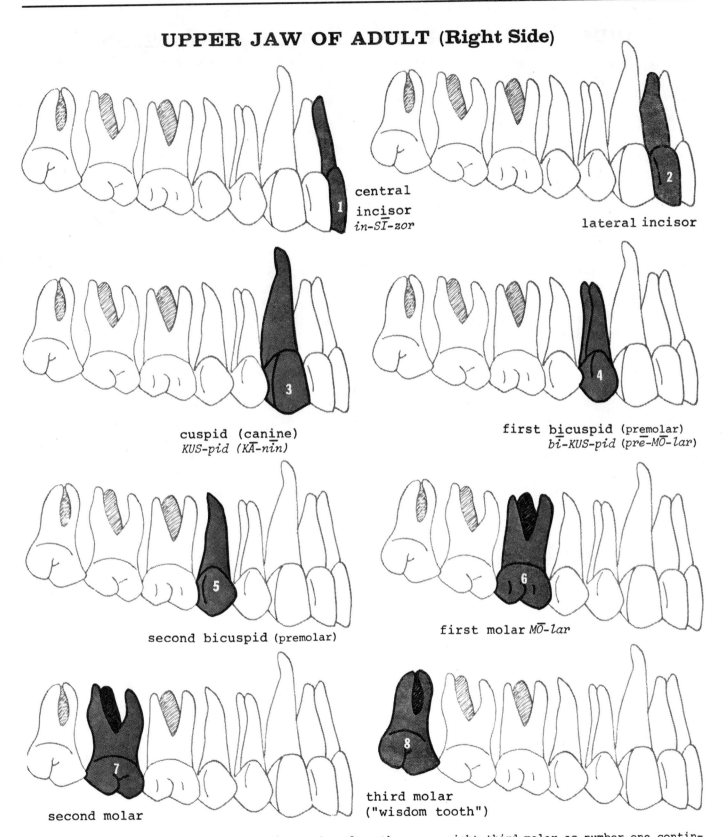

central incisor
in-SĪ-zor

lateral incisor

cuspid (canine)
KUS-pid (KĀ-nin)

first bicuspid (premolar)
bī-KUS-pid (prē-MŌ-lar)

second bicuspid (premolar)

first molar *MŌ-lar*

second molar

third molar
("wisdom tooth")

Alternate numbering: insurance companies number from the upper right third molar as number one contin-uing to the upper left third molar as number 16. The lower left third molar is then number 17 and pro-ceeds to the lower right molar as number 32. The drawing shown is based on a mirror image. This is an exception as anatomical drawings usually show the right side as on the reader's left.

LEFT ←															RIGHT
16	15	14	13	12	11	10	9	8	7	6	5	4	3	2	1
17	18	19	20	21	22	23	24	25	26	27	28	29	30	31	32

SIDE VIEW OF CHILD'S TEETH (Deciduous Teeth)

dē-SID-ū-us

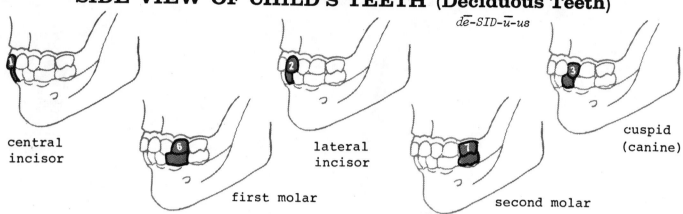

central
incisor

first molar

lateral
incisor

second molar

cuspid
(canine)

As a child's jaws grow it allows more room for teeth. When the deciduous teeth fall out and are replaced by the permanent teeth room is now available for the eruption of the bicuspid and second bicuspid. As continued growth occurs space is provided for the third molars or "wisdom teeth." However, these latter teeth do not always appear in every individual.

SIDE VIEW OF ADULT TEETH

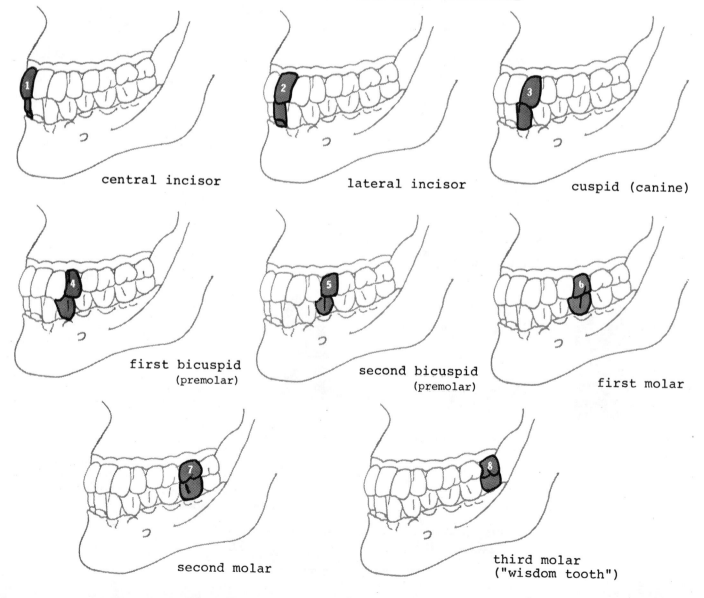

central incisor

lateral incisor

cuspid (canine)

first bicuspid
(premolar)

second bicuspid
(premolar)

first molar

second molar

third molar
("wisdom tooth")

INTERIOR OF STOMACH

cardiac orifice
KAR-dē-ak OR-i-fis

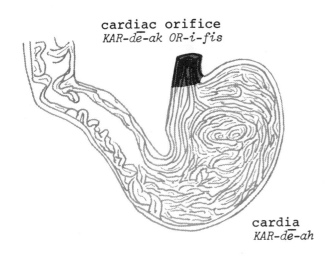

cardia
KAR-dē-ah

fundus
FUN-dus

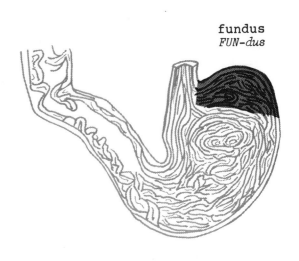

lesser
curvature

rugae
ROO-gē

greater
curvature

body

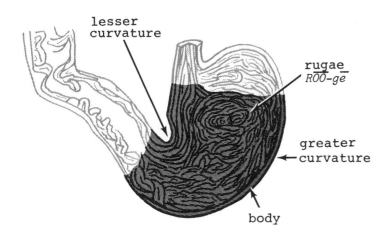

pyloric portion

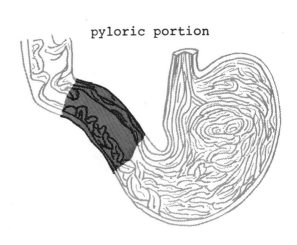

pylorus
pī-LŌ-rus

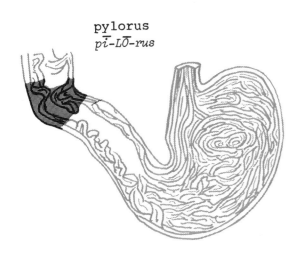

duodenum

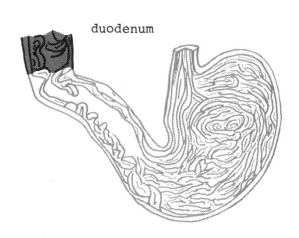

RELATIONSHIP OF STOMACH, DUODENUM AND PANCREAS

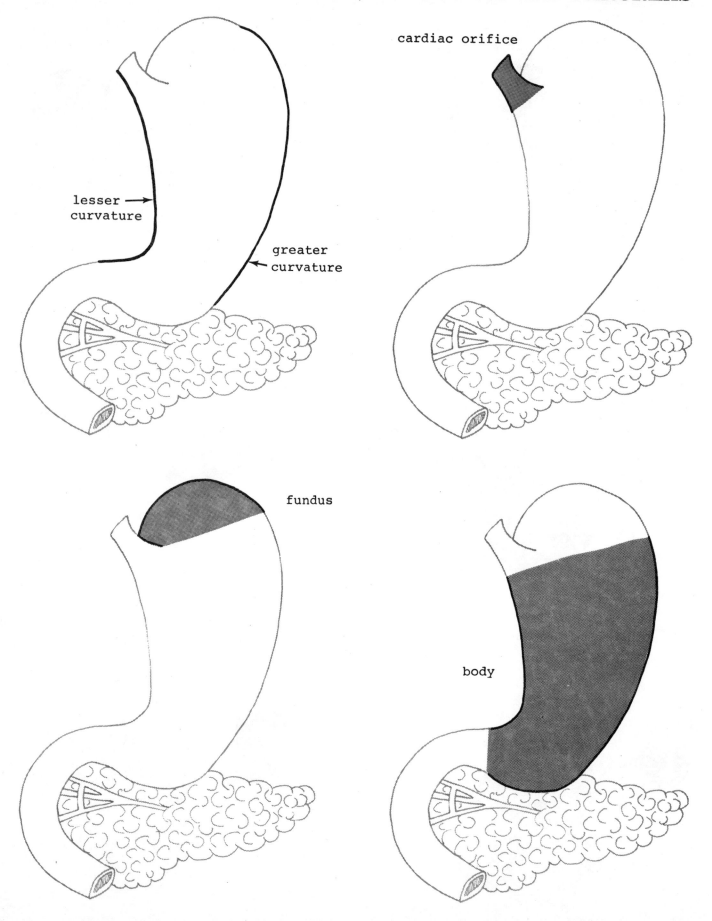

lesser
curvature

greater
curvature

cardiac orifice

fundus

body

RELATIONSHIP OF STOMACH, DUODENUM AND PANCREAS

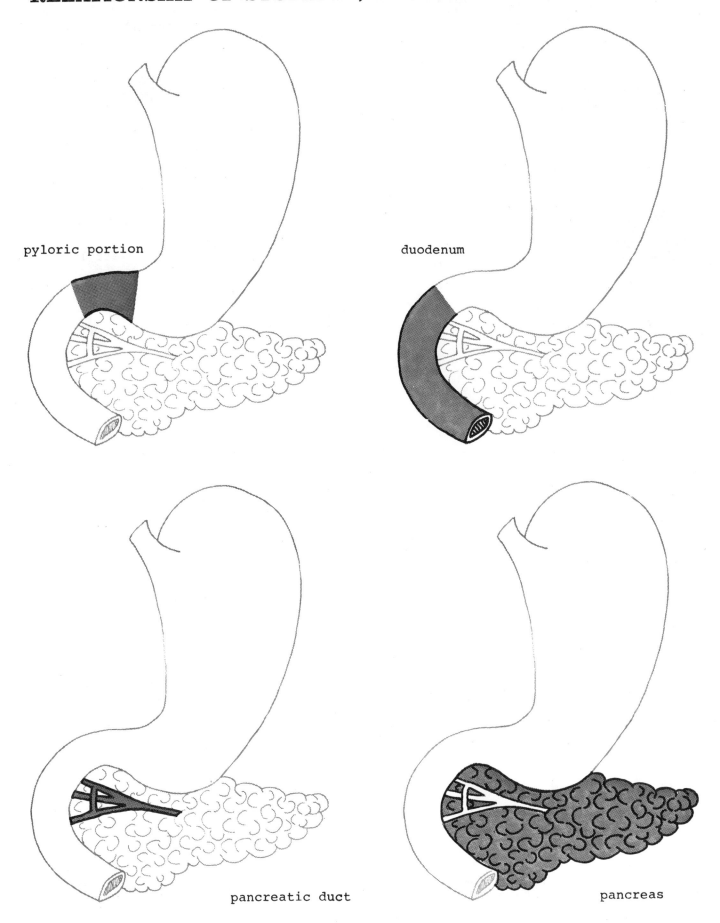

pyloric portion

duodenum

pancreatic duct

pancreas

ANTERIOR SURFACE OF LIVER

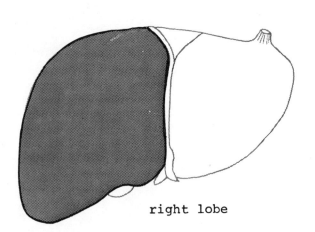

right lobe

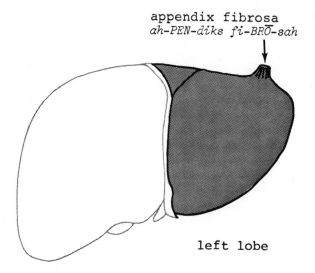

appendix fibrosa
ah-PEN-diks fi-BRŌ-sah

left lobe

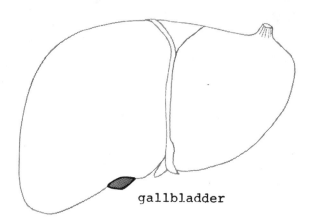

gallbladder

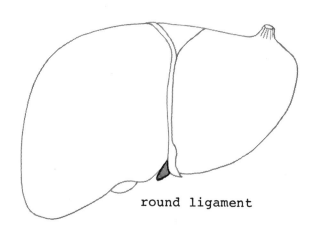

round ligament

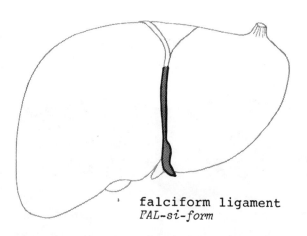

falciform ligament
FAL-si-form

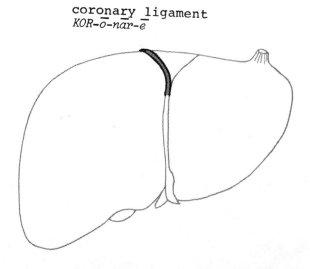

coronary ligament
KOR-ō-nār-ē

DORSAL (VISCERAL) SURFACE OF LIVER

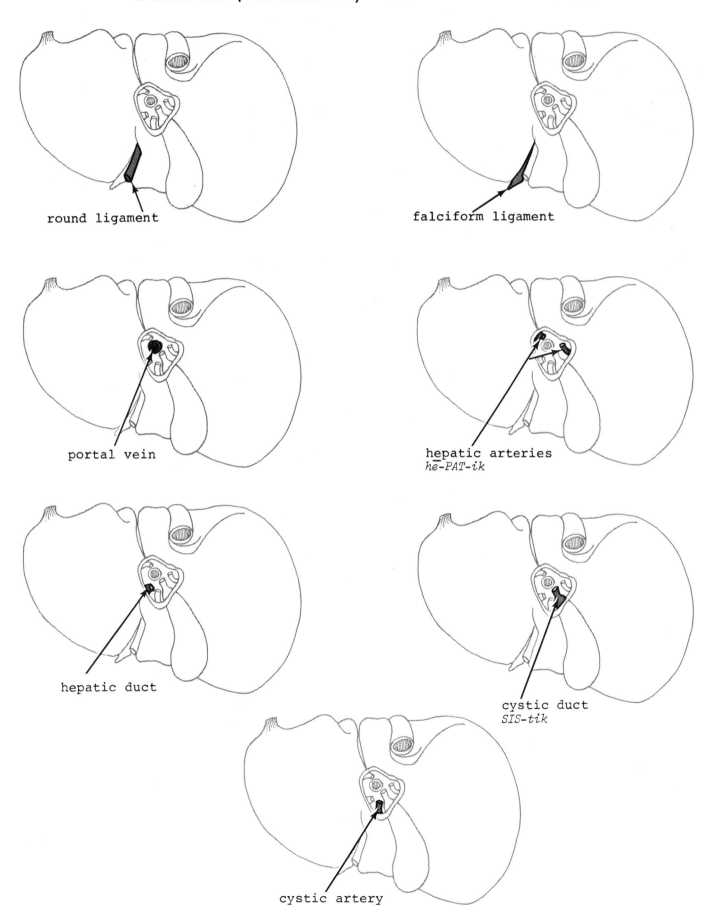

round ligament

falciform ligament

portal vein

hepatic arteries
hē-PAT-ik

hepatic duct

cystic duct
SIS-tik

cystic artery

DORSAL (VISCERAL) SURFACE OF LIVER

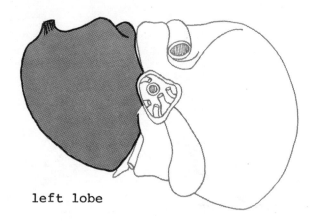

left lobe

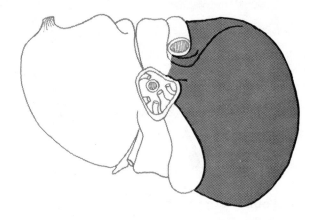

right lobe

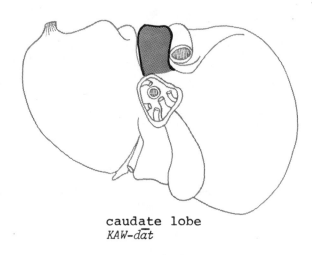

caudate lobe
KAW-dāt

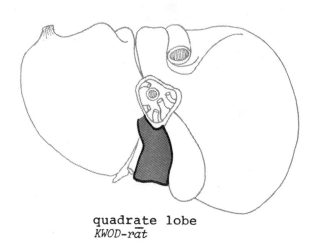

quadrate lobe
KWOD-rāt

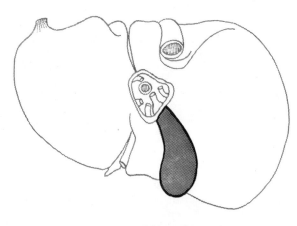

gallbladder

inferior vena cava
VĒ-nah KĀ-vah

VISCERAL IMPRESSIONS ON DORSAL SURFACE OF LIVER

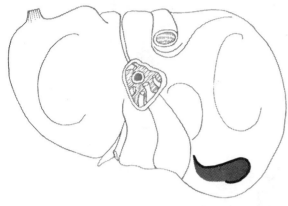

colic impression
KOL-ik

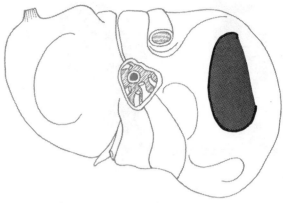

renal impression
RĒ-nal

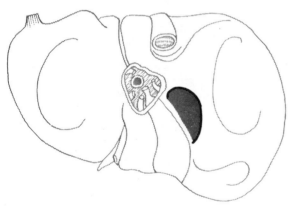

duodenal impression
dū-OD-i-nal or
dū-ō-DĒ-nal

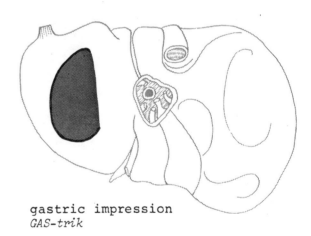

gastric impression
GAS-trik

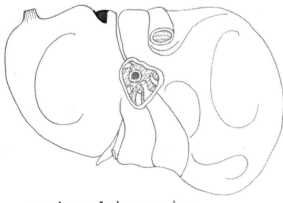

esophageal impression
ē-SOF-ah-jē-al

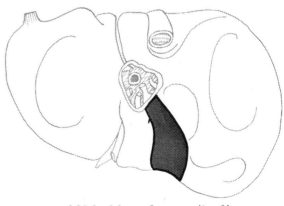

gallbladder fossa (bed)
FOS-ah

GALLBLADDER AND DUCTS

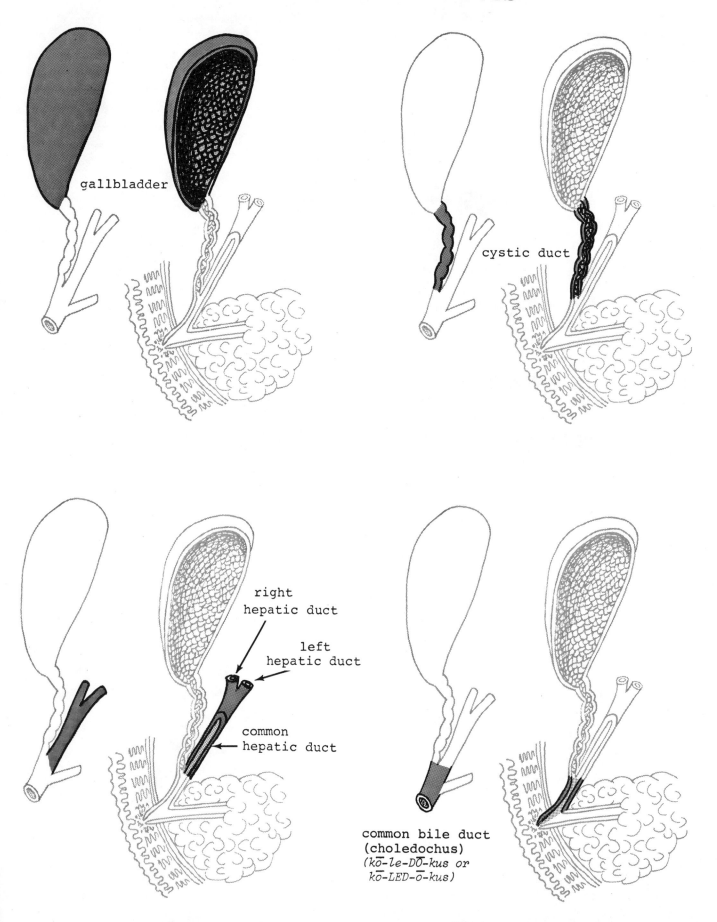

gallbladder

cystic duct

right
hepatic duct

left
hepatic duct

common
hepatic duct

common bile duct
(choledochus)
*(kō-le-DŌ-kus or
kō-LED-ō-kus)*

GALLBLADDER AND DUCTS

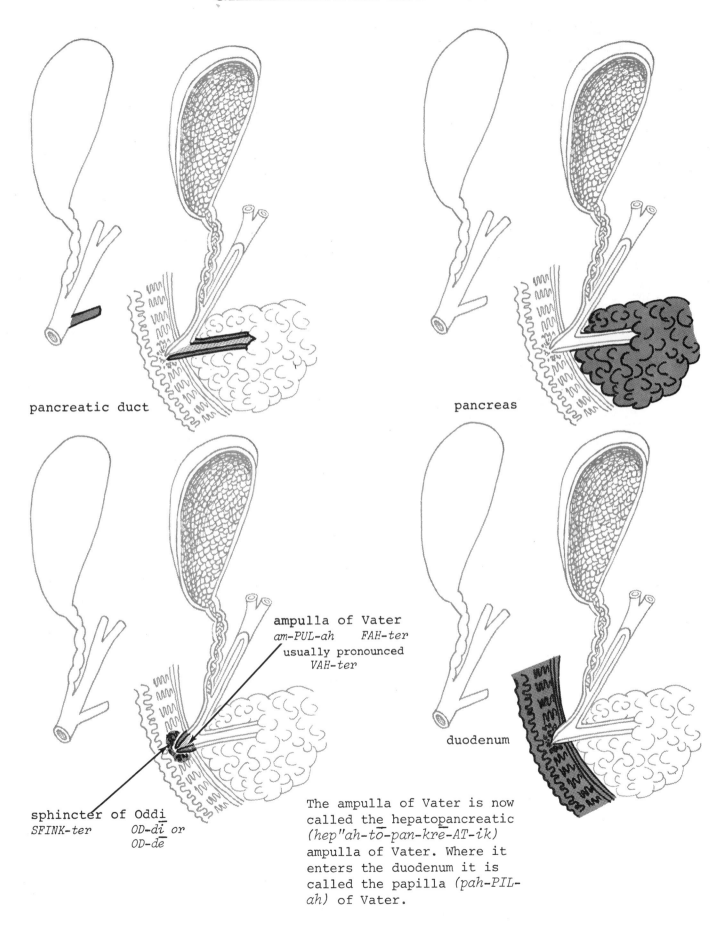

pancreatic duct

pancreas

ampulla of Vater
am-PUL-ah   FAH-ter
usually pronounced
VAH-ter

sphincter of Oddi
SFINK-ter   OD-dī or
OD-dē

duodenum

The ampulla of Vater is now called the hepatopancreatic *(hep"ah-tō-pan-krē-AT-ik)* ampulla of Vater. Where it enters the duodenum it is called the papilla *(pah-PIL-ah)* of Vater.

THE BILIARY SYSTEM

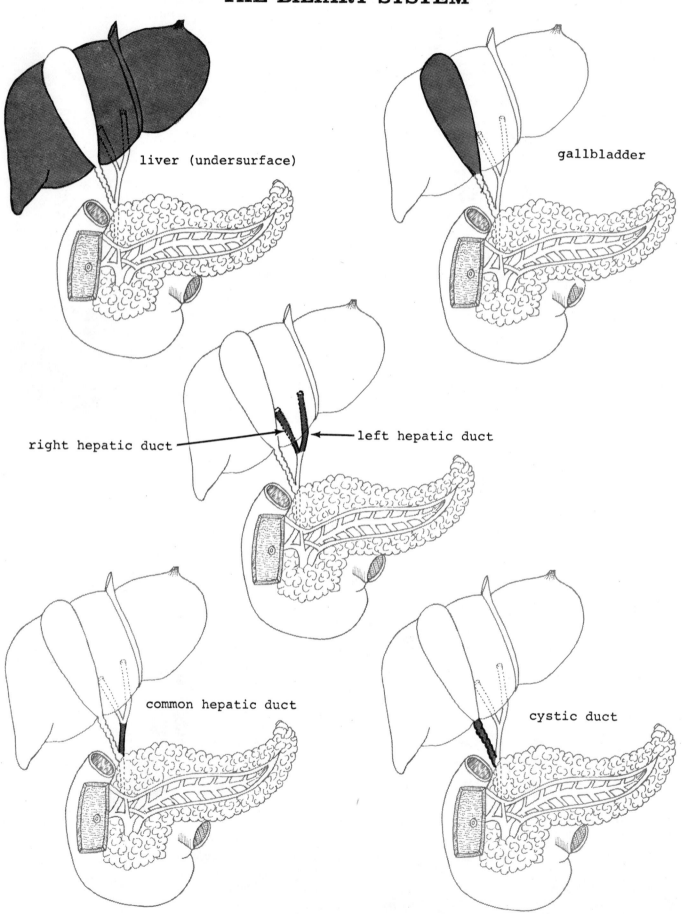

liver (undersurface)

gallbladder

right hepatic duct

left hepatic duct

common hepatic duct

cystic duct

THE BILIARY SYSTEM

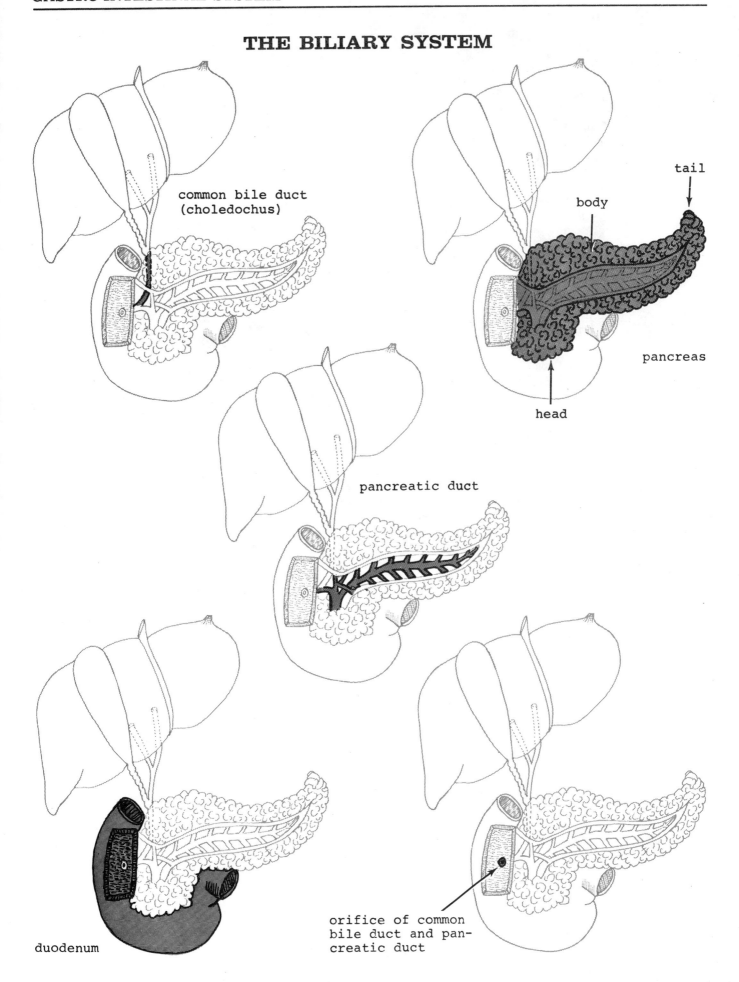

common bile duct
(choledochus)

tail

body

pancreas

head

pancreatic duct

duodenum

orifice of common
bile duct and pan-
creatic duct

A VILLUS

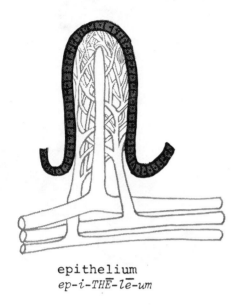

epithelium
ep-i-THĒ-lē-um

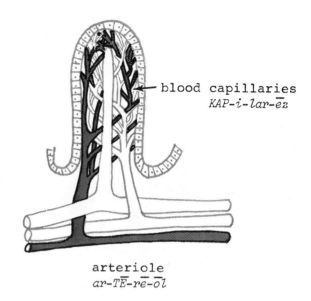

blood capillaries
KAP-i-lar-ēz

arteriole
ar-TĒ-rē-ōl

The villi are tiny finger-like projections which cover the entire length of the small intestine giving its inner lining (the mucosa) a velvety appearance. They number between four and five million in the human. Their size is just at the borderline of visibility with the naked eye. They are quite irregular in size and shape. Some are larger in certain parts of the intestine than in other parts.

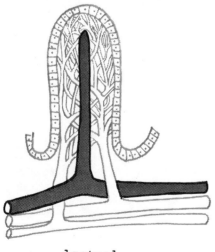

lacteal
LAK-tē-al

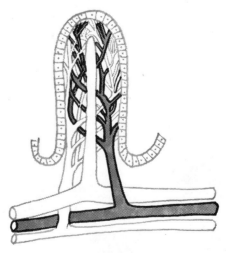

venule
VĒ-nul

COATS OF THE INTESTINE (Cross Section)

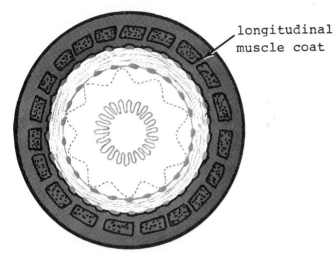

serous coat (serosa)
S$\overline{E}$-rus   *(se-R$\overline{O}$-sah)*

longitudinal
muscle coat

nerve plexuses
PLEK-se-sez

circular muscle coat

submucous coat (submucosa)
sub-MY$\overline{OO}$-kus   *(sub-MY$\overline{OO}$-k$\overline{o}$-sah)*

mucous coat (mucosa)
MY$\overline{OO}$-kus   *(my$\overline{oo}$-K$\overline{O}$-sah)*

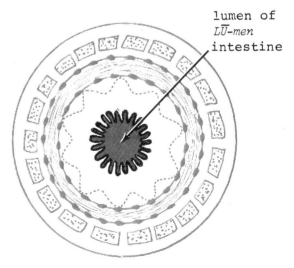

lumen of
L$\overline{U}$-men
intestine

villi
VIL-l$\overline{i}$

COATS OF THE INTESTINE SHOWN IN LAYERS

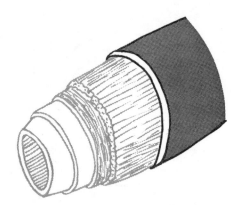

serous coat (serosa)

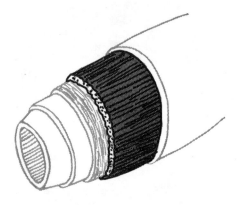

longitudinal muscle coat

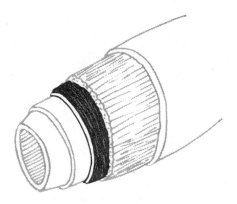

circular muscle coat

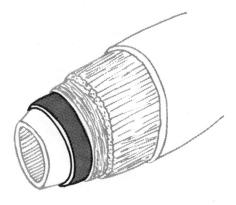

submucous coat (submucosa)

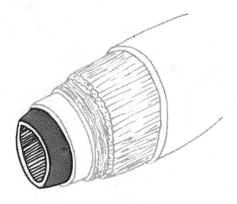

mucous coat (mucosa)

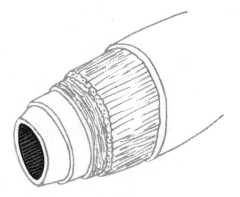

lumen of intestine
LŪ-men

CECUM AND ACCESSORY ORGANS

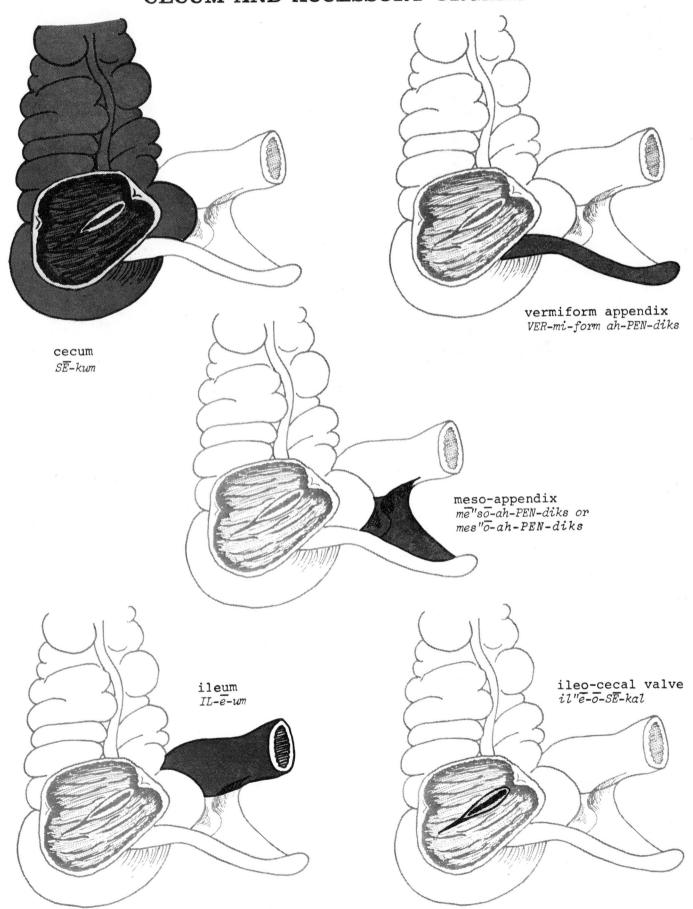

cecum
SĒ-kum

vermiform appendix
VER-mi-form ah-PEN-diks

meso-appendix
*mē"sō-ah-PEN-diks or
mes"ō-ah-PEN-diks*

ileum
IL-ē-um

ileo-cecal valve
il"ē-ō-SĒ-kal

THE OMENTUM

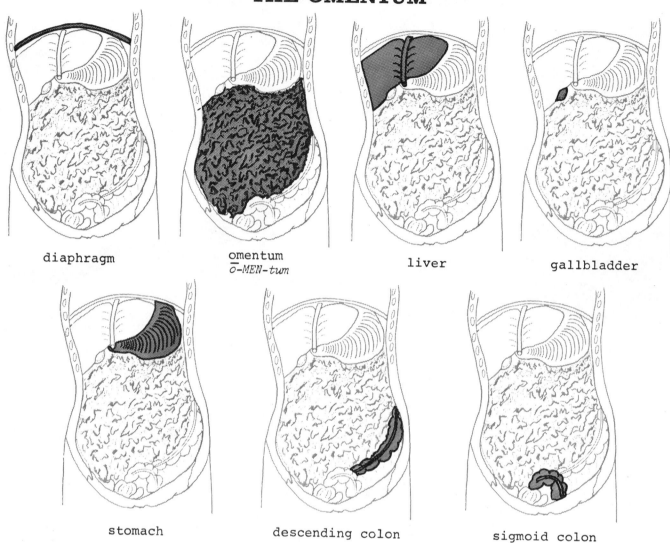

diaphragm

omentum
Ō-MEN-tum

liver

gallbladder

stomach

descending colon

sigmoid colon

The omentum is called "the apron of the abdomen." It is a continuation of the peritoneum per"i-to-NĒ-um and hangs over the intestines like an apron.

SECTION OF SMALL INTESTINE SHOWING MESENTERY AND ARTERIAL SUPPLY

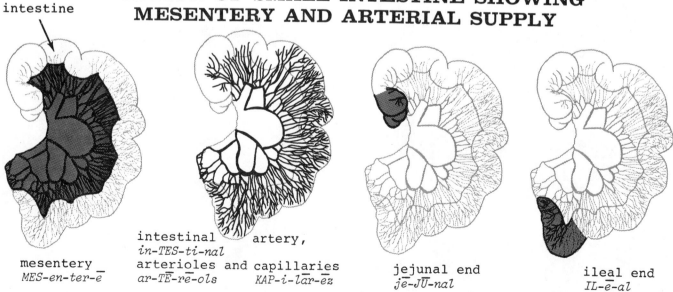

intestine

mesentery
MES-en-ter-ē

intestinal artery,
in-TES-ti-nal
arterioles and capillaries
ar-TĒ-rē-ols   KAP-i-lar-ēz

jejunal end
je-JŪ-nal

ileal end
IL-ē-al

THE LARGE INTESTINE

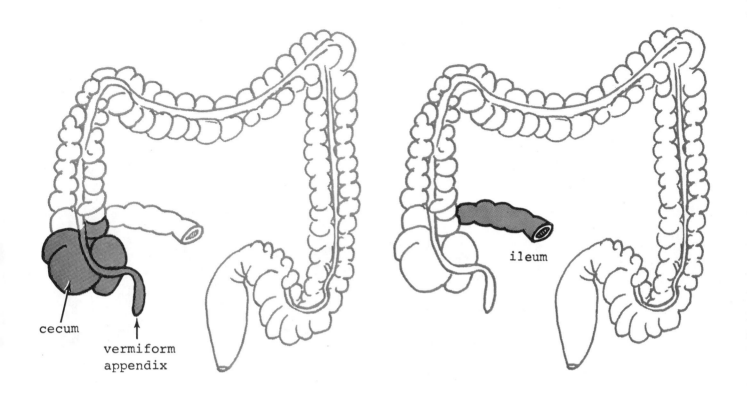

cecum

vermiform
appendix

ileum

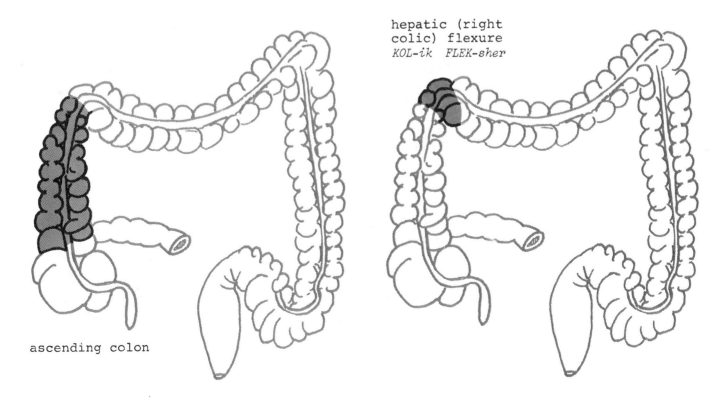

hepatic (right
colic) flexure
KOL-ik FLEK-sher

ascending colon

THE LARGE INTESTINE

transverse colon

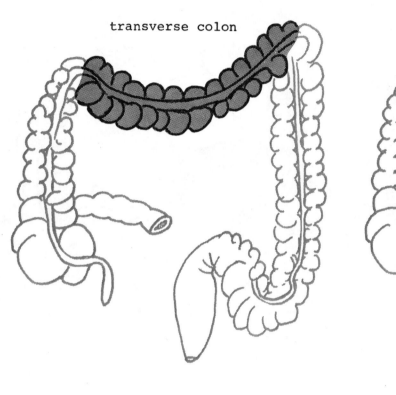

splenic (left
SPLEN-ik
colic) flexure

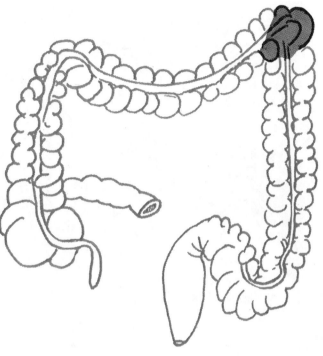

descending colon

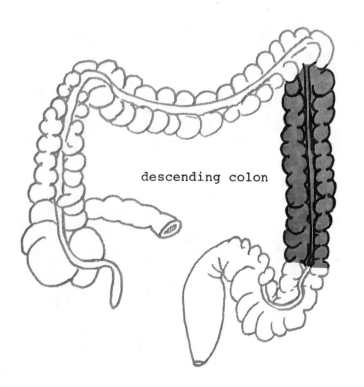

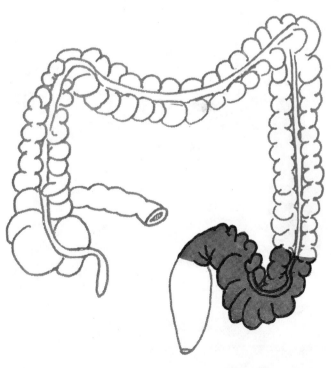

sigmoid colon

THE LARGE INTESTINE

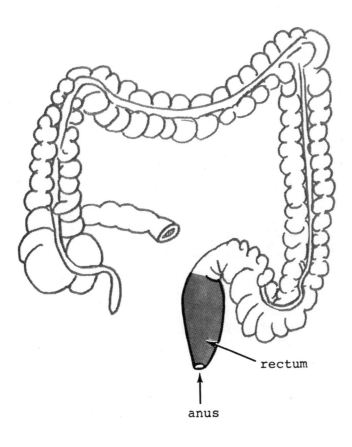

rectum

anus

INDEX OF ENGLISH TRANSLATION

aliment/ary
food

alveol/ar
cavity

ampulla
small jar

anus
to sit

appendix
to hang upon

arteri/ole
artery   small

bi/cuspid
two   point

canine
dog

capillaries
hairlike

cardia
heart
(esophageal end of stomach)

caud/ate
tail

cecum
blind

cementum
substance holding cells or tissues together

chole/dochus
bile   receptacle

circum/vall/ate
around   to wall

col/ic
colon

colon
large intestine

coron/ary
crown

cuspid
point

cyst/ic
sac

deciduous
to fall off

dentin
tooth

diaphragm
a partition

duoden/al
twelve

duodenum
twelve
(refers to its length of about 12″)

epi/glottis
on, upon   mouth of windpipe

epi/thelium
on, upon   nipple

esophage/al
gullet

esophagus
gullet

falci/form
sickle   shape

fascia
band

fibrosa
fiber

fili/form
thread   shape

flexure
bend or fold

fossa
ditch

fundus
bottom

fungi/form
fungus   shape

gall/bladder
bile   sac

gastr/ic
stomach

gastro/-intestin/al
stomach   intestines

genio/hyoid
chin   hyoid bone

gingiva
gums

gingiv/al
gums

glosso/palatine
tongue   palate

hepat/ic
liver

hepato/pancreat/ic
liver   pancreas

ile/al
ileum

ileo/cec/al
ileum   cecum

ileum
portion of small intestine

incisor
to cut into

intestin/al
intestine

jejun/al
empty

jejunum
empty

lacte/al
milk

larynx
windpipe

lumen
light

mandible
a jaw

masseter
masticator

mes/enter/y
middle   intestine

meso/-appendix
middle   appendix

molar
mass

mucosa
mucous membrane

muc/ous
mucus

oculi
eye

Oddi
after a 19th Century Italian physician

omentum
covering membrane

omo/hyoid
shoulder   hyoid bone

orbicul/aris
orbit

orifice
opening

palate
roof of mouth

palatine
palate

palato/gloss/al
palate   tongue

palato/pharynge/al
palate   pharynx

pan/creas
all   flesh

pancreat/ic
pancreas

papillae (pl.)
nipple

par/otid
beside   ear

peri/dont/al
around   tooth

peri/toneum
around   to stretch

pharyngo/palatine
pharynx   palate

pharynx
throat

plexuses (pl.)
network

pre/molar
before   mass

pylor/ic
pylorus

pylorus
gate

quadr/ate
four, square

rectum
straight

ren/al
kidney

rugae
ridge or fold

saliv/ary
spittle

serosa
whey or serum

ser/ous
whey or serum

sigm/oid
letter "S"

sphincter
to bind fast

splen/ic
spleen

sterno/cleido/mastoid
sternum   clavicle   mastoid process

sterno/hyoid
sternum   hyoid bone

sub/lingu/al
beneath   tongue

sub/mandibul/ar
beneath   mandible

sub/maxill/ary
beneath   maxilla

sub/mucosa
beneath mucous membrane

sub/muc/ous
beneath mucus (mucous membrane)

tempor/al
time, temple

tonsil
a stake

trachea
rough

uvula
little grape

Vater
after a 17th Century German anatomist

vena cava
vein   hollow

ven/ule
vein   small

vermi/form
worm   shape

villi
tuft of hair

CHAPTER IX

GENITO-URINARY SYSTEM

The Genito-urinary or Urogenital system consists of the urinary organs which form and discharge the urine. The organs involved are the kidneys, ureters, bladder and urethra.

The genital organs of the system are concerned with the process of reproduction. They include the ovaries, oviducts, uterus and vagina of the female and the scrotum (containing the testes), penis, ductus deferens, seminal vesicles and prostate of the male.

The urinary organs are closely related to the reproductive organs in both the male and female. In the male the two systems have the same termination.

In the male, urinary and reproductive disturbances are treated by a urologist. The urologist also treats conditions of the urinary system in the female. Reproductive disturbances of the female are treated by a gynecologist.

RELATIONSHIP OF URINARY ORGANS, FEMALE REPRODUCTIVE ORGANS AND RECTUM

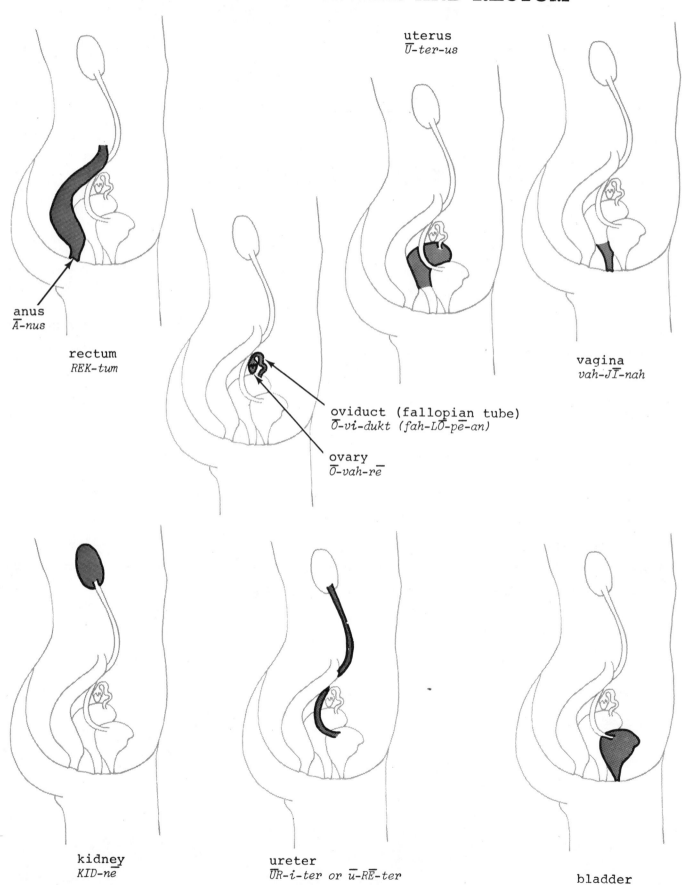

uterus
Ū-ter-us

anus
Ā-nus

rectum
REK-tum

oviduct (fallopian tube)
Ō-vi-dukt (fah-LŌ-pē-an)

ovary
Ō-vah-rē

vagina
vah-JĪ-nah

kidney
KID-nē

ureter
ŪR-i-ter or ū-RĒ-ter

bladder

FEMALE REPRODUCTIVE SYSTEM
AND RELATED STRUCTURES (Sagittal View)

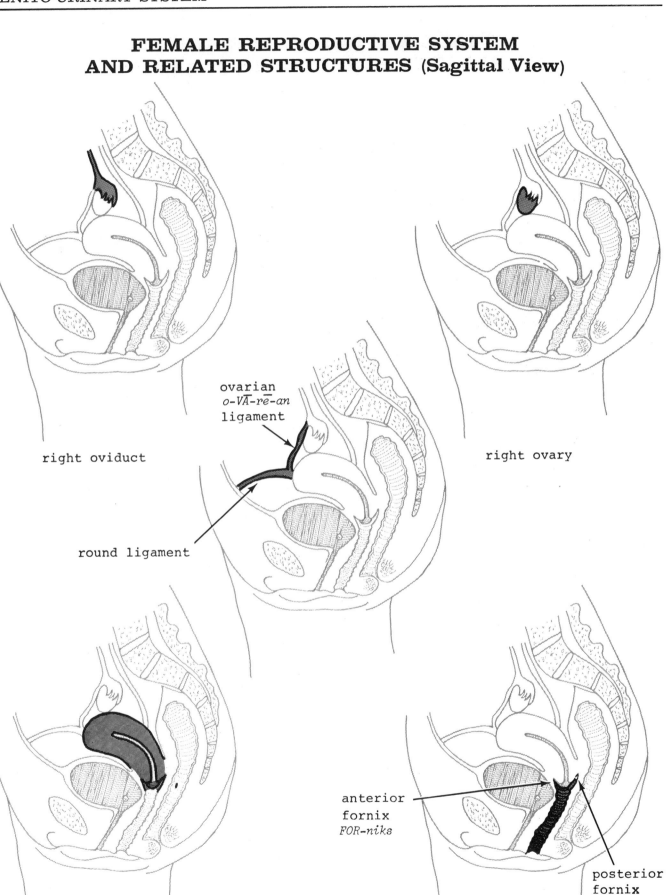

right oviduct

ovarian
o-VĀ-rē-an
ligament

right ovary

round ligament

uterus

anterior
fornix
FOR-niks

posterior
fornix

vagina

FEMALE REPRODUCTIVE SYSTEM
AND RELATED STRUCTURES (Sagittal View)

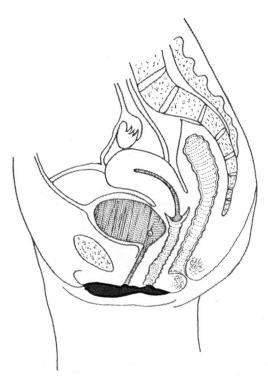

labium minorum
LĀ-bē-um mi-NŌ-rum

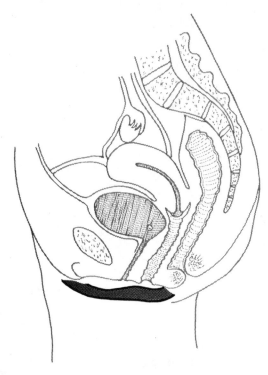

labium majorum
mah-JŌ-rum

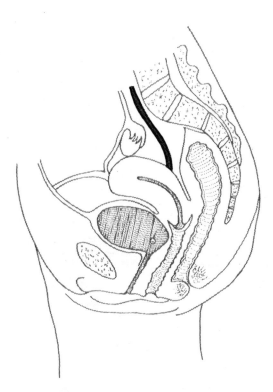

right ureter
ŪR-i-ter (ū-RĒ-ter)

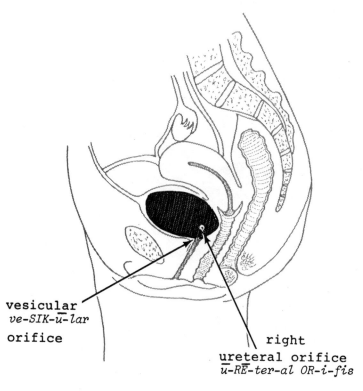

vesicular
ve-SIK-ū-lar
orifice

right
ureteral orifice
ū-RĒ-ter-al OR-i-fis

bladder

FEMALE REPRODUCTIVE SYSTEM
AND RELATED STRUCTURES (Sagittal View)

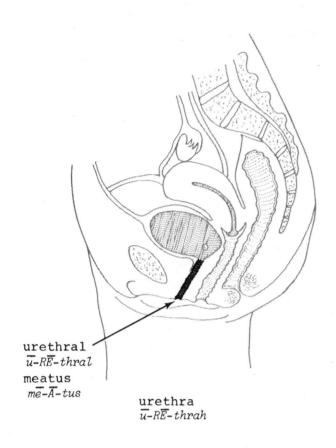

urethral
ū-RĒ-thral
meatus
mē-Ā-tus

urethra
ū-RĒ-thrah

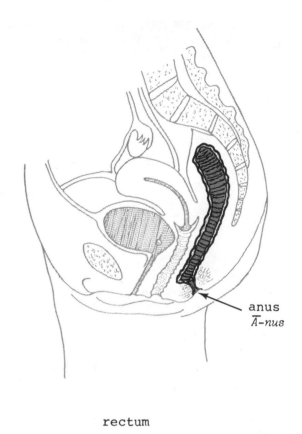

anus
Ā-nus

rectum

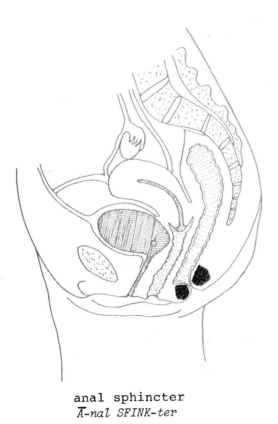

anal sphincter
Ā-nal SFINK-ter

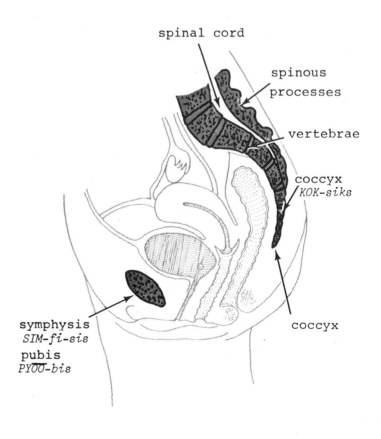

spinal cord

spinous
processes

vertebrae

coccyx
KOK-siks

symphysis
SIM-fi-sis
pubis
PYOO-bis

coccyx

EXTERNAL GENITALIA OF FEMALE

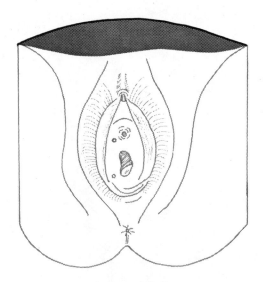

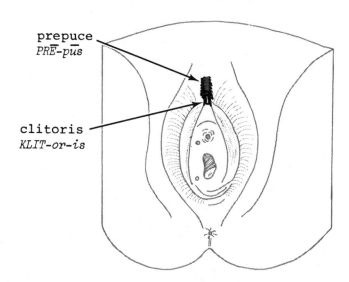

prepuce
PRĒ-pūs

clitoris
KLIT-or-is

mons pubis
monz PYOO-bis

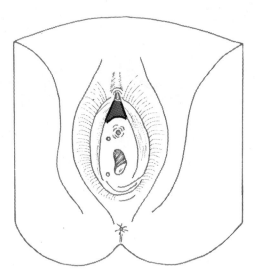

introitus or vestibule
in-TRŌ-ĭ-tus   *VES-ti-būl*

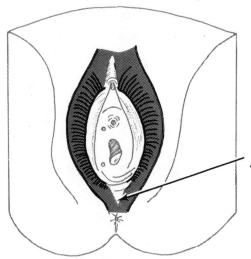

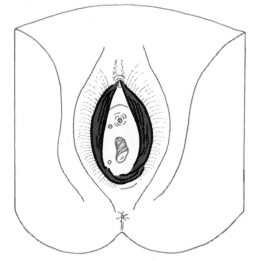

fourchet
fōor-SHET

labia majora
LĀ-bē-ah mah-JŌ-rah

labia minora
mi-NŌ-rah

← vulva →

EXTERNAL GENITALIA OF FEMALE

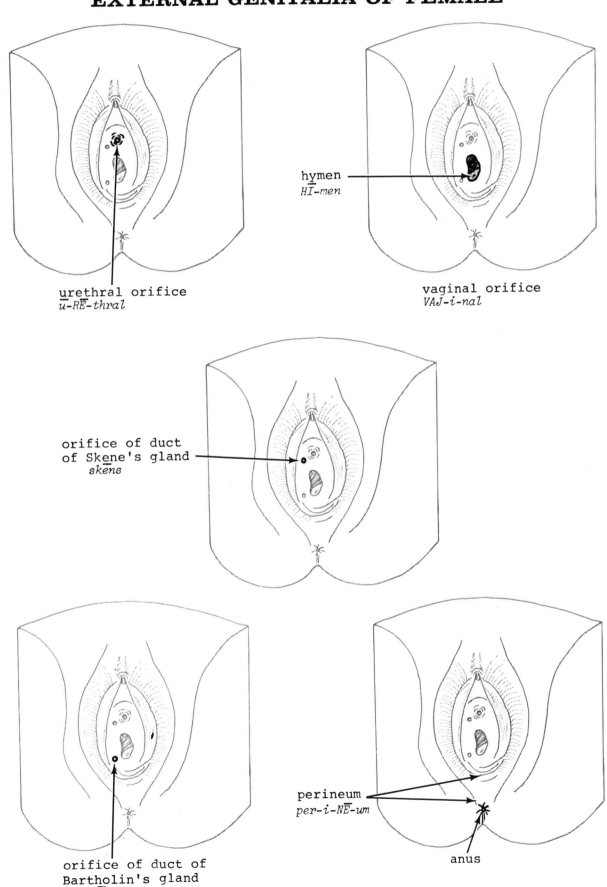

urethral orifice
u-RĒ-thral

hymen
HĪ-men

vaginal orifice
VAJ-i-nal

orifice of duct
of Skene's gland
skēns

orifice of duct of
Bartholin's gland
BAR-tō-linz or
BAR-thō-linz

perineum
per-i-NĒ-um

anus

SECTION THROUGH UTERUS AND VAGINA

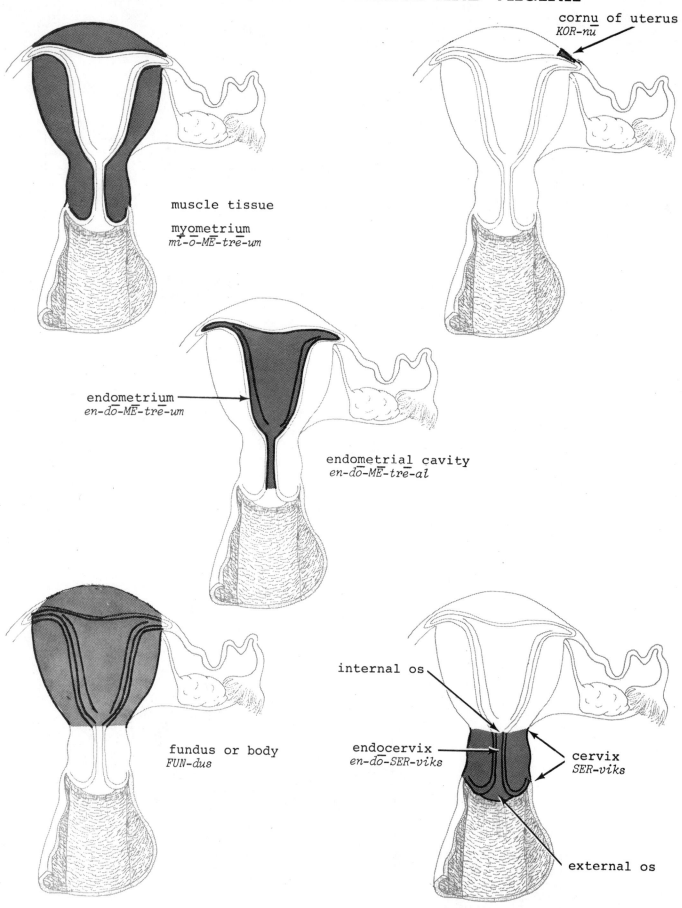

cornu of uterus
KOR-nū

muscle tissue

myometrium
mī-ō-MĒ-trē-um

endometrium
en-dō-MĒ-trē-um

endometrial cavity
en-dō-MĒ-trē-al

fundus or body
FUN-dus

internal os

endocervix
en-dō-SER-viks

cervix
SER-viks

external os

SECTION THROUGH UTERUS AND VAGINA

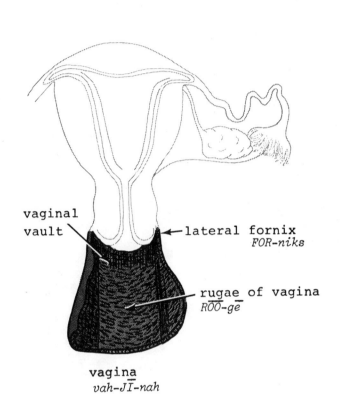

vaginal
vault

lateral fornix
FOR-niks

rugae of vagina
R͞OO-g͞e

vagina
vah-J͞I-nah

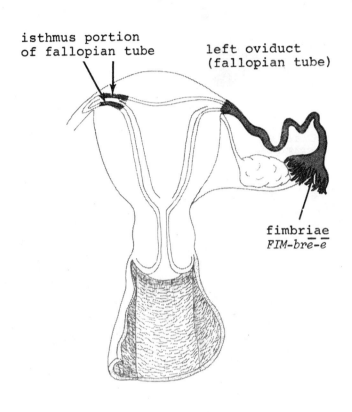

isthmus portion
of fallopian tube

left oviduct
(fallopian tube)

fimbriae
FIM-br͞e-͞e

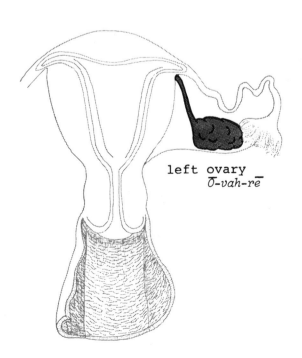

left ovary
͞O-vah-r͞e

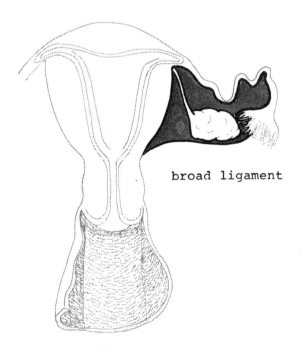

broad ligament

OVARY SECTIONED

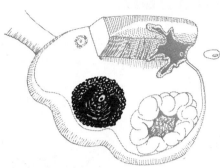

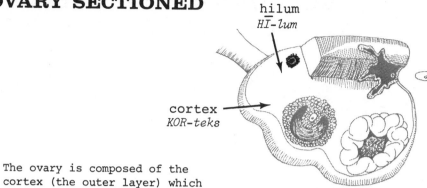

Graafian follicle
GRAF-ē-an FOL-i-kl
(a mature vesicular
follicle) *ve-SIK-ū-lar*

The ovary is composed of the
cortex (the outer layer) which
contains the follicles, and
the hilum which does not contain follicles. The blood
vessels enter through the
hilum. The cavity of a col-
lapsed follicle which has
ruptured to release an ovum,
contains some fluid tinged
with blood and is given the
name of corpus hemorrhagicum.

hilum
HĪ-lum

cortex
KOR-teks

immature follicle

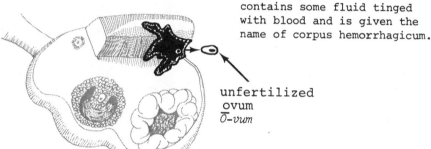

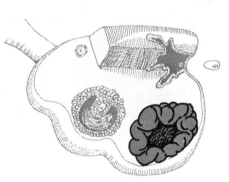

unfertilized
ovum
Ō-vum

corpus hemorrhagicum
hem-ō-RAJ-i-kum

corpus luteum
KOR-pus LŪ-tē-um

LOWER HALF OF BREAST DISSECTED
TO SHOW LACTIFEROUS DUCTS

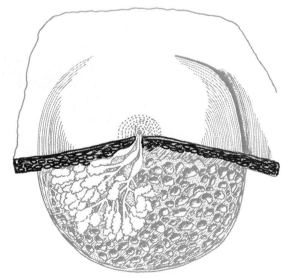

fat

loculi in connective tissue
LOK-ū-lī

LOWER HALF OF BREAST DISSECTED
TO SHOW LACTIFEROUS DUCTS

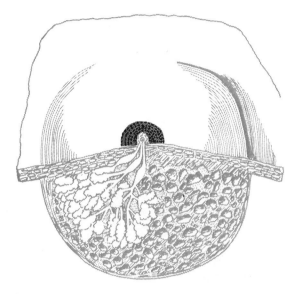

areola
ah-RĒ-ō-lah

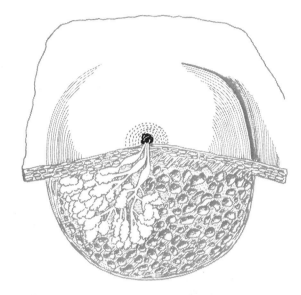

papilla
pah-PIL-ah
(nipple)

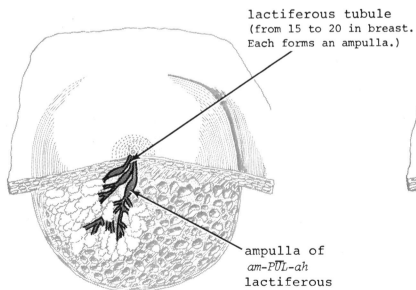

lactiferous tubule
(from 15 to 20 in breast.
Each forms an ampulla.)

ampulla of
am-PŪL-ah
lactiferous
duct

lactiferous ducts
lak-TIF-er-us

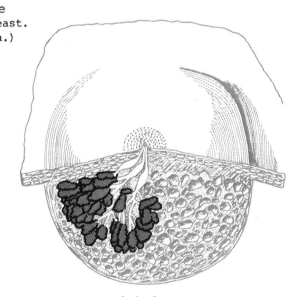

lobules
LOB-ūlε

SAGITTAL SECTION OF MALE REPRODUCTIVE
AND URINARY ORGANS

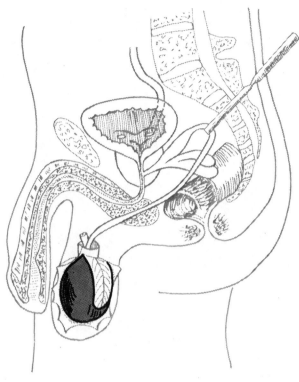

testis
TES-tis

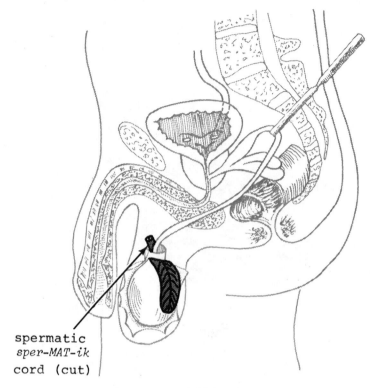

spermatic
sper-MAT-ik
cord (cut)

epididymis
ep-i-DID-i-mis

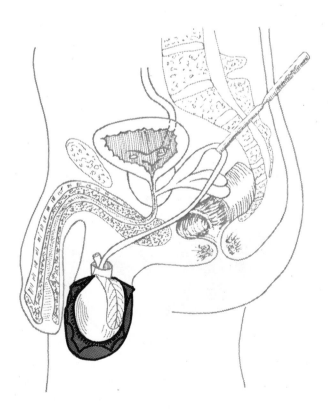

scrotum
SKRŌ-tum

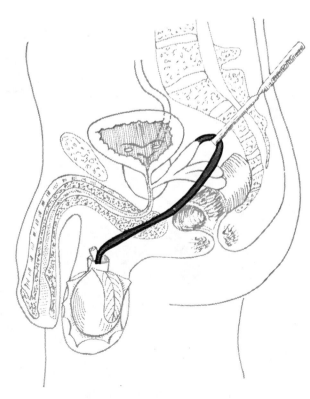

left ductus deferens
DUK-tus DEF-er-enz

SAGITTAL SECTION OF MALE REPRODUCTIVE
AND URINARY ORGANS

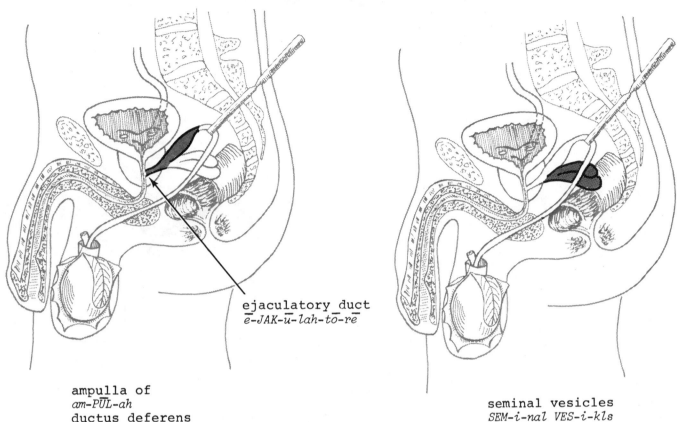

ejaculatory duct
$\overline{e}$-JAK-$\overline{u}$-lah-t$\overline{o}$-r$\overline{e}$

ampulla of
am-P$\overline{UL}$-ah
ductus deferens

seminal vesicles
SEM-i-nal VES-i-kls

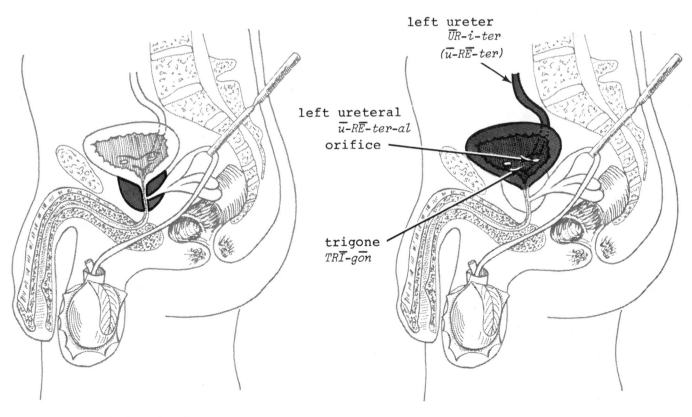

left ureter
$\overline{UR}$-i-ter
($\overline{u}$-R$\overline{E}$-ter)

left ureteral
$\overline{u}$-R$\overline{E}$-ter-al
orifice

trigone
TR$\overline{I}$-g$\overline{o}$n

prostate gland
PROS-t$\overline{a}$t

bladder

SAGITTAL SECTION OF MALE REPRODUCTIVE
AND URINARY ORGANS

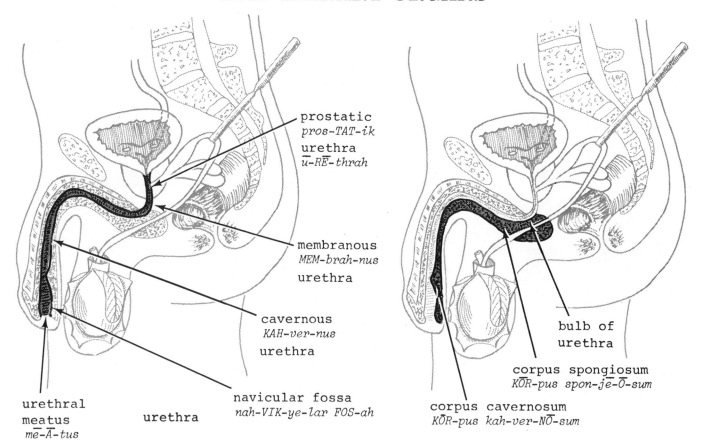

prostatic
pros-TAT-ik
urethra
ū-RĒ-thrah

membranous
MEM-brah-nus
urethra

cavernous
KAH-ver-nus
urethra

navicular fossa
nah-VIK-ye-lar FOS-ah

urethral
meatus
mē-Ā-tus

urethra

bulb of
urethra

corpus spongiosum
KŌR-pus spon-jē-Ō-sum

corpus cavernosum
KŌR-pus kah-ver-NŌ-sum

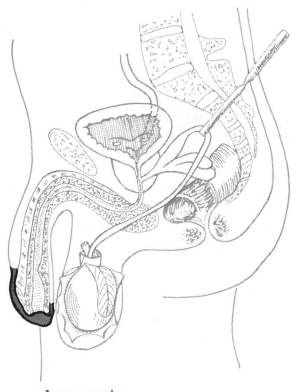

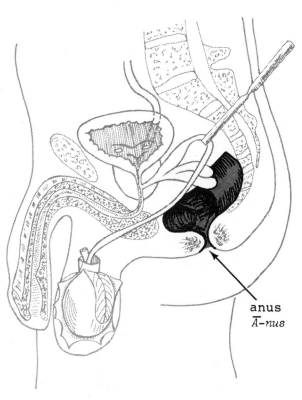

glans penis
glanz PĒ-nis

rectum
REK-tum

anus
Ā-nus

SAGITTAL SECTION OF MALE REPRODUCTIVE AND URINARY ORGANS

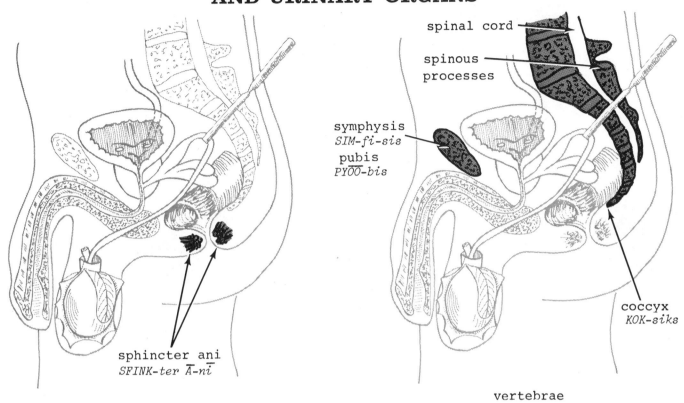

spinal cord

spinous processes

symphysis
SIM-fi-sis
pubis
PYOO-bis

coccyx
KOK-siks

sphincter ani
SFINK-ter A-ni

vertebrae

DORSAL VIEW OF BLADDER AND ACCESSORY MALE ORGANS

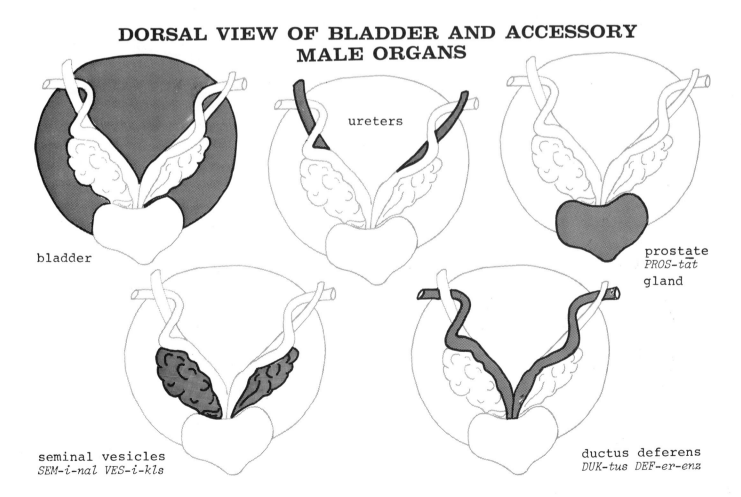

bladder

ureters

prostate
PROS-tat
gland

seminal vesicles
SEM-i-nal VES-i-kls

ductus deferens
DUK-tus DEF-er-enz

MALE ORGANS OF REPRODUCTION

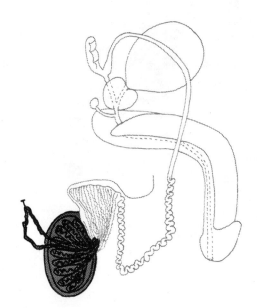

testis
TES-tis

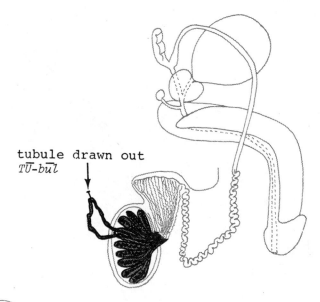

tubule drawn out
TŪ-būl

seminiferous tubules
se-mi-NIF-er-us

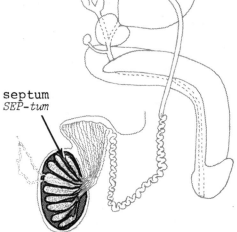

septum
SEP-tum

tunica vaginalis
TŪ-nik-ah vaj-i-NAL-is

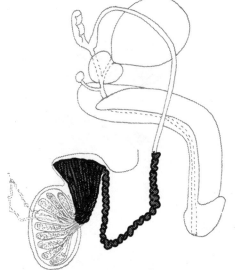

epididymis
ep-i-DID-i-mis

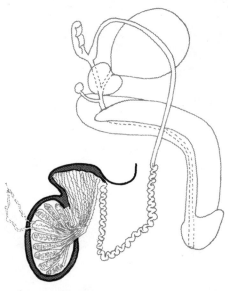

tunica albuginea
al-bū-JIN-ē-ah

MALE ORGANS OF REPRODUCTION

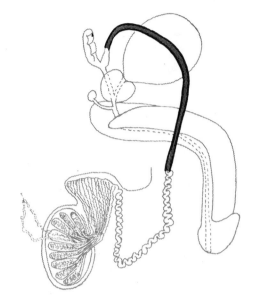

ductus deferens
DUK-tus DEF-er-enz
(vas deferens)
(vaz)

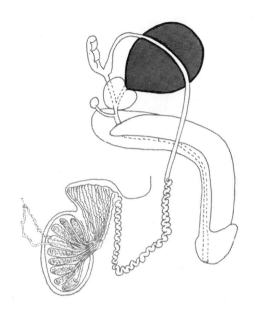

bladder

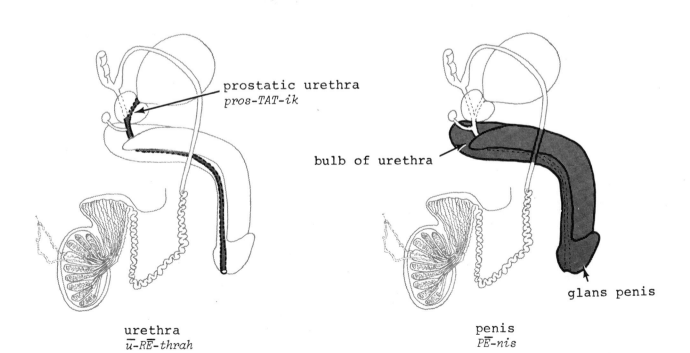

prostatic urethra
pros-TAT-ik

bulb of urethra

glans penis

urethra
ū-RĒ-thrah

penis
PĒ-nis

MALE ORGANS OF REPRODUCTION

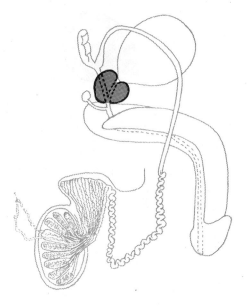

prostate gland

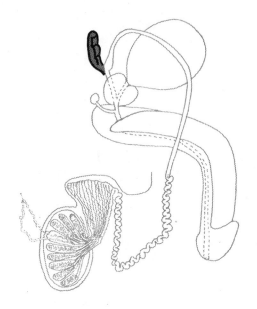

seminal vesicle

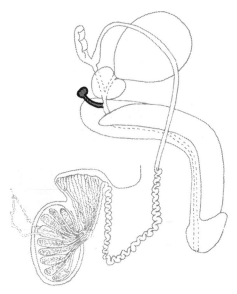

Cowper's gland
KOW-perz

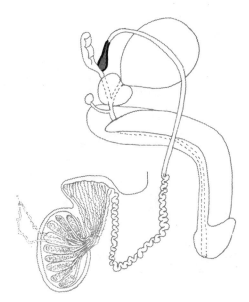

ampulla of vas
am-PŪL-ah   vaz

ANATOMICAL LOCATION OF URINARY SYSTEM

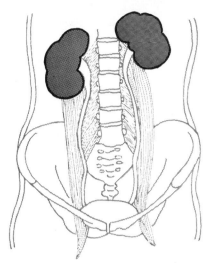

right and left kidneys
The right kidney is usually
lower than the left kidney.

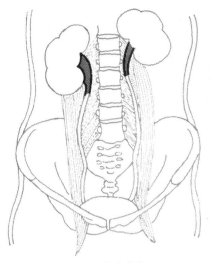

pelves of kidneys

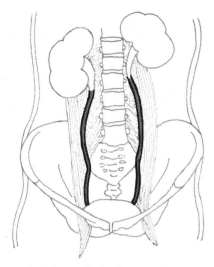

right and left ureters

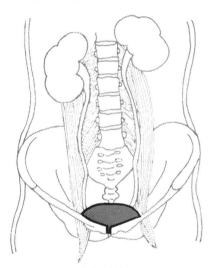

bladder

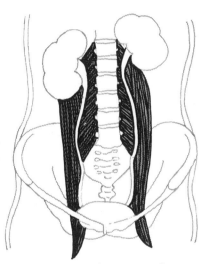

psoas major muscle
SŌ-as

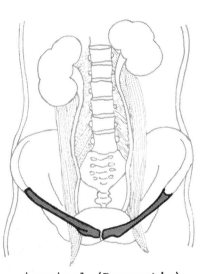

inguinal (Poupart's)
IN-gwin-al (POO-parts)
ligament

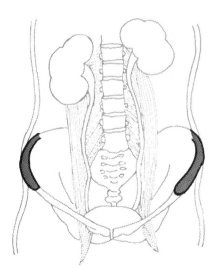

iliac crest

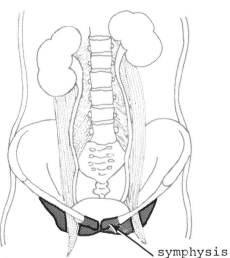

ischium   symphysis
                     pubis

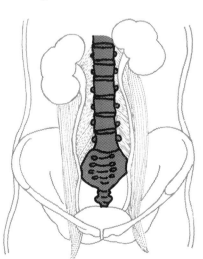

vertebrae

URINARY SYSTEM – MALE

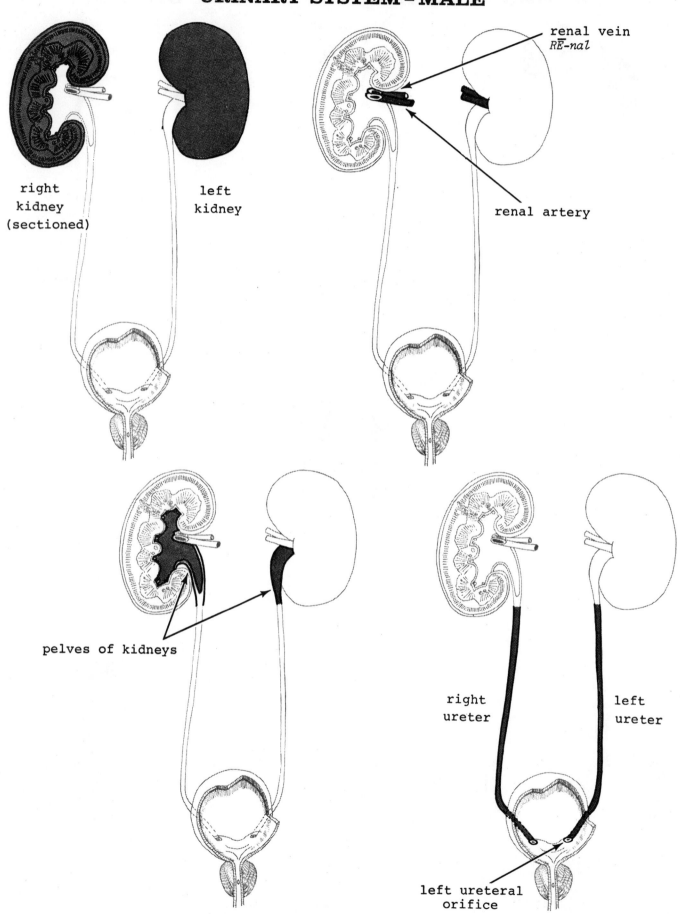

right
kidney
(sectioned)

left
kidney

renal vein
RĒ-nal

renal artery

pelves of kidneys

right
ureter

left
ureter

left ureteral
orifice

URINARY SYSTEM – MALE

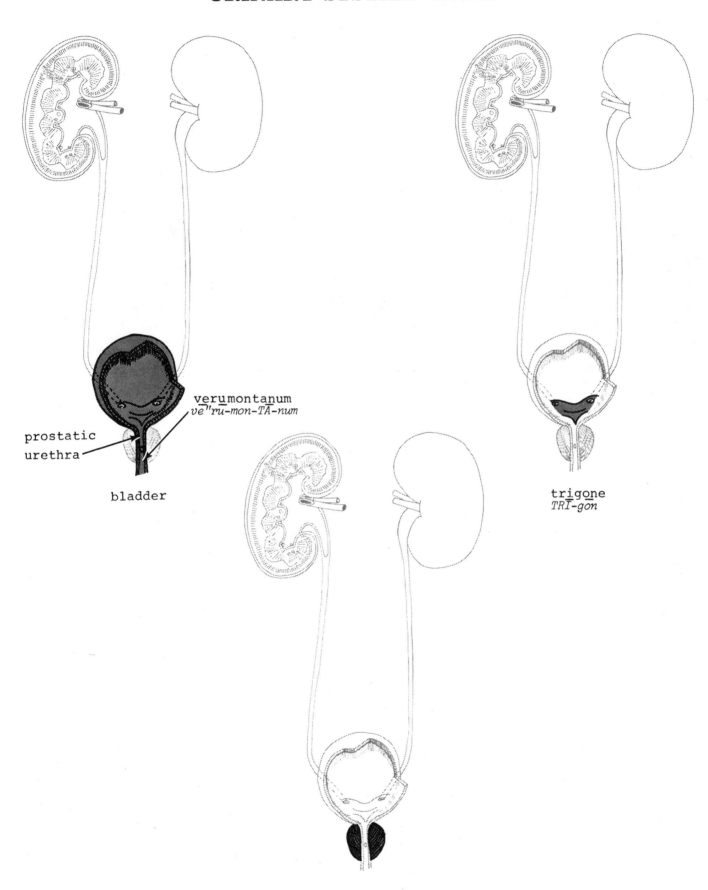

verumontanum
ve"ru-mon-TA-num

prostatic
urethra

bladder

trigone
TRI-gon

prostate gland

RIGHT KIDNEY SECTIONED

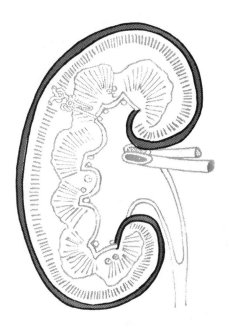

renal capsule

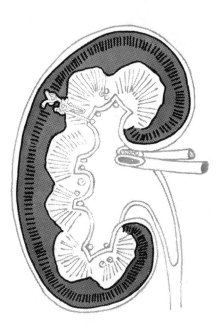

cortex

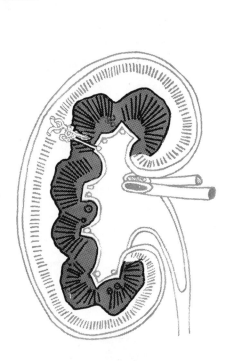

medulla
me-DUL-ah

orifice of nephron
in papilla
pah-PIL-ah

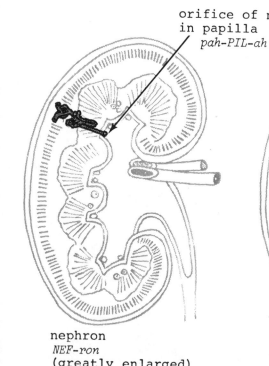

nephron
NEF-ron
(greatly enlarged)

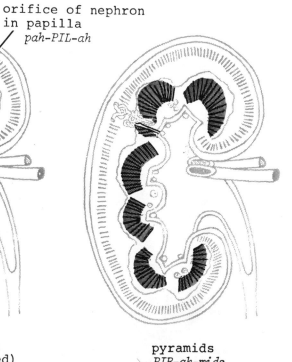

pyramids
PIR-ah-mids

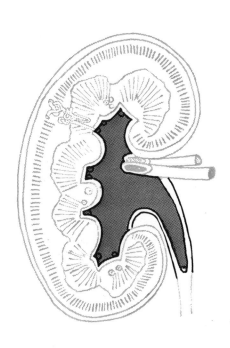

pelvis of kidney

RIGHT KIDNEY SECTIONED

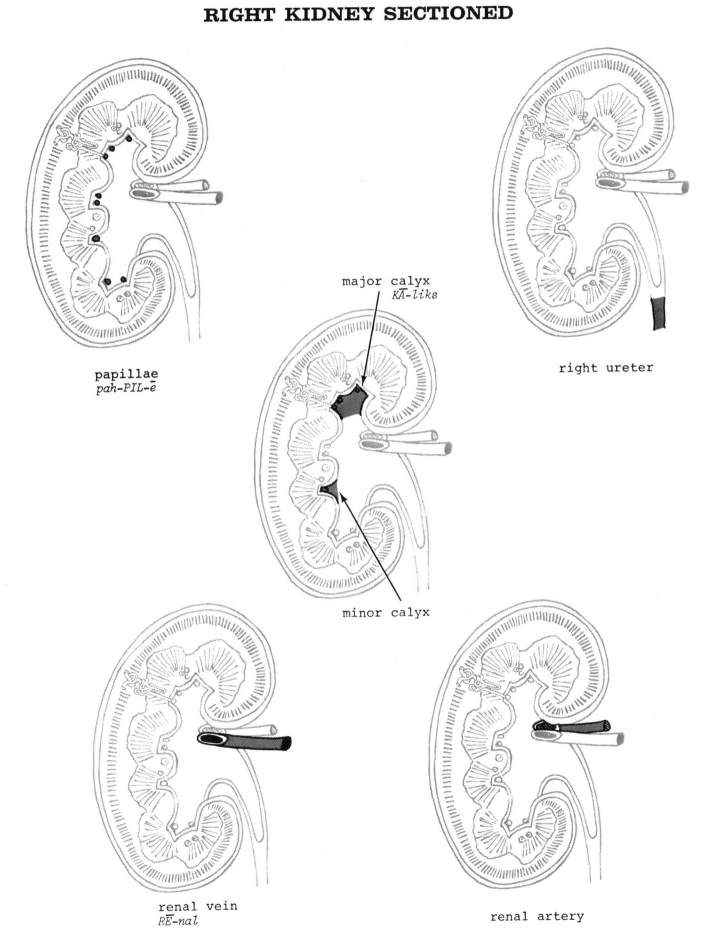

papillae
pah-PIL-ē

major calyx
KĀ-liks

minor calyx

right ureter

renal vein
RĒ-nal

renal artery

NEPHRON UNIT OF KIDNEY

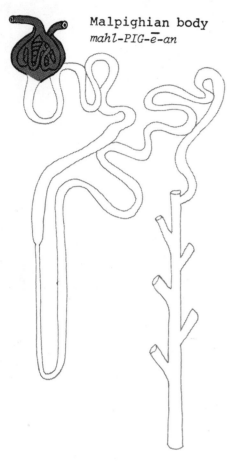

Malpighian body
mahl-PIG-ē-an

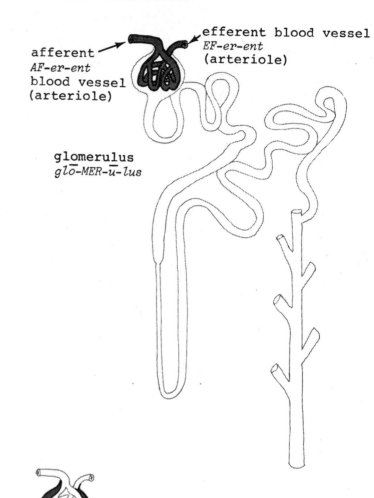

afferent
AF-er-ent
blood vessel
(arteriole)

efferent blood vessel
EF-er-ent
(arteriole)

glomerulus
glō-MER-ū-lus

The Malpighian body is made up of
the glomerulus and Bowman's capsule.

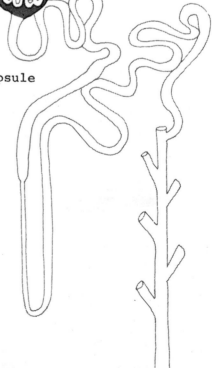

Bowman's capsule
BŌ-manz

The Malpighian body, the proximal and distal
convoluted tubes are located in the cortex.
Henle's loop and the collecting tubule are
continued into the medulla. The urine empties
from the collecting tubule into the papilla
of the calyx. Actually, many tubules join together to form a central duct (the duct of
Bellini) which then empties into the papilla.
The urine then goes into the pelvis of the
kidney from where it goes into the ureter
to the bladder and exists through the urethra.

NEPHRON UNIT OF KIDNEY

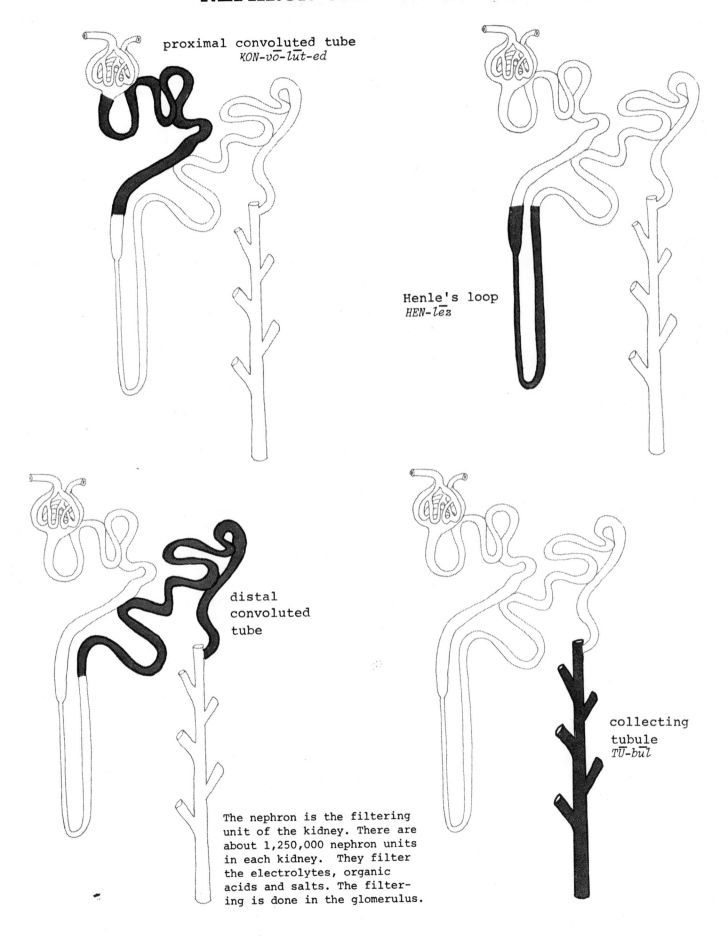

proximal convoluted tube
KON-vō-lūt-ed

Henle's loop
HEN-lēz

distal
convoluted
tube

collecting
tubule
TŪ-būl

The nephron is the filtering
unit of the kidney. There are
about 1,250,000 nephron units
in each kidney. They filter
the electrolytes, organic
acids and salts. The filtering is done in the glomerulus.

INDEX OF ENGLISH TRANSLATION

af/ferent
toward   to carry

albuginea
white

ampulla
small jar

an/al
anus

ani
anus

anus
from "as to sit" or "annus"- "a ring"

areola
small space

arteri/ole
artery   small

Bartholin's
after a 17th Century Danish anatomist

Bellini
after a 17th Century Italian anatomist

Bowman's
after a 19th Century English physician

calyx
cup of flower

cavern/ous
hollow

cavernosum
hollow

cervix
neck

clitoris
key to door

coccyx
cuckoo

con/voluted
together   to roll

cornu
horn

corpus
body

cortex
bark or rind

Cowper's
after a 17th Century English anatomist

deferens
to carry down

ductus
duct

ef/ferent
out   to carry

ejaculat/ory
to throw out

endo/cervix
within   neck

endo/metri/al
within   uterus

endo/metrium
within   uterus

epi/didymis
on, upon   twin

fallopian
after a 16th Century Italian anatomist

fimbriae
a fringe

foll/icle
bag   little

fornix
arch, vault

fossa
ditch

fourchet
little   fork

fundus
bottom (the point furthest removed
from the opening of an organ)

glans
gland

glomer/ulus
ball   small

Graafian
after a 17th Century Dutch physiologist

hemo/rrhagicum
blood   to burst forth

Henle's
after a 19th Century German anatomist

hilum
a small bit or trifle

hymen
membrane

ili/ac
hip bone

inguin/al
groin

intro/itus
into   to go

ischium
hip, haunch

kidney
a structure near the womb

labia (pl.)
lip

labium (sing.)
lip

lacti/fer/ous
milk   to bear

lob/ules
lobes   small

loc/uli
place   small

luteum
yellow

majora (pl.)
large

majorum (sing.)
large

Malpighian
after a 17th Century Italian anatomist

meatus
a passage

medulla
marrow

membran/ous
membrane

minora (pl.)
small

minorum (sing.)
small

mons
mountain

myo/metrium
muscle   uterus

navic/ul/ar
ship   small

nephron
kidney

orifice
mouth or opening

os
mouth

ovari/an
egg bearer

ovary
egg bearer

ovi/duct
egg   to draw

papilla
nipple

penis
to hang down

perineum
floor of pelvis

Poupart's
after a 17th Century French anatomist

pre/puce
before   penis

pros/tate
before   to stand

prostat/ic
prostate

psoas
loin

pubis
pubic bone

pyramid
a cone shaped eminence

rectum
straight

ren/al
kidney

rugae (pl.)
wrinkle or crease

scrotum
bag

semin/al
seed

semini/fer/ous
seed   to bear

septum
wall off

Skene's
after a Brooklyn gynecologist

spermat/ic
seed

sphincter
to bind fast

spongiosum
sponge

sym/physis
together   to grow

testis
a witness (testicle)

trigone
triangle

tub/ule
tube   small

tunica
coat

ureter
urinary canal

ureter/al
ureter

urethra
canal leading from bladder

urethr/al
urethra

uterus
womb

vagina
hollow

vagin/al
vagina

vagin/alis
vagina

vas
vessel

veru/montanum
a spit,   mountain
dart or ridge

vertebrae (pl.)
to turn

ves/icles
bladder   small

vesic/ul/ar
bladder   small

vestibule
an entrance court

vulva
a covering

CHAPTER X

SENSES

Taste (the Mouth) is included in the Gastro-Intestinal System.

Smell (the Nose) is included in the Respiratory System.

The three remaining Senses are Touch (the Skin) or Integumentary System, Hearing (the Ear) and Sight (the Eye). These appear in the following pages.

Abnormal conditions of the skin are treated by a dermatologist, of the ear by an otologist and of the eye by an ophthalmologist.

Otologists may also include in their practice rhinology (the nose), laryngology (the larynx and throat). They are then known as ENT (Ear, Nose, Throat) practitioners. Some may also include in their practice the treatment of the eye for which they receive the initials EENT (Eye, Ear, Nose, Throat). There are other medical titles given these specialities such as otorhinolaryngologist but for the layman the initials prove to be the easiest to understand.

TOUCH – THE SKIN (INTEGUMENTARY SYSTEM)
HAIR FOLLICLE

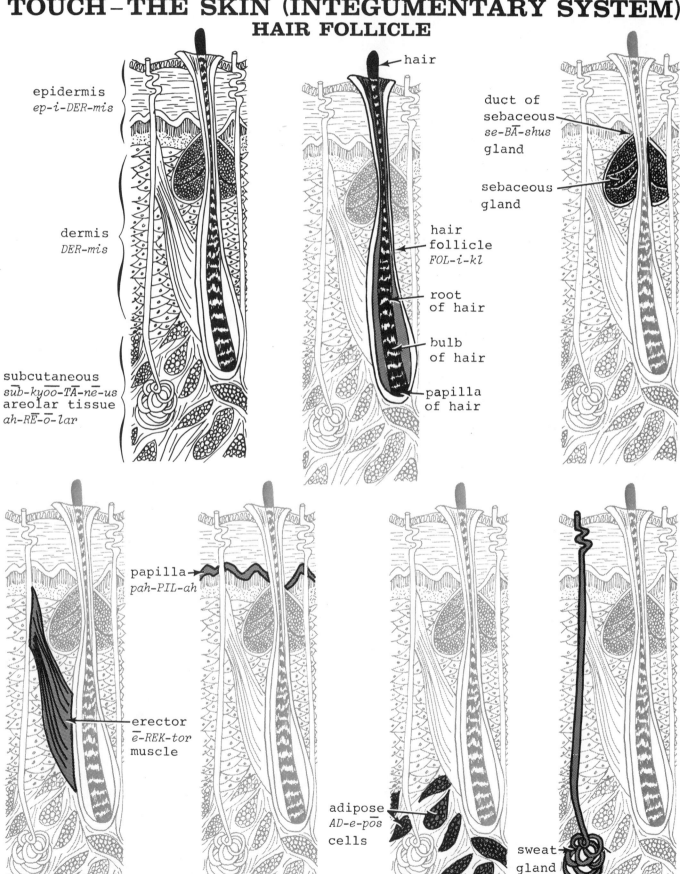

epidermis
ep-i-DER-mis

dermis
DER-mis

subcutaneous
sŭb-kyōō-TĀ-nē-us
areolar tissue
ah-RĒ-ō-lar

hair

hair
follicle
FOL-i-kl

root
of hair

bulb
of hair

papilla
of hair

duct of
sebaceous
se-BĀ-shus
gland

sebaceous
gland

papilla
pah-PIL-ah

erector
ē-REK-tor
muscle

adipose
AD-e-pōs
cells

sweat
gland

STRUCTURE OF SKIN

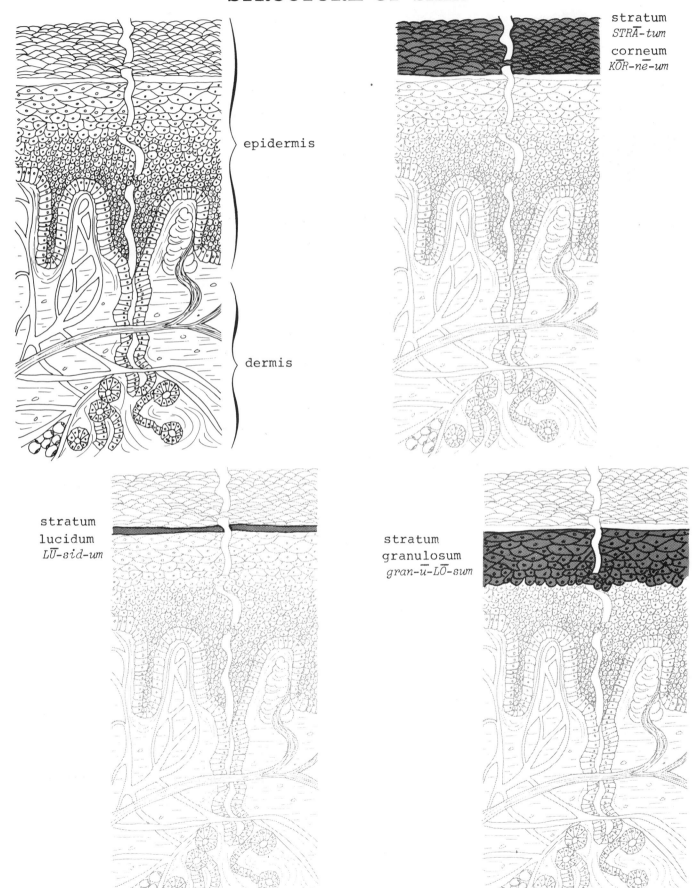

epidermis

dermis

stratum
STRĀ-tum

corneum
KŌR-nē-um

stratum
lucidum
LŪ-sid-um

stratum
granulosum
gran-ū-LŌ-sum

STRUCTURE OF SKIN

stratum
germinatum
jer-mi-NĀ-tum

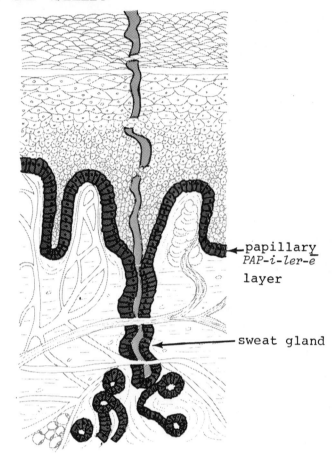

papillary
PAP-i-ler-ē
layer

sweat gland

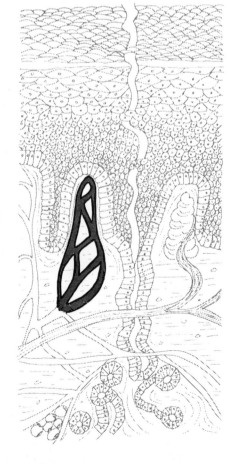

papilla

nerve

STRUCTURE OF SKIN

tactile corpuscle
TAK-til KŌR-pus-l

fat cells

STRUCTURE OF FINGERNAIL

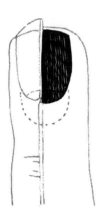

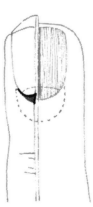

cuticle
KYOO-ti-kl

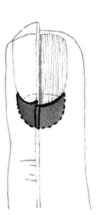

buried part
of nail

nail bed

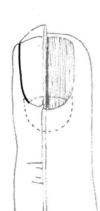

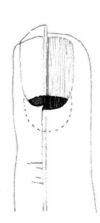

lateral nail groove

lunula
LŪ-nu-lah

HEARING–THE EAR
AURICLE (PINNA) OF EXTERNAL EAR

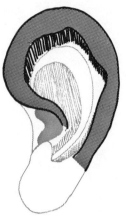

helix
HĒ-liks

antihelix
an-te-HĒ-liks

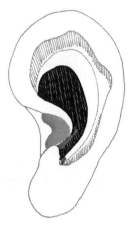

concha
KONG-kah

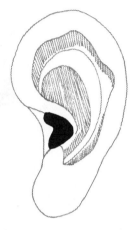

external auditory meatus
AW-de-tō-rē mē-Ā-tus

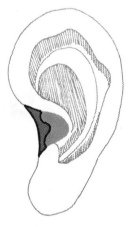

tragus
TRĀ-gus

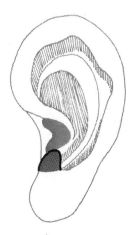

antitragus
an-te-TRĀ-gus

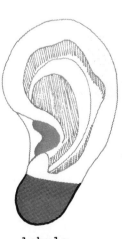

lobule
LOB-ūl

DIVISIONS OF THE EAR

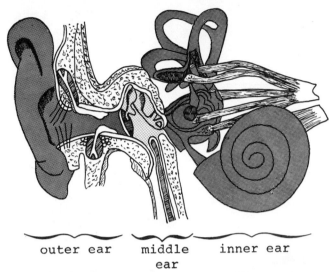

outer ear   middle   inner ear
             ear

STRUCTURE OF THE EAR

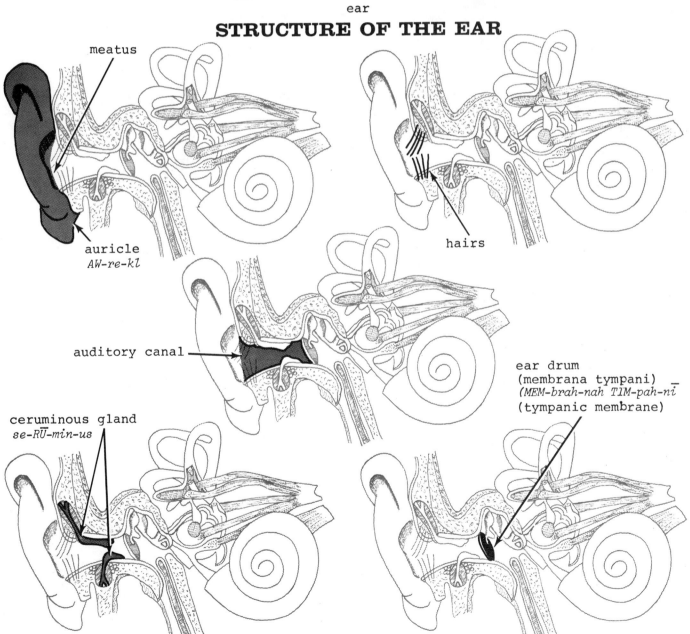

meatus

auricle
AW-re-kl

hairs

auditory canal

ear drum
(membrana tympani)
(MEM-brah-nah TIM-pah-ni
(tympanic membrane)

ceruminous gland
se-RŪ-min-us

STRUCTURE OF THE EAR

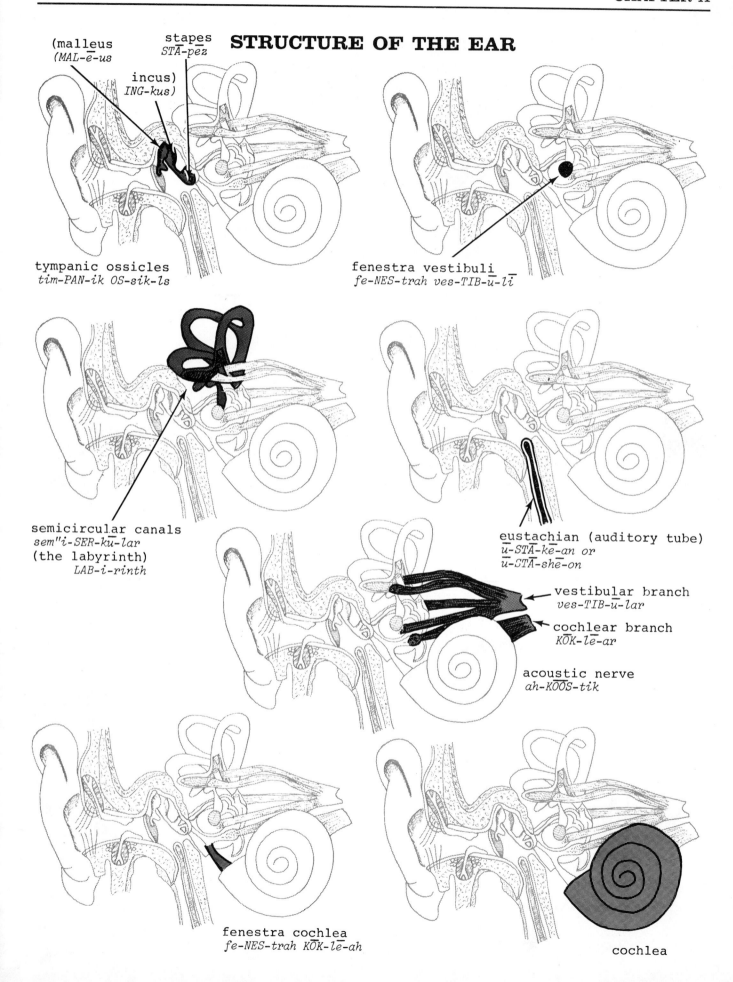

(malleus
(MAL-ē̄-us

stapes
STĀ-pēz

incus)
ING-kus)

tympanic ossicles
tim-PAN-ik OS-sik-ls

fenestra vestibuli
fe-NES-trah ves-TIB-ū-lī̄

semicircular canals
sem"i-SER-kū-lar
(the labyrinth)
LAB-i-rinth

eustachian (auditory tube)
ū-STĀ-ke-an or
ū-STĀ-she-on

vestibular branch
ves-TIB-ū-lar

cochlear branch
KŌK-le-ar

acoustic nerve
ah-KŌŌS-tik

fenestra cochlea
fe-NES-trah KŌK-le-ah

cochlea

SIGHT – THE EYE
FRONT VIEW OF RIGHT EYE

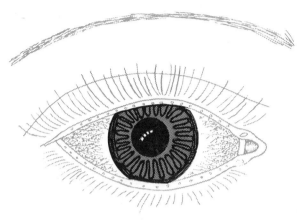

cornea
KŌR-nē-ah

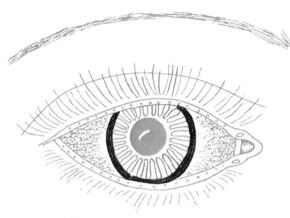

limbus
LIM-bus

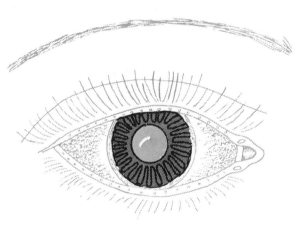

iris
Ī-ris

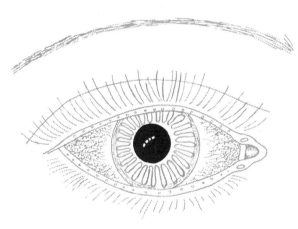

lens
lenz

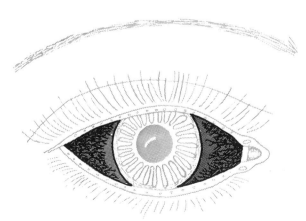

sclera
SKLĒ-rah or SKLER-ah

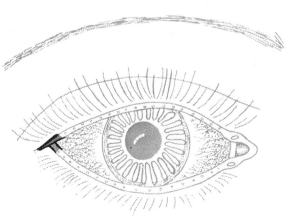

lateral (external) canthus
KAN-thus

FRONT VIEW OF RIGHT EYE

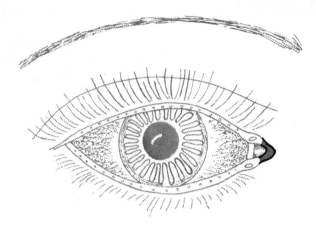

medial(internal)canthus

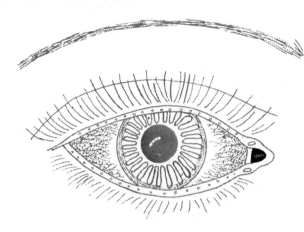

caruncle
KAR-ung-kl

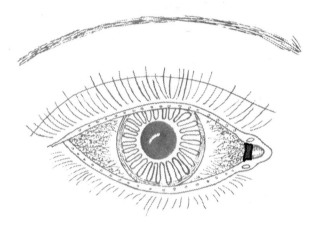

plica semilunaris
PLĪ-kah sem"ē-lū-NAR-is

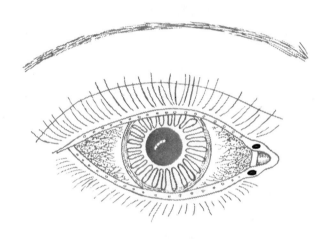

puncta lacrimalia
PUNK-tah lak-ri-MĀ-lē-ah

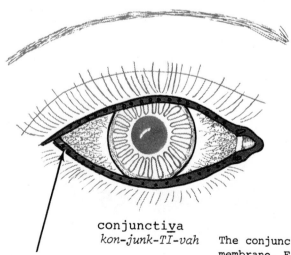

conjunctiva
kon-junk-TĪ-vah

lid margin of
conjunctiva

The conjunctiva is a continuous
membrane. From the lid margin of
the upper lid it lines the lid,
then continues over the sclera,
lines the lower lid and ends at
the lower lid margin.

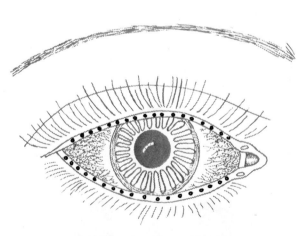

orifices of meibomian
OR-i-fis-es   mi-BŌ-me-an
or tarsal glands
TAHR-sal

STRUCTURE OF LID AND TEAR APPARATUS

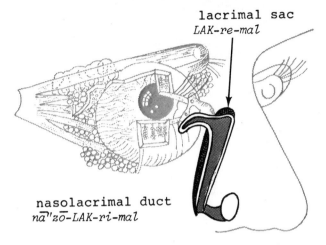

lacrimal sac
LAK-re-mal

nasolacrimal duct
nā"zō-LAK-ri-mal

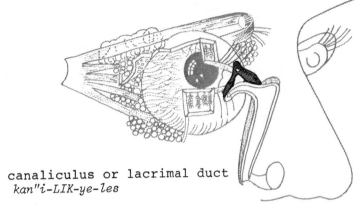

canaliculus or lacrimal duct
kan"i-LIK-ye-les

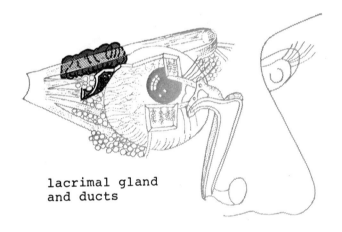

lacrimal gland
and ducts

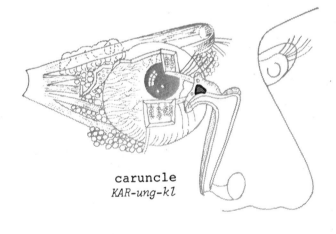

caruncle
KAR-ung-kl

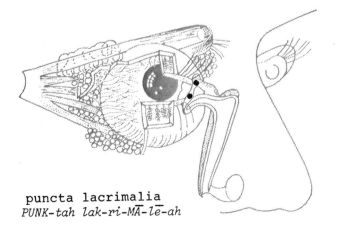

puncta lacrimalia
PUNK-tah lak-ri-MĀ-lē-ah

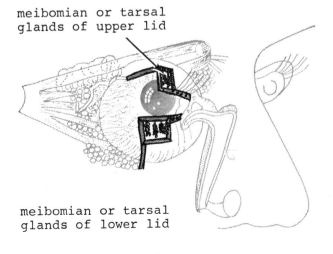

meibomian or tarsal
glands of upper lid

meibomian or tarsal
glands of lower lid

STRUCTURE OF LID AND TEAR APPARATUS

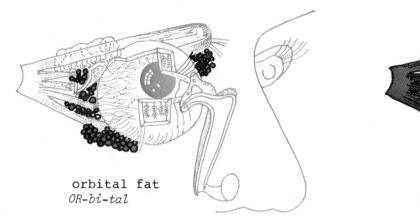

orbital fat
OR-bi-tal

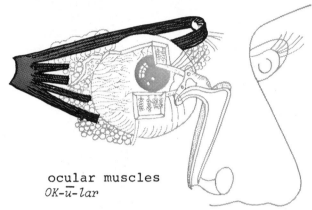

ocular muscles
OK-u-lar

HORIZONTAL SECTION OF LEFT EYE
VIEWED FROM ABOVE

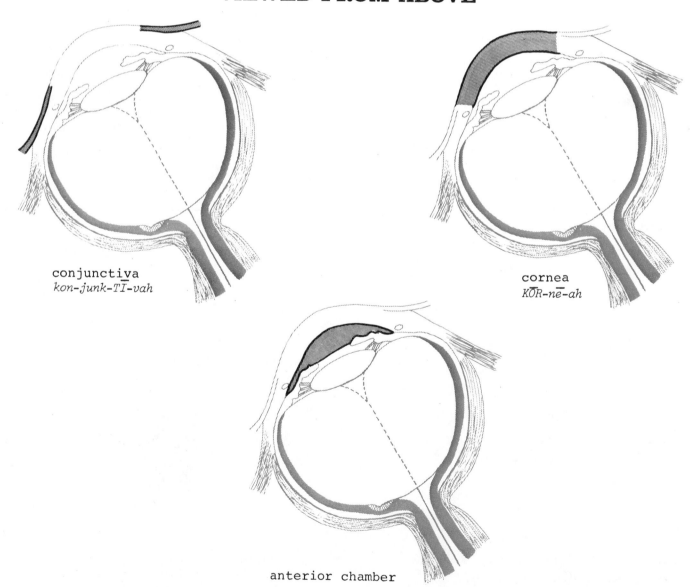

conjunctiva
kon-junk-TI-vah

cornea
KOR-ne-ah

anterior chamber

HORIZONTAL SECTION OF LEFT EYE
VIEWED FROM ABOVE

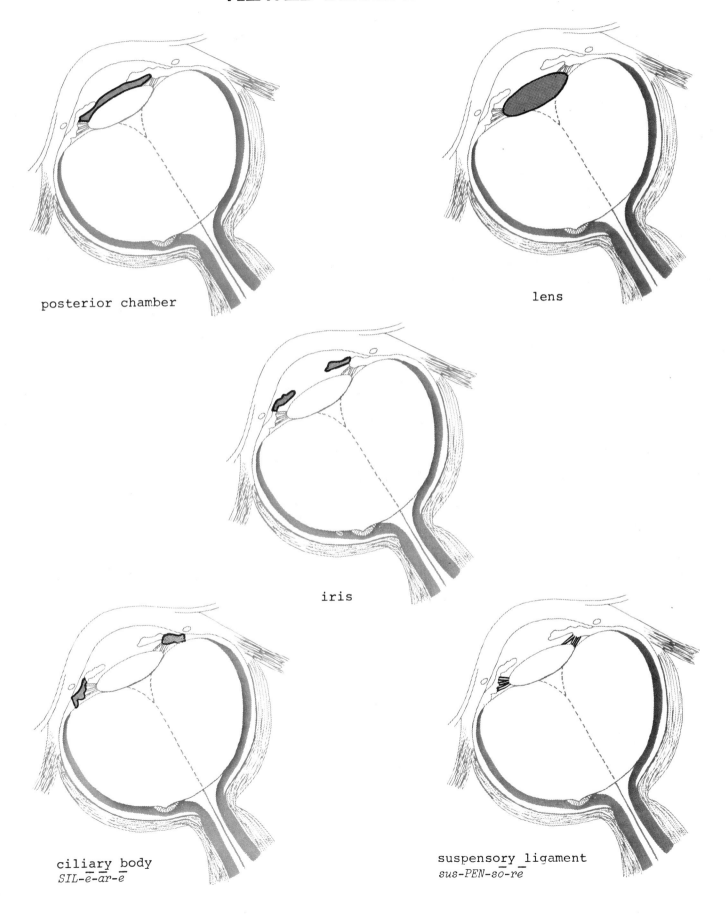

posterior chamber

lens

iris

ciliary body
SIL-ē-ar-ē

suspensory ligament
sus-PEN-sō-rē

HORIZONTAL SECTION OF LEFT EYE
VIEWED FROM ABOVE

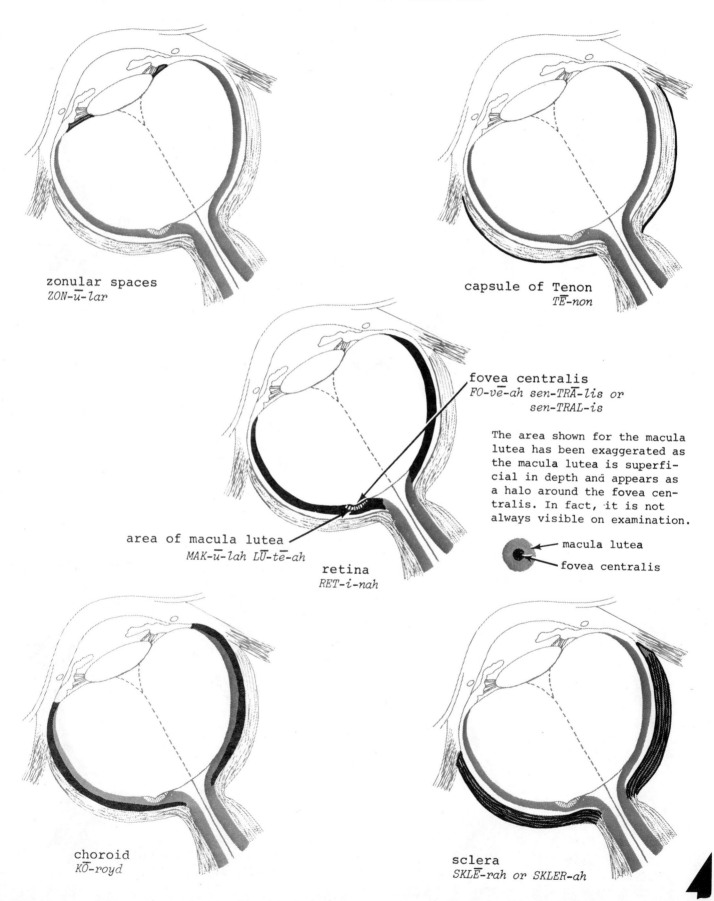

zonular spaces
ZON-ū-lar

capsule of Tenon
TĒ-non

fovea centralis
*FO-vē-ah sen-TRĀ-lis or
sen-TRAL-is*

The area shown for the macula
lutea has been exaggerated as
the macula lutea is superficial in depth and appears as
a halo around the fovea centralis. In fact, it is not
always visible on examination.

macula lutea
fovea centralis

area of macula lutea
MAK-ū-lah LŪ-tē-ah

retina
RET-i-nah

choroid
KŌ-royd

sclera
SKLĒ-rah or SKLER-ah

HORIZONTAL SECTION OF LEFT EYE
VIEWED FROM ABOVE

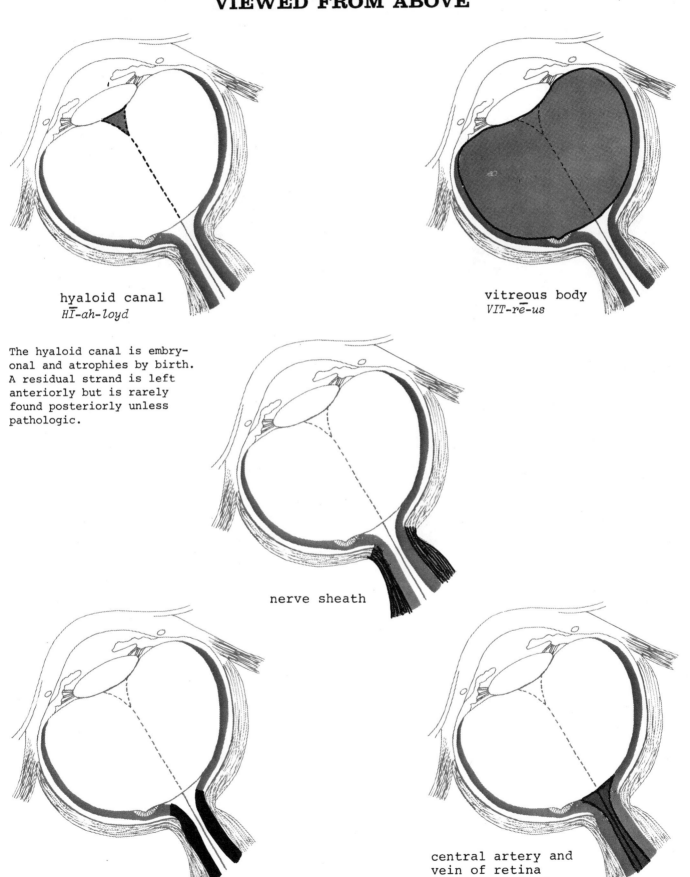

hyaloid canal
HĪ-ah-loyd

vitreous body
VIT-rē-us

The hyaloid canal is embry-
onal and atrophies by birth.
A residual strand is left
anteriorly but is rarely
found posteriorly unless
pathologic.

nerve sheath

optic nerve

central artery and
vein of retina
RET-i-nah

HORIZONTAL SECTION OF LEFT EYE
VIEWED FROM ABOVE

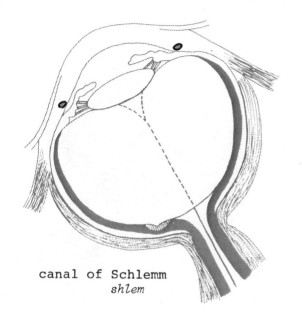

canal of Schlemm
shlem

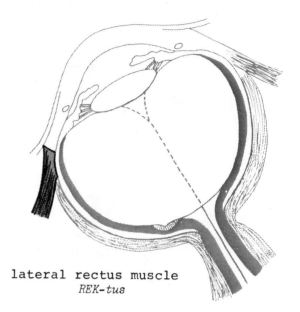

lateral rectus muscle
REK-tus

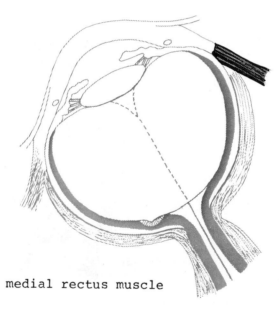

medial rectus muscle

INDEX OF ENGLISH TRANSLATION

acoust/ic
to hear

adip/ose
fat

anti/helix
against   helix (coil)

anti/tragus
against   tragus (goat)

areol/ar
areola

audit/ory
hearing

aur/icle
ear   little

canali/culus
canal   small

canthus
corner of eye

caruncle
small flesh

centr/alis
center

cerumin/ous
wax

chor/oid
skin

cili/ary
eyelash

cochlea (sing.)
snail shell

cochle/ar
cochlea

concha
snail

con/junctiva
together   to join

cornea
horny

corneum
horny

corpus/cle
body   little

cut/icle
skin   small

dermis
skin

epi/dermis
on, upon   skin

erect/or
to make rigid

eustachian
after a 16th Century Italian anatomist

fenestra
window

foll/icle
bag   little

fovea
pit

germinatum
seed

granulosum
granule

helix
coil

hyal/oid
glass

incus
anvil

labyrinth
a maze

lacrim/al
tear

lacrimalia
tear

limbus
a border

lob/ule
lobe   small

lucidum
clear

lun/ula
moon   small

lutea
yellow

macula
spot

malleus
hammer

meatus
a passage

meibomian
After a 17th century German anatomist.

membrana
membrane

naso/lacrim/al
nose   tear

ocul/ar
eye

orbit/al
orbit

opt/ic
eye

orifices (pl.)
mouth, opening

oss/icles (pl.)
bone   little

papilla
nipple

pinna
wing

plica
a fold

puncta
point

rectus
straight

zon/ul/ar
zone   little

retina
a net

Schlemm
after a 19th Century German physiologist

sclera
hard

sebace/ous
oily

semi /lun/aris
half   moon

stapes
stirrup

stratum
layer

sub/cutane/ous
beneath   skin

suspens/ory
to hold up a part

tactile
touch

tars/al
**a broad,
flat surface**

Tenon
after an 18th Century French surgeon

tragus
goat

tympani
drum

tympan/ic
drum

vestibul/ar
vestibule

vestibuli
vestibule

vitre/ous
glassy

INDEX

The pages listed in different styles of type indicate the following:

italic signifies the pronunciation is given for the term on that page.
standard signifies only that the term appears on given pages.
gothic signifies the page on which the term is translated.

A comparison of these types in number form would appear as follows:

(italic) 1, (standard) 1, (gothic) 1.

STUDY SECTION

for

ILLUSTRATED
PROGRAMMED
GROSS ANATOMY

by

Verlee E. Gross

NOTE: For convenience the pages of the Study Section are perforated so they can be removed from the book. They are also drilled for use in a loose-leaf binder.

TO REMOVE PAGES: Fold each page at the perforation before tearing it out. This will prevent tearing of the paper outside of the perforation at the binding.

CHAPTER II

SKELETAL SYSTEM

DIVISIONS OF SKELETAL SYSTEM

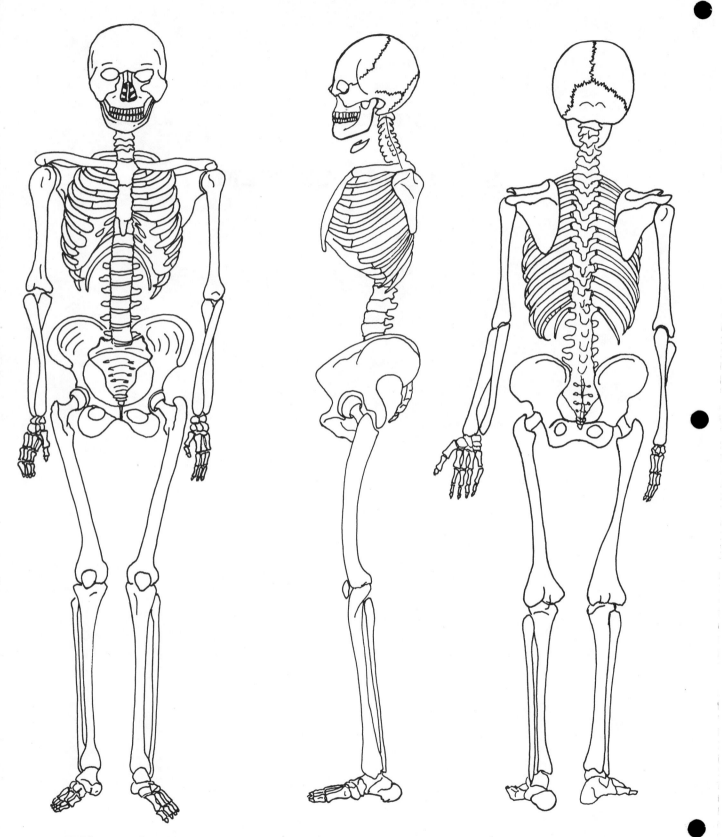

Indicate the divisions of the skeleton by coloring in the areas as follows:

CRANIAL and FACIAL BONES — RED   UPPER EXTREMITIES — PURPLE
TRUNK — BLUE   LOWER EXTREMITIES — GREEN

CRANIAL AND FACIAL BONES

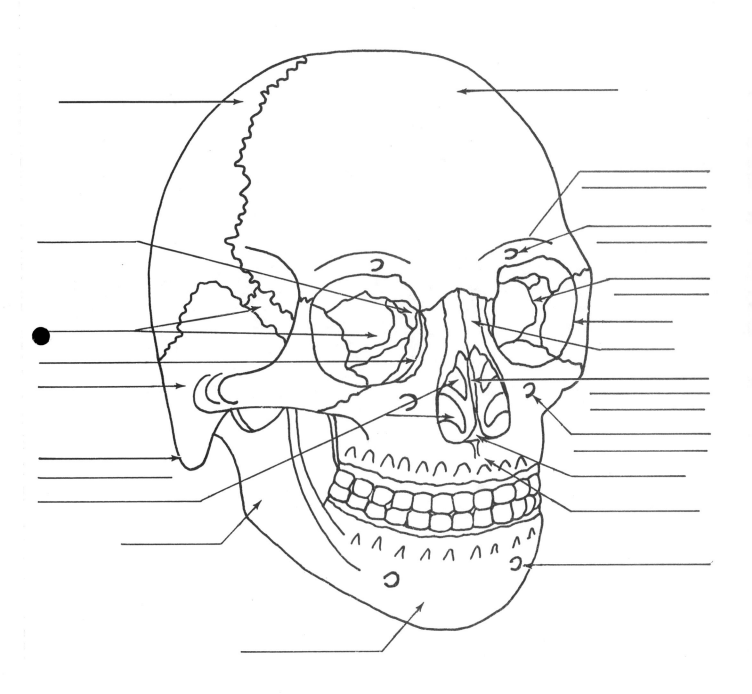

CRANIAL AND FACIAL BONES (Lateral View)

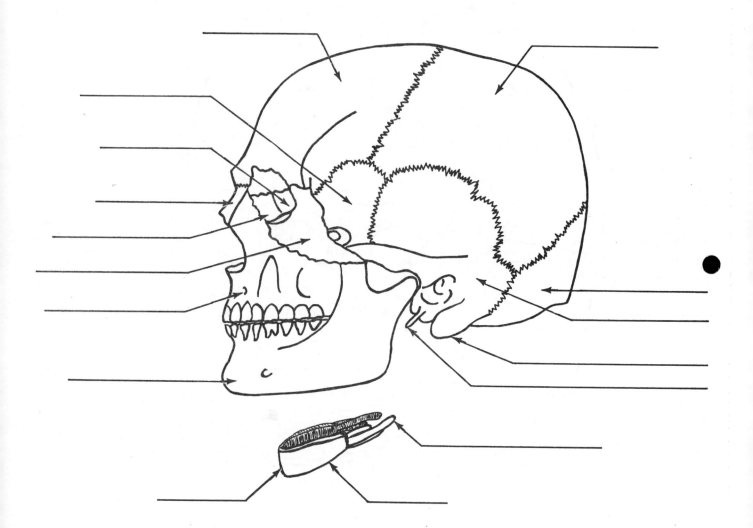

CROSS SECTION OF SKULL

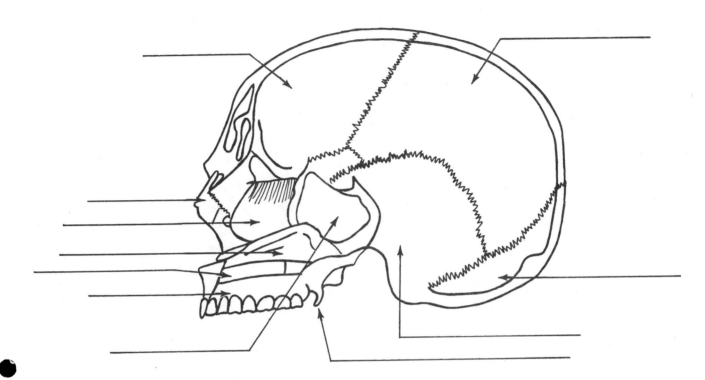

FLOOR OF CRANIAL CAVITY

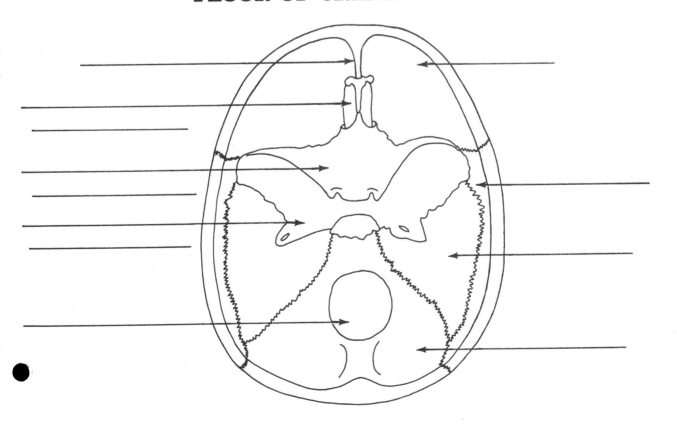

OSSICLES OF THE EAR

RIGHT SIDE OF MANDIBLE

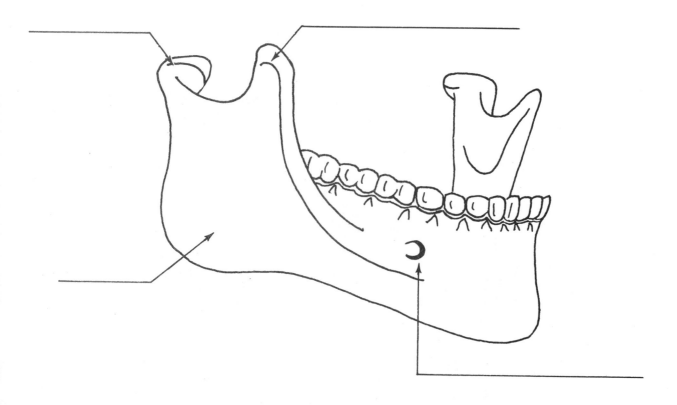

UPPER EXTREMITY

RIGHT
(Palmer View)

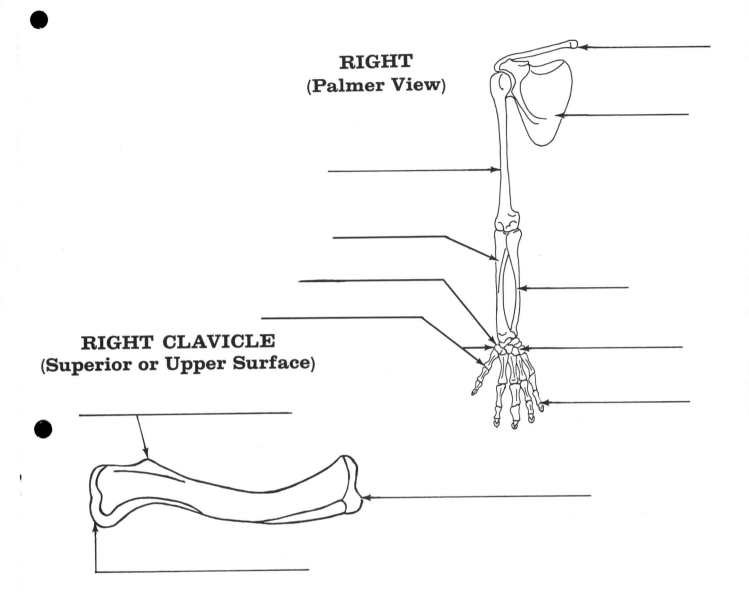

RIGHT CLAVICLE
(Superior or Upper Surface)

RIGHT SCAPULA
(Posterior or Dorsal Surface)

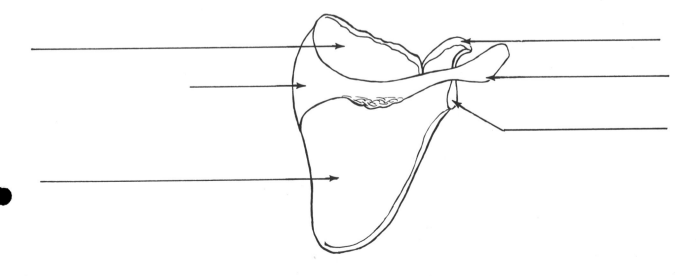

RIGHT HUMERUS

(Anterior View)

(Posterior View)

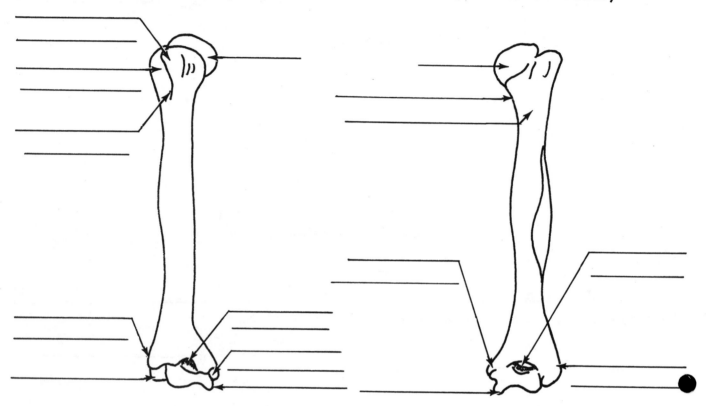

RADIUS AND ULNA (Posterior View)

UPPER PART OF ULNA

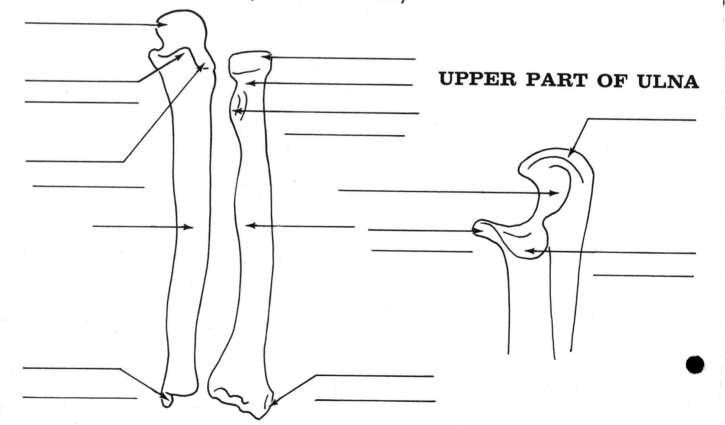

RIGHT HAND (Palmer Surface)

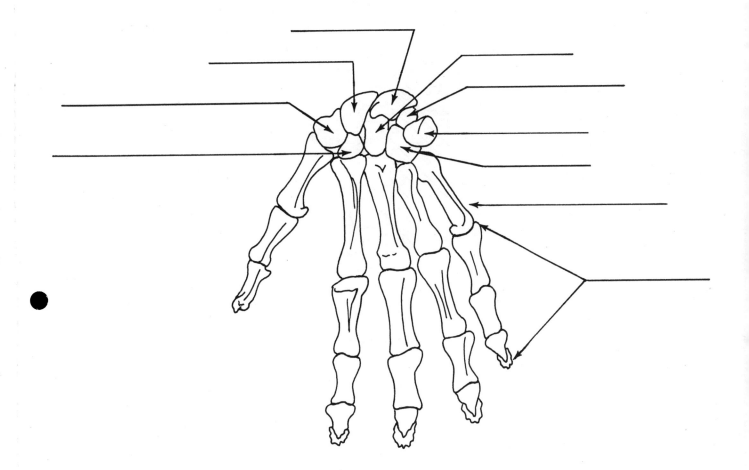

TRUNK
(Anterior View)

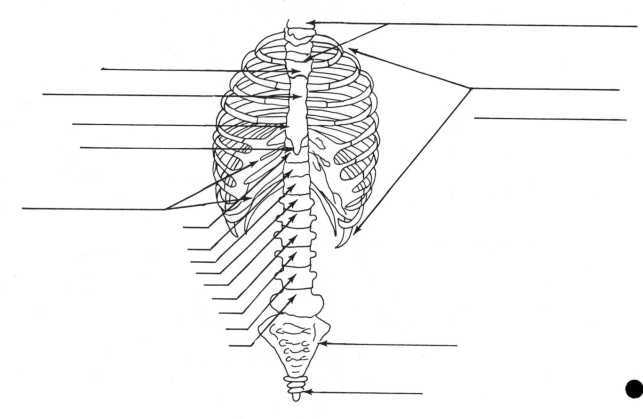

(Dorsal View)

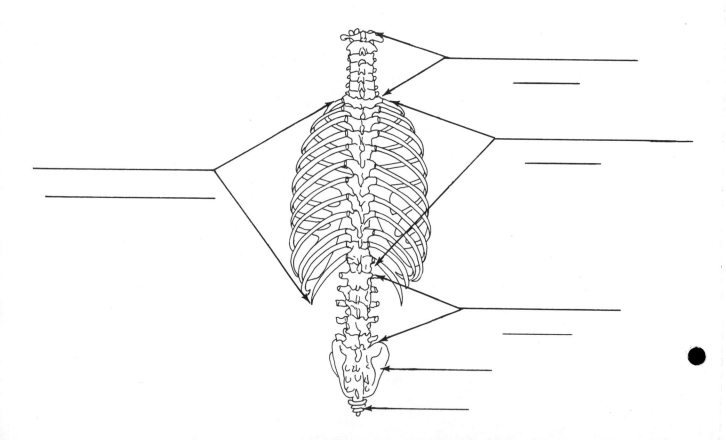

LUMBAR VERTEBRAE (Lateral View)

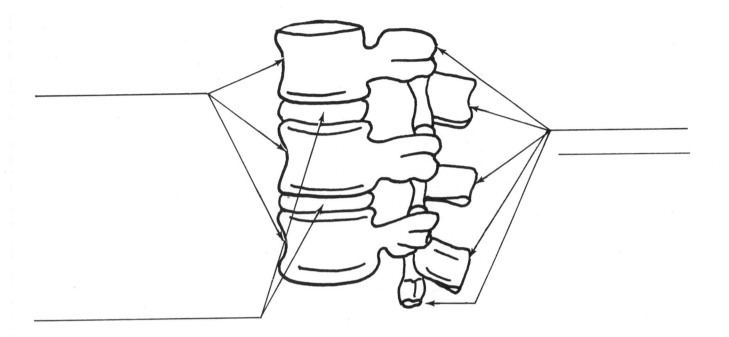

THORACIC VERTEBRA

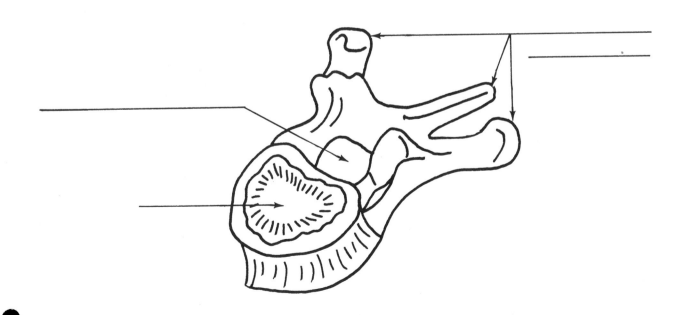

LUMBAR VERTEBRA (Viewed from Above)

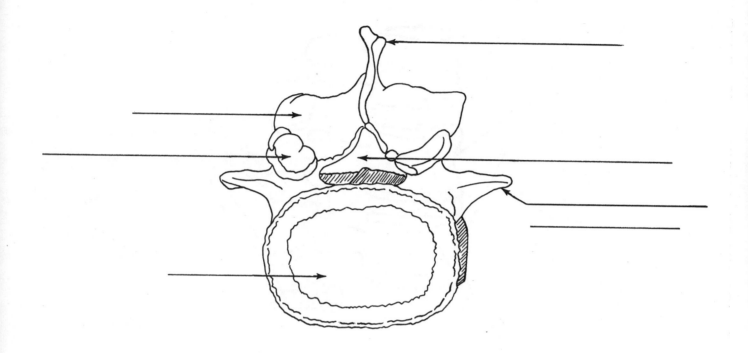

LUMBAR VERTEBRA (Lateral View)

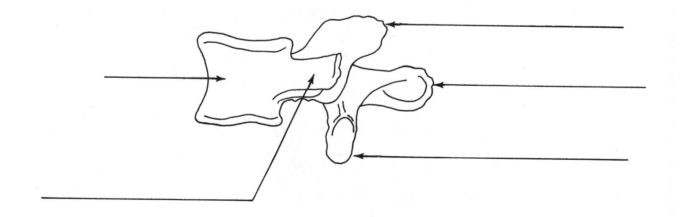

LOWER EXTREMITY

MALE PELVIS

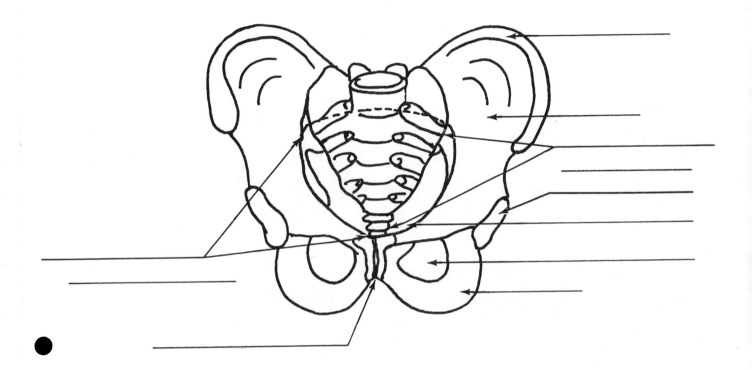

FEMALE PELVIS

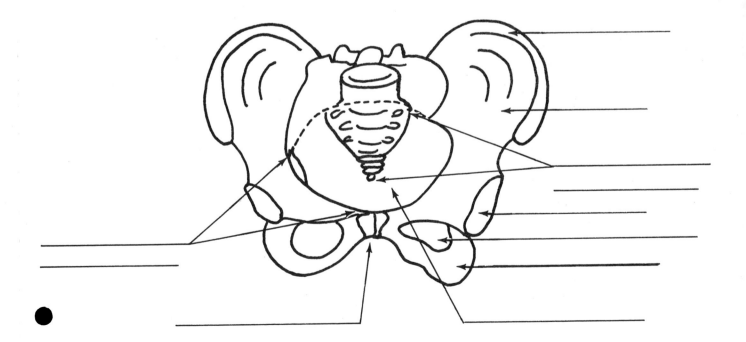

LOWER EXTREMITY

RIGHT
(Anterior View)

GROWTH AREAS
OF A LONG BONE

THE TIBIA AND FIBULA
(Posterior View)

THE FEMUR

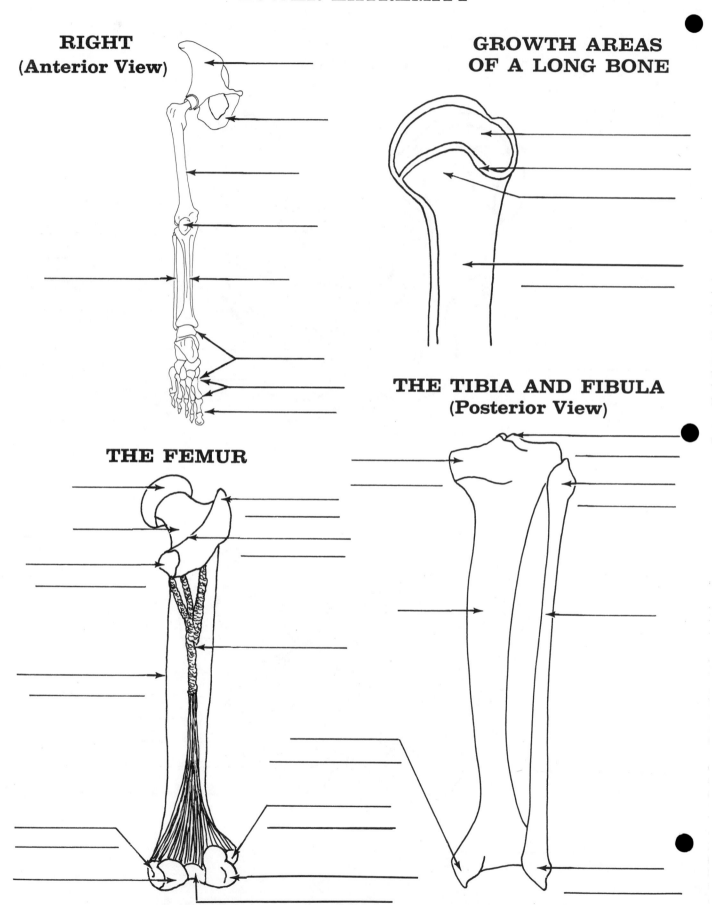

THE TIBIA AND FIBULA (Anterior View)

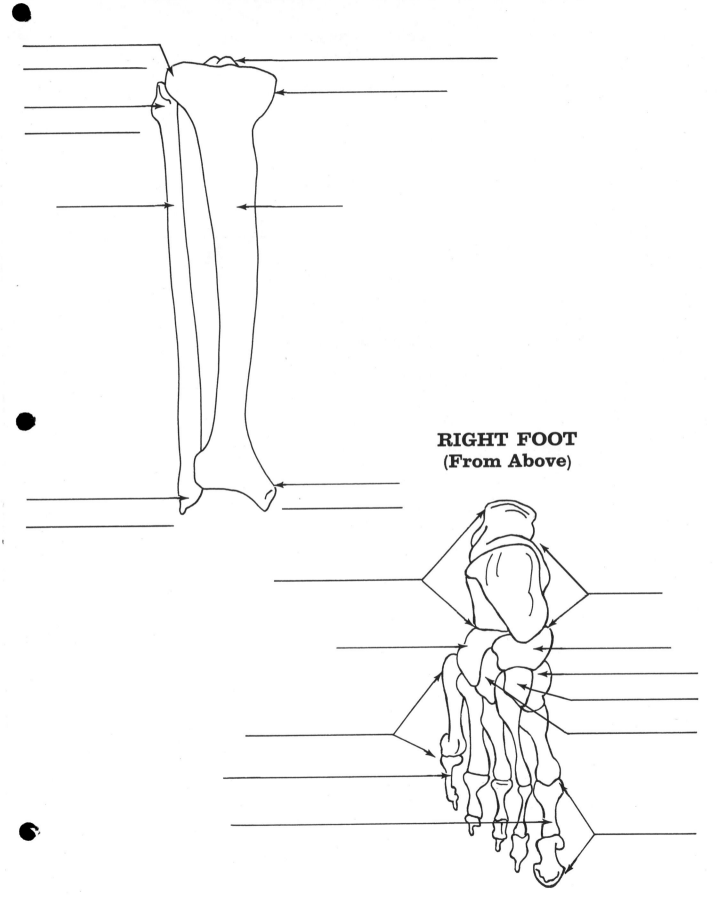

RIGHT FOOT
(From Above)

JOINTS AND ARTICULATIONS

SYNARTHROSES
(Immovable Joints)

SKULL

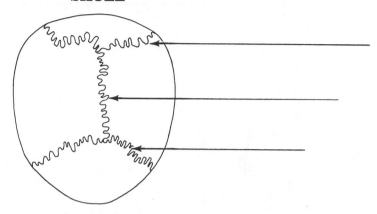

AMPHIARTHROSES
(Slightly Movable Joints)

FEMALE PELVIS

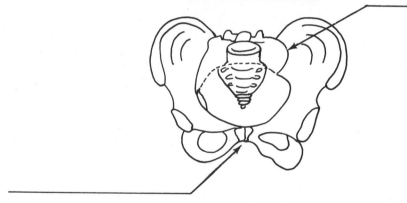

SYNDESMOSIS

TIBIOFIBULAR JOINT

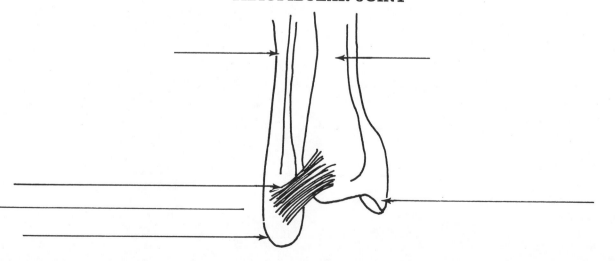

JOINTS AND ARTICULATIONS

DIARTHROSES (Freely Movable Joints)

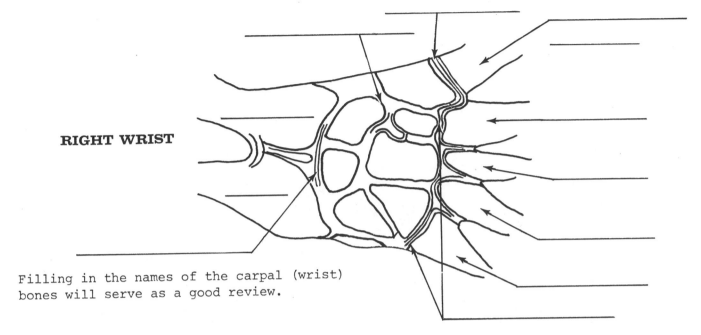

RIGHT WRIST

Filling in the names of the carpal (wrist)
bones will serve as a good review.

HINGE JOINT
RIGHT KNEE JOINT

BALL-AND-SOCKET JOINT
RIGHT HIP JOINT

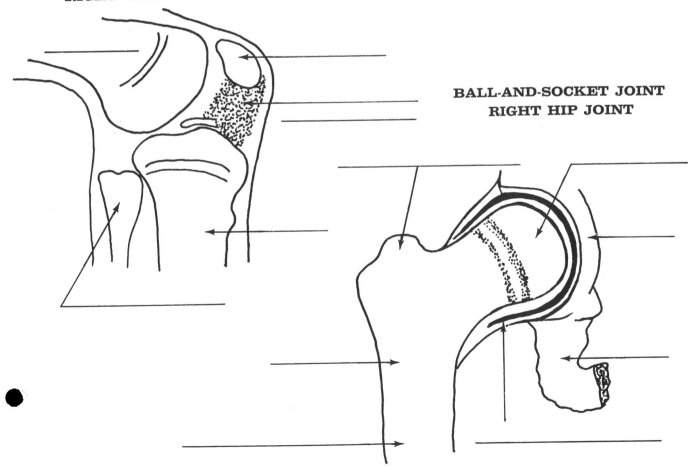

JOINTS AND ARTICULATIONS

RIGHT KNEE JOINT CAPSULE

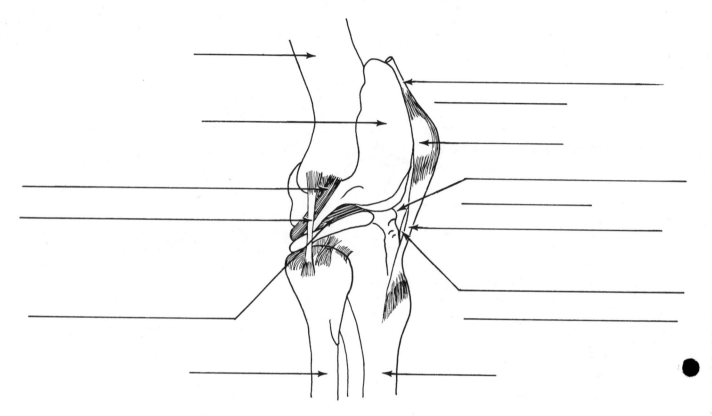

LIGAMENTS OF LEFT SHOULDER

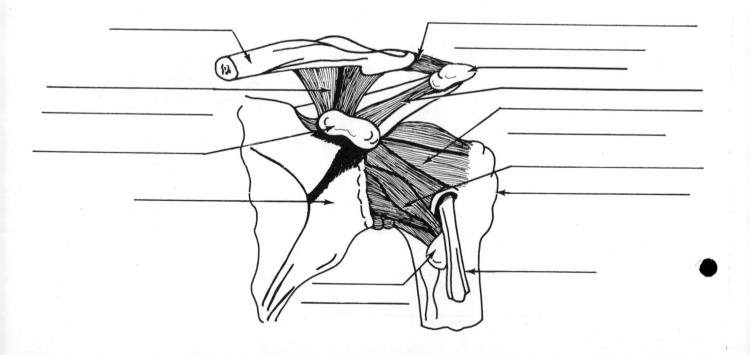

JOINTS AND ARTICULATIONS

RIGHT SHOULDER JOINT CAPSULE

(Anterior View)

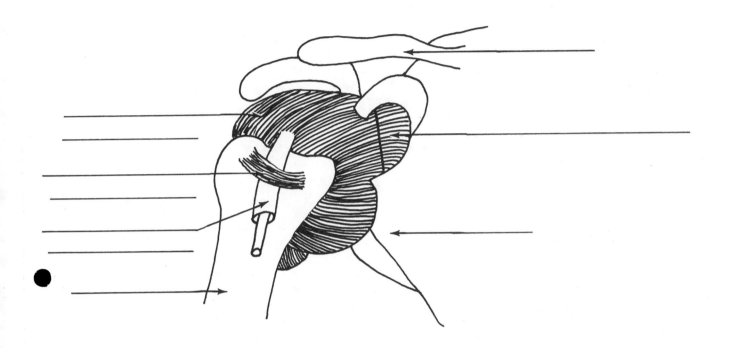

CHAPTER III

MUSCULAR SYSTEM

SUPERFICIAL MUSCLES OF HEAD AND NECK

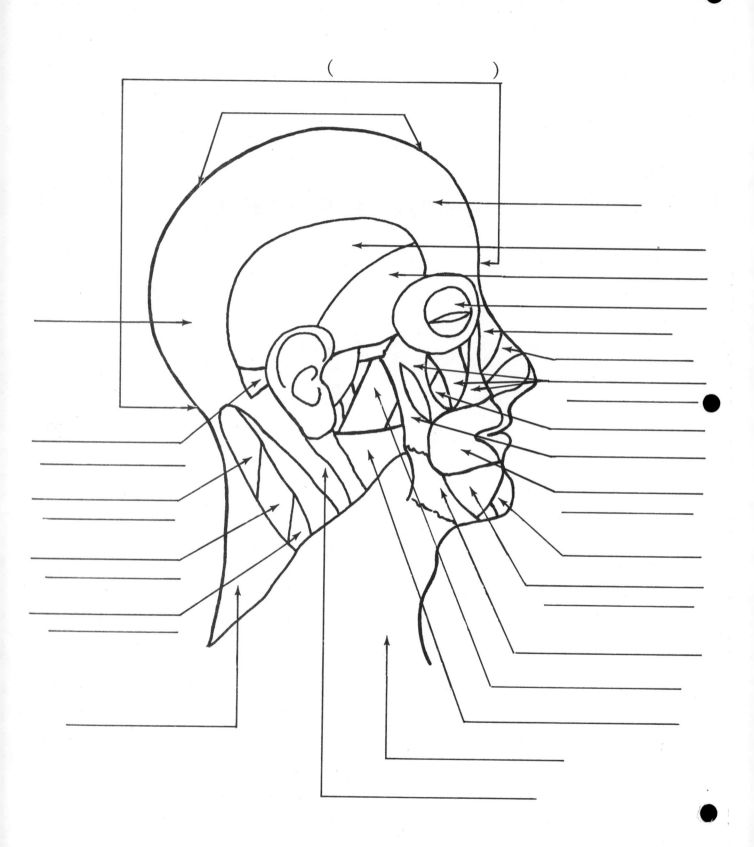

(   )

MUSCLES OF THE EYE

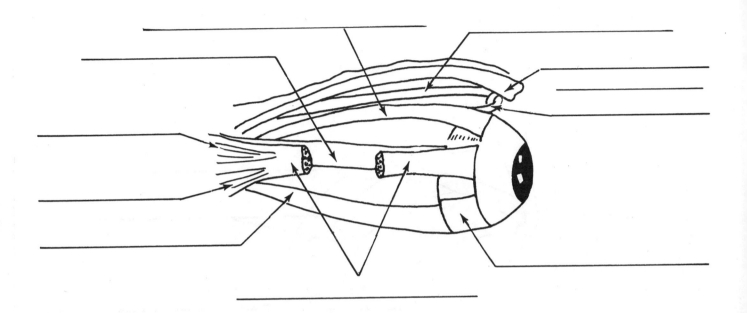

MUSCLES OF THE NECK

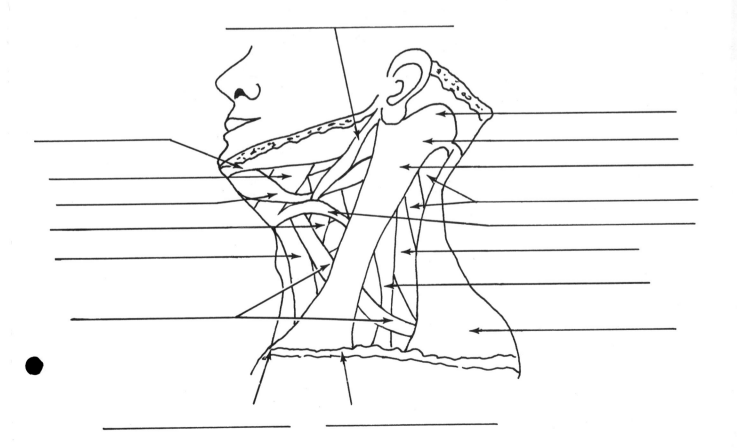

SUPERFICIAL MUSCLES OF CHEST AND UPPER ARM
(Anterior View)

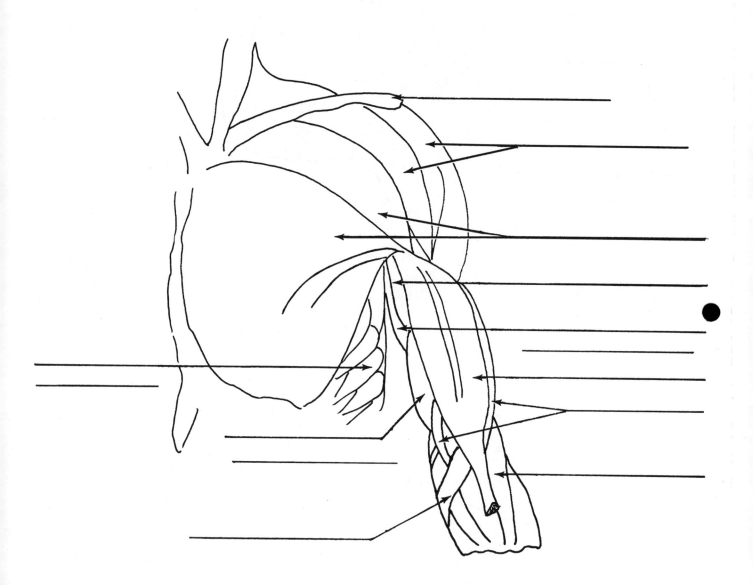

DEEP MUSCLES OF CHEST AND UPPER ARM
(Anterior View)

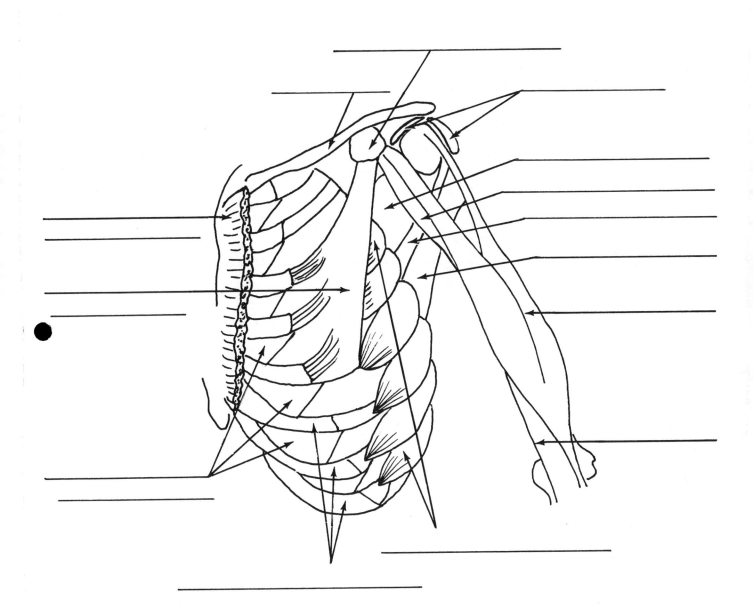

SUPERFICIAL MUSCLES, LEFT FOREARM (Palmer View)

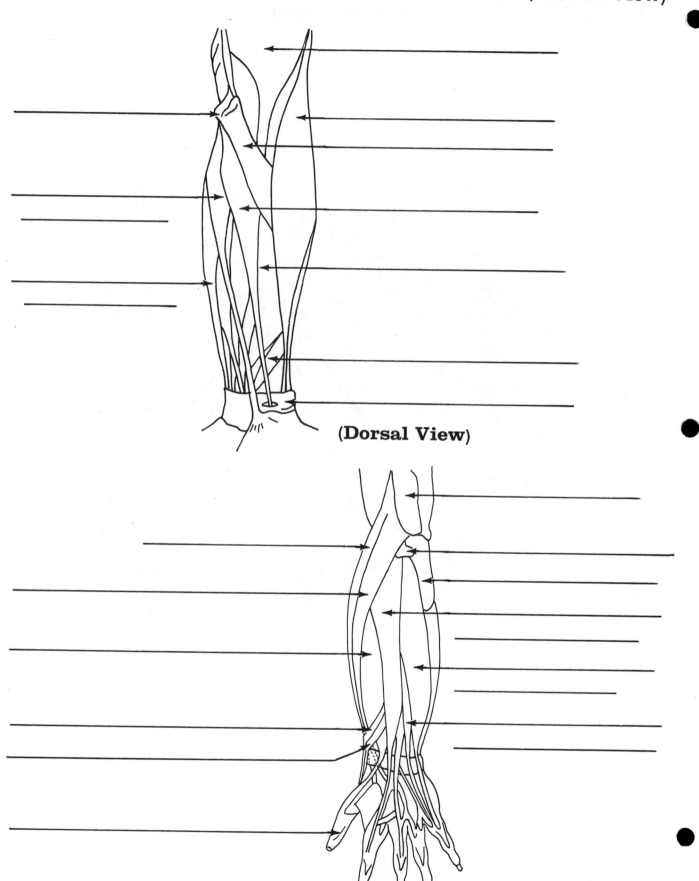

(Dorsal View)

DEEP MUSCLES, LEFT FOREARM (Palmer View)

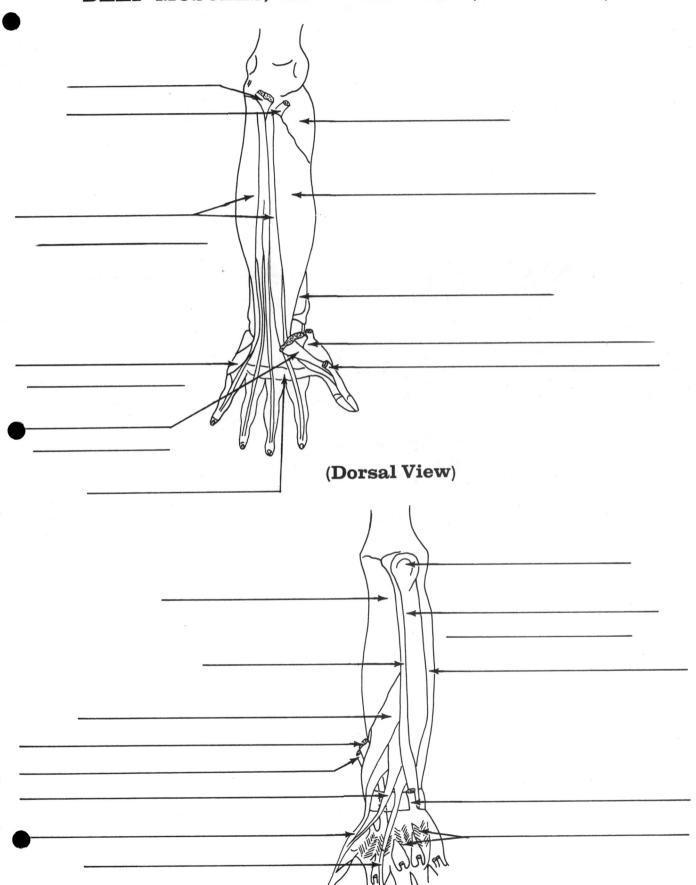

(Dorsal View)

MUSCLES OF RIGHT HAND (Palmer View)

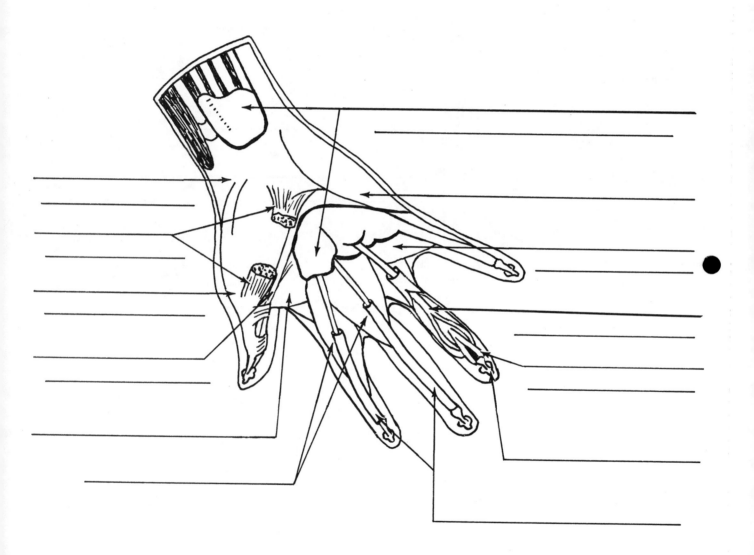

ABDOMINAL MUSCLES

Superficial   **Deep**

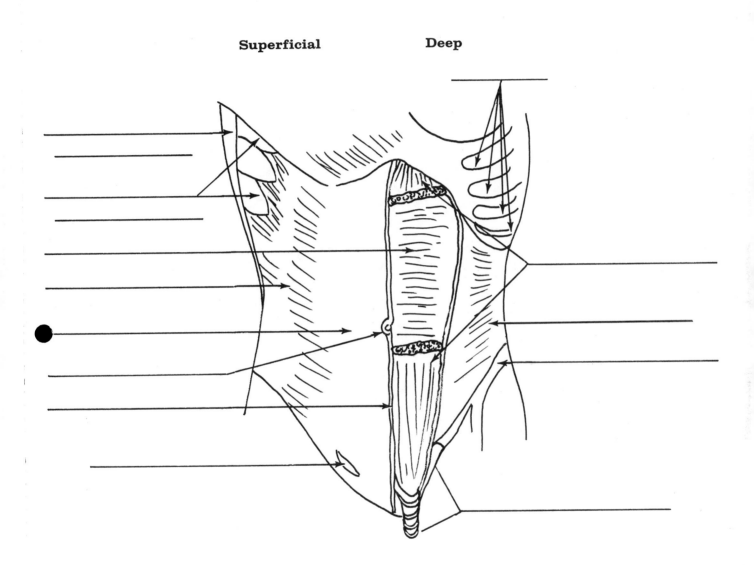

MUSCLES OF BACK

Deep   Superficial

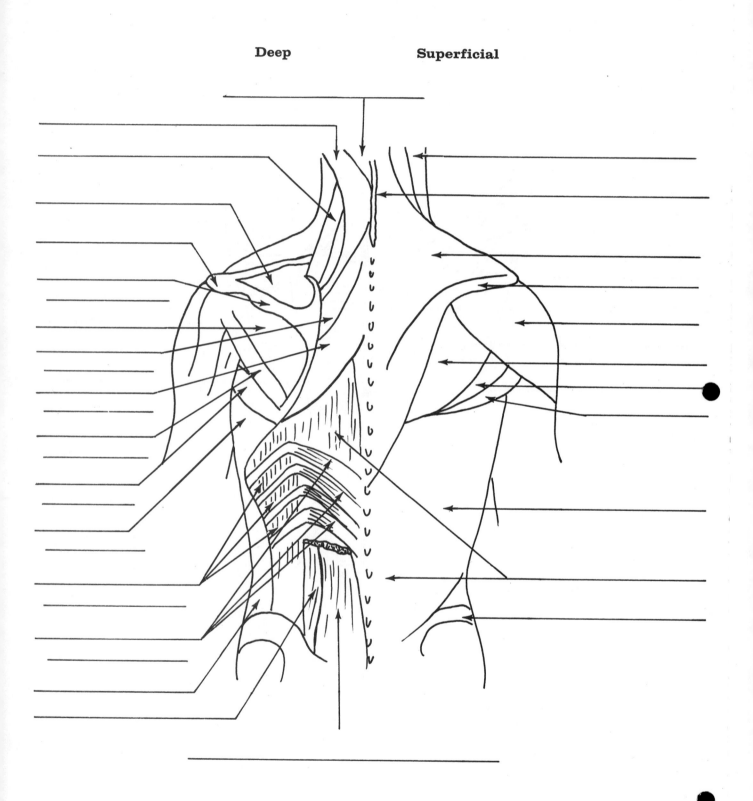

DEEP MUSCLES OF BACK

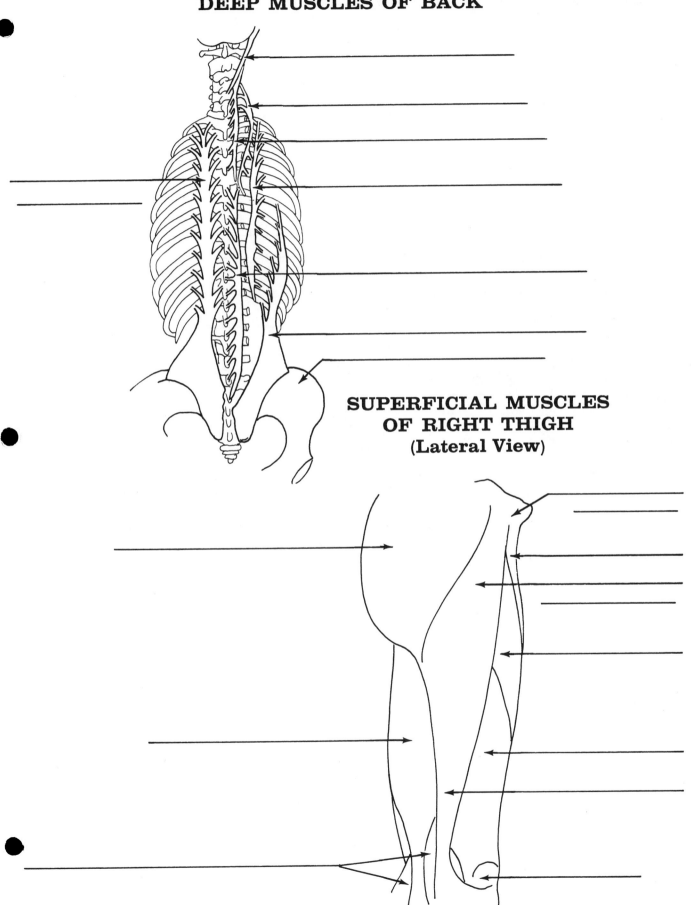

SUPERFICIAL MUSCLES
OF RIGHT THIGH
(Lateral View)

SUPERFICIAL MUSCLES, RIGHT THIGH (Lateral View)

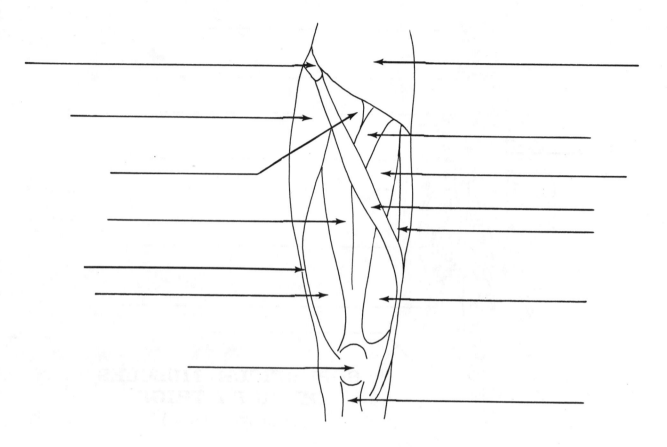

(Dorsal View)

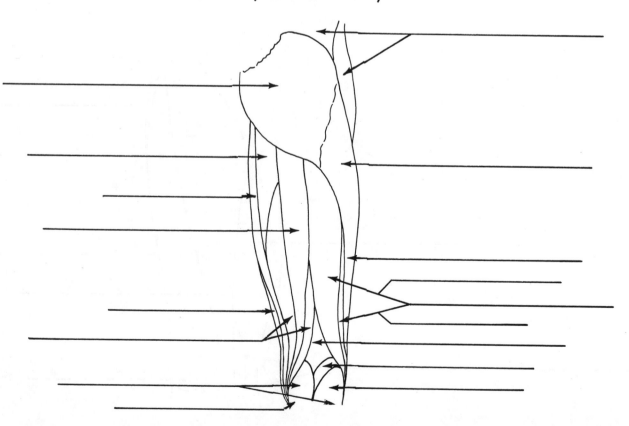

SUPERFICIAL MUSCLES, LOWER RIGHT LEG (Anterior View)

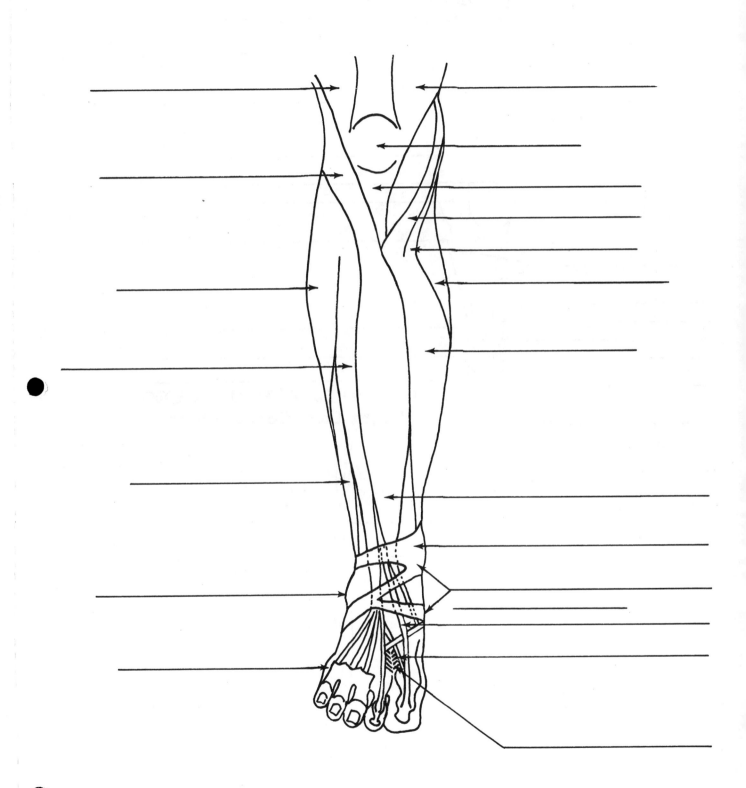

SUPERFICIAL MUSCLES, LOWER RIGHT LEG (Dorsal View)

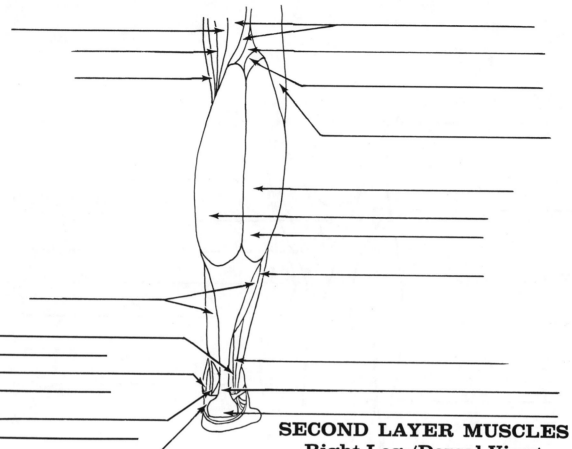

SECOND LAYER MUSCLES
Right Leg (Dorsal View)

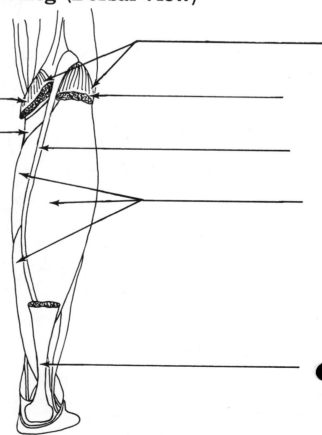

CHAPTER IV

NERVOUS SYSTEM

LOBES AND FISSURES OF CEREBRUM

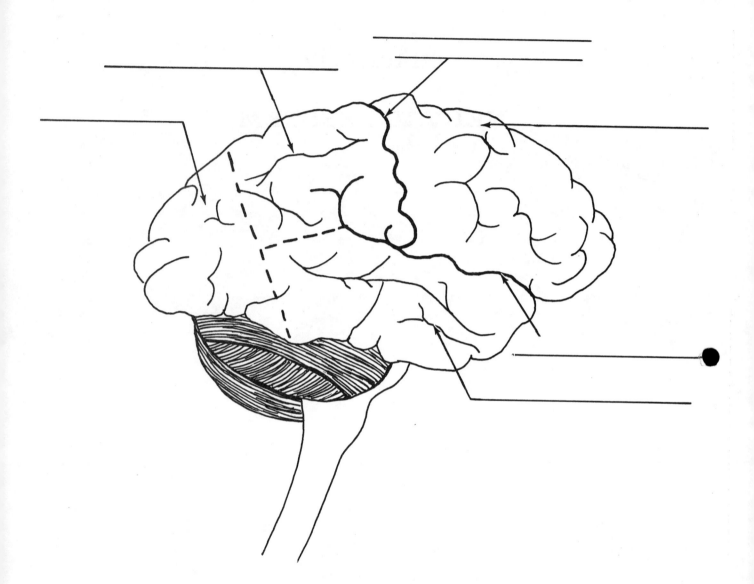

COMPONENTS OF BRAIN, RIGHT SIDE (Separated)

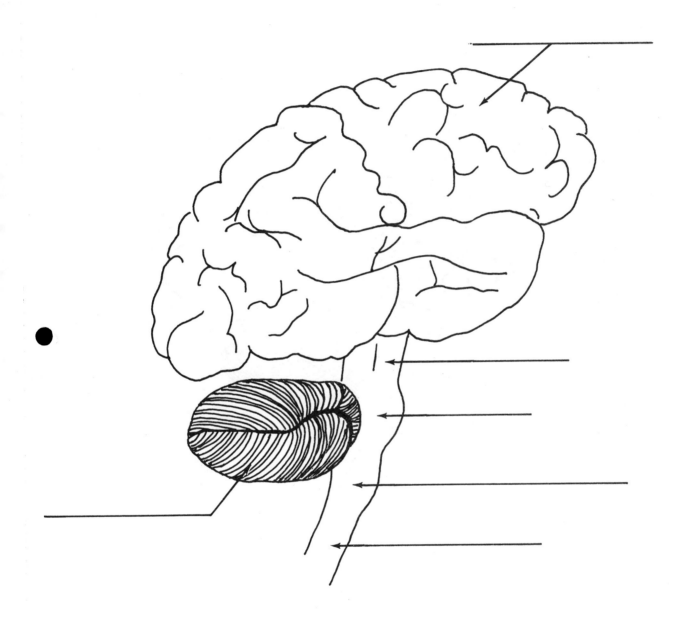

VENTRICLES OF CEREBRUM

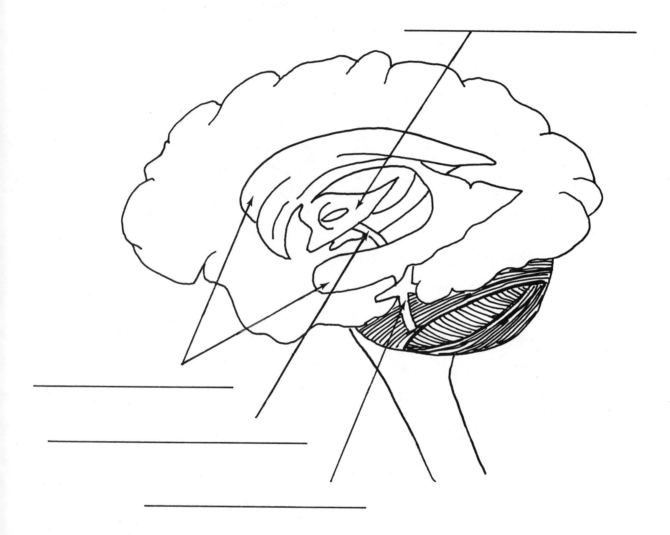

THE CEREBRUM

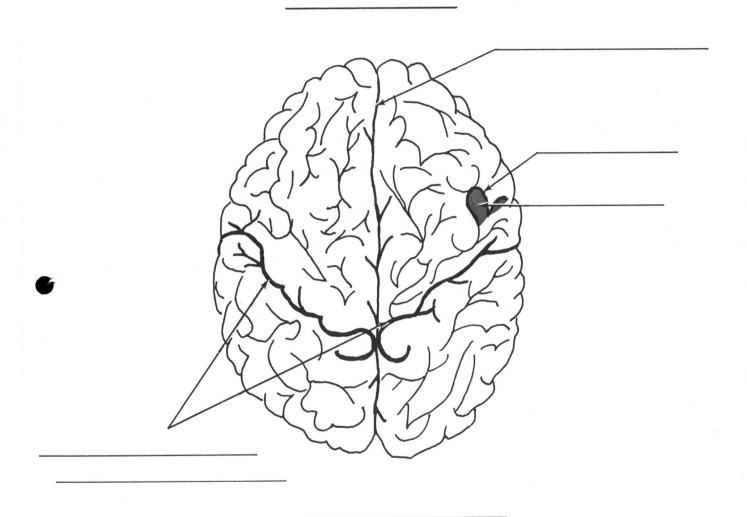

LEFT HALF OF BRAIN (Sagittal Section)

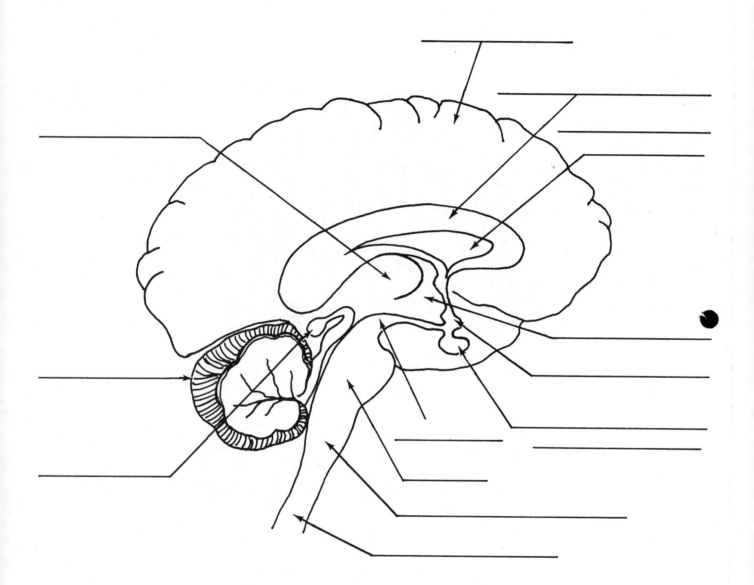

SPINAL CORD, VERTEBRAE AND MENINGES

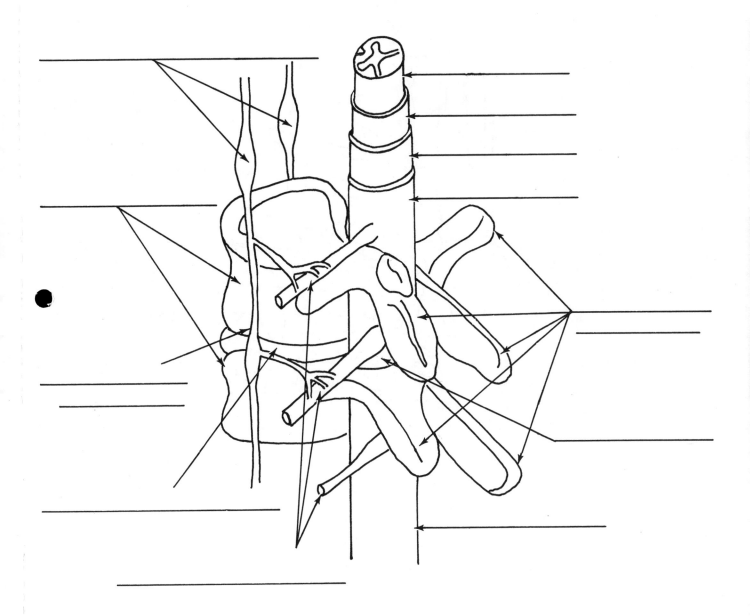

CRANIAL NERVES

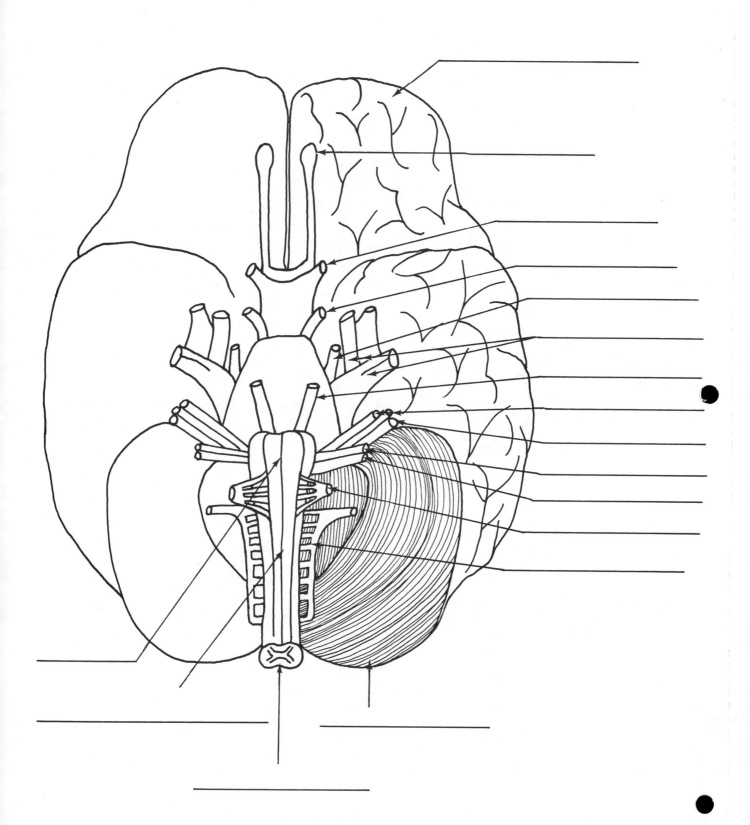

DORSAL VIEW OF SPINAL CORD AND EMERGING NERVES
CENTRAL NERVOUS SYSTEM – (Voluntary Nerves)

SYMPATHETIC NERVOUS SYSTEM

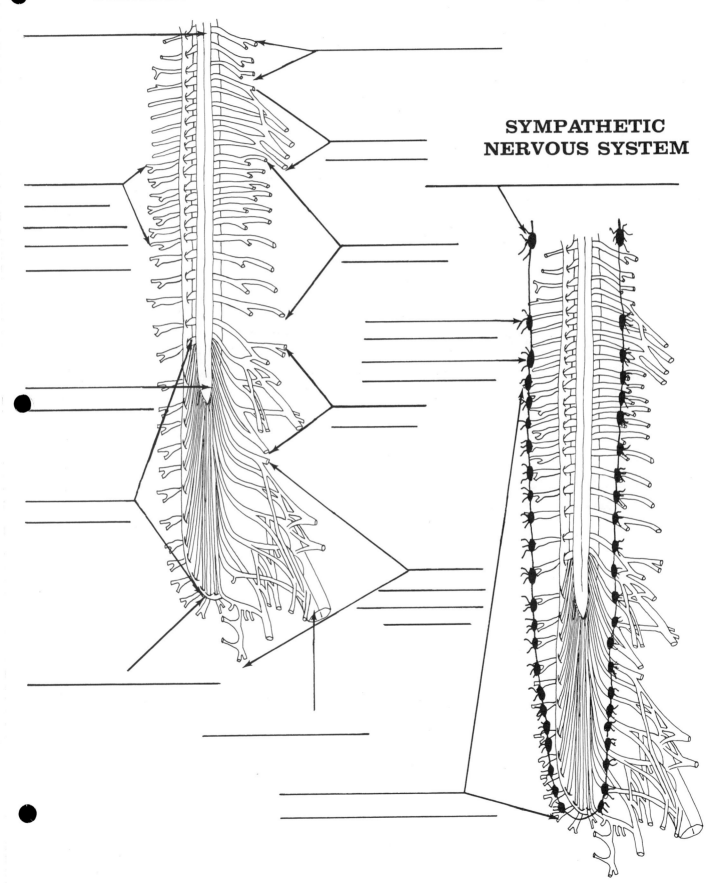

AUTONOMIC NERVOUS SYSTEM
(Involuntary Nerves)

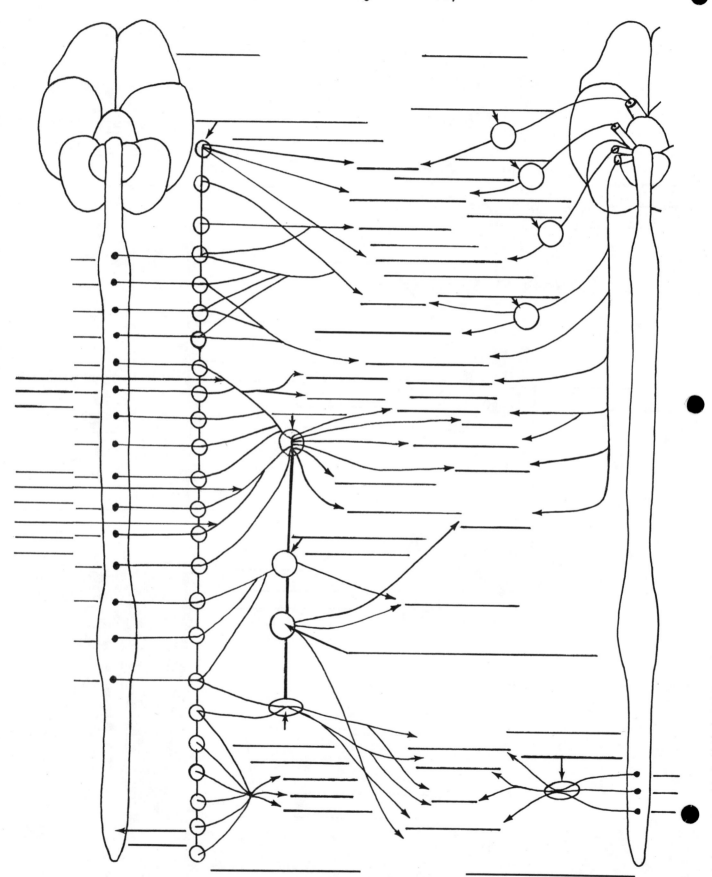

CHAPTER V

CIRCULATORY SYSTEM

PRINCIPAL ARTERIES OF HEAD, NECK AND BODY

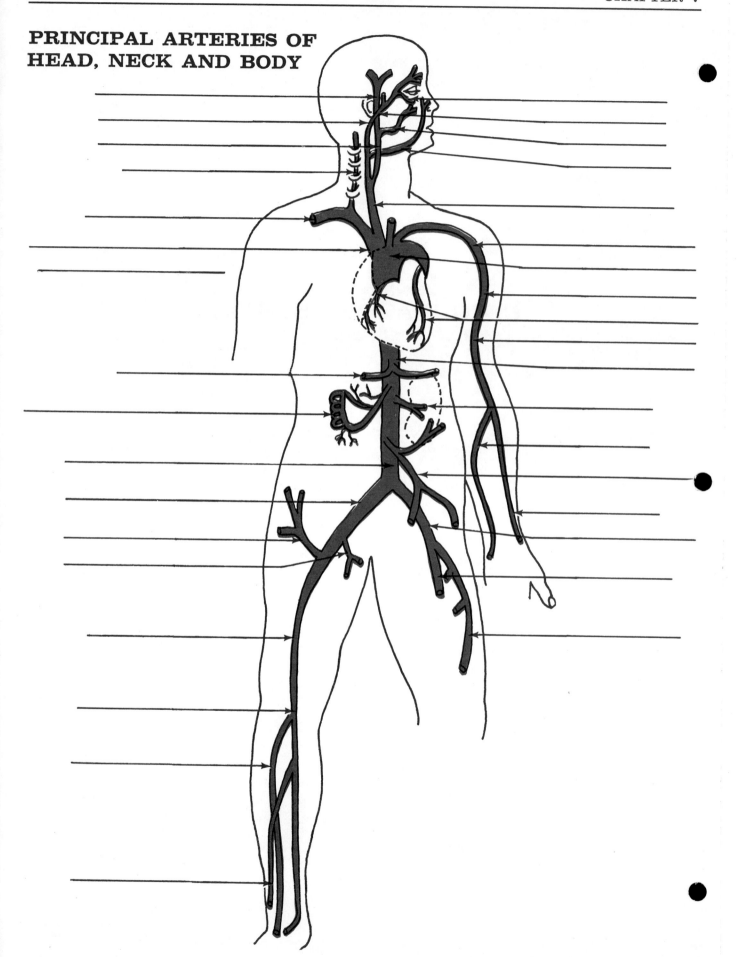

PRINCIPAL VEINS OF HEAD, NECK AND BODY

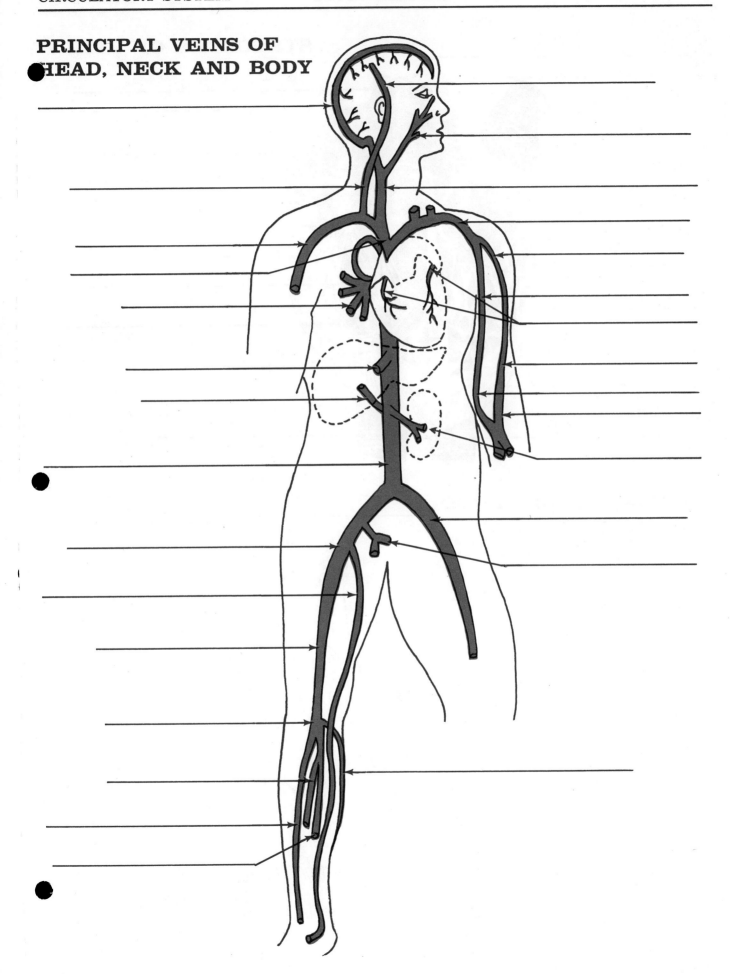

STRUCTURE OF HEART

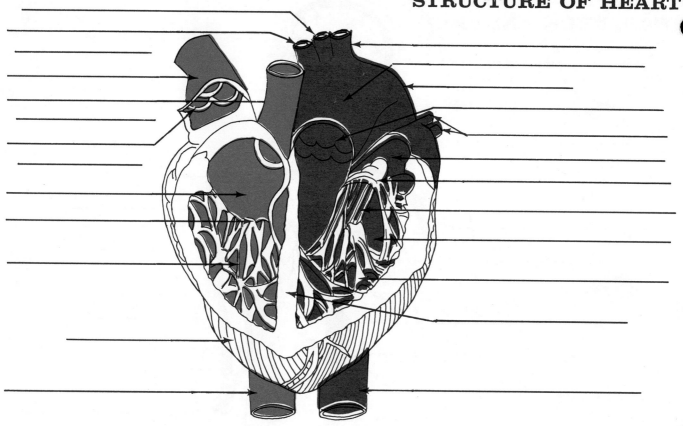

CIRCULATION OF BLOOD
THROUGH THE HEART

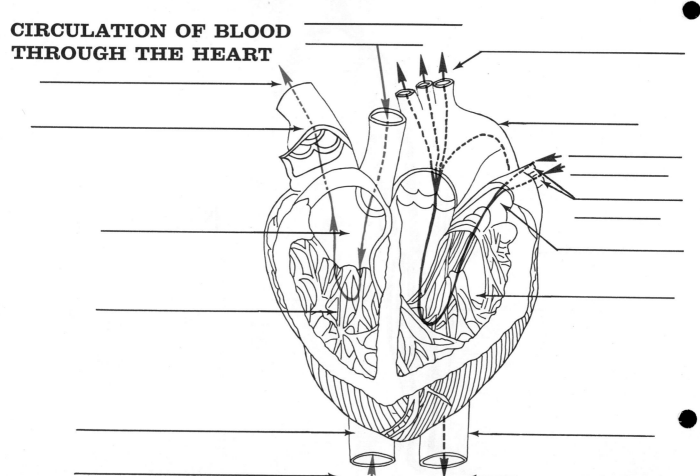

SIMPLIFIED VERSION OF CIRCULATION

Put in arrows to indicate blood flow.

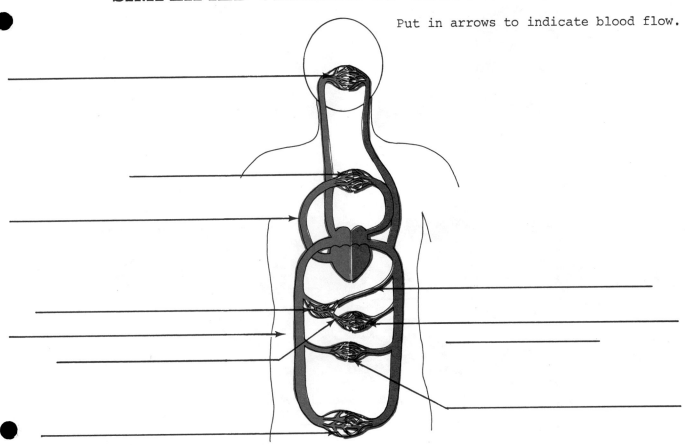

CORONARY CIRCULATION

ARTERIAL

VENOUS

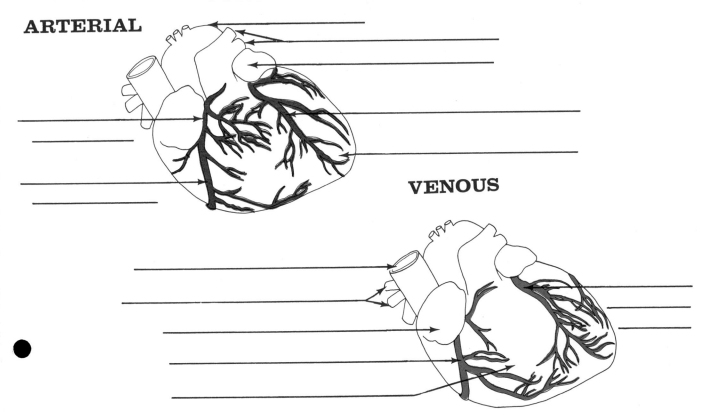

NORMAL BLOOD CELLS

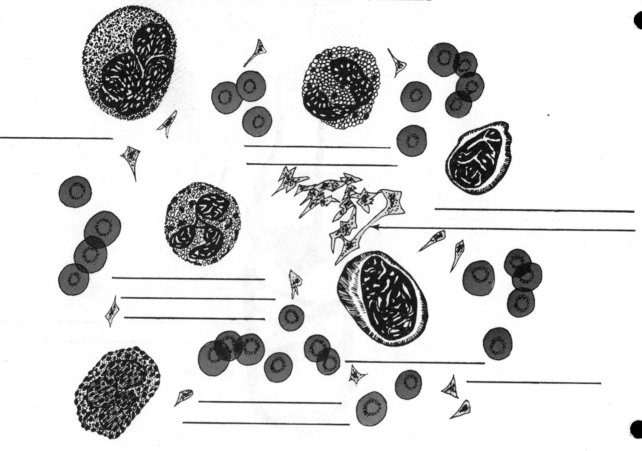

LYMPHATIC SYSTEM

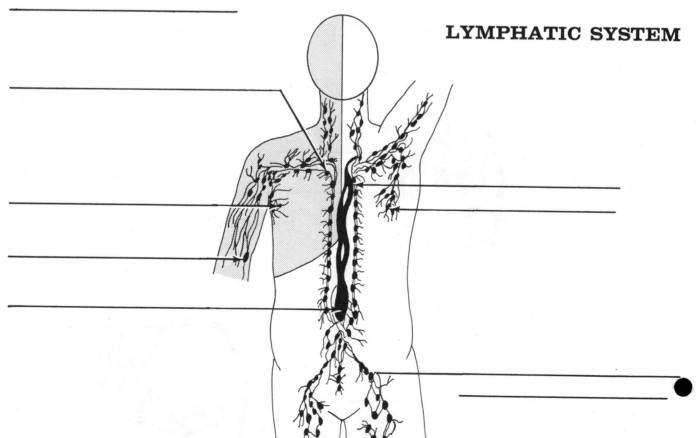

CHAPTER VI

ENDOCRINE SYSTEM

ENDOCRINE SYSTEM SHOWING LOCATION OF GLANDS

CHAPTER VII

RESPIRATORY SYSTEM

RESPIRATORY ORGANS

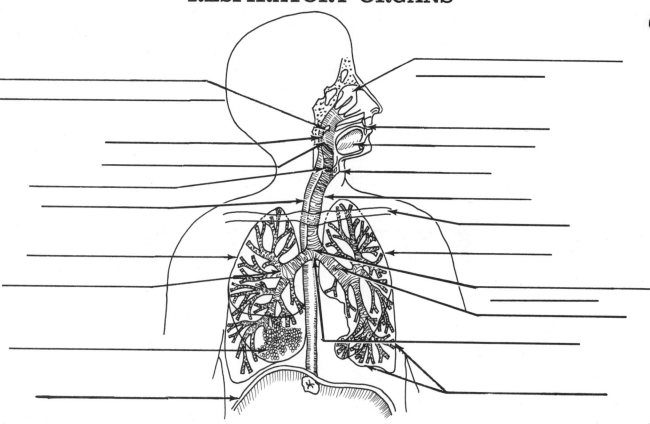

UPPER RESPIRATORY ORGANS

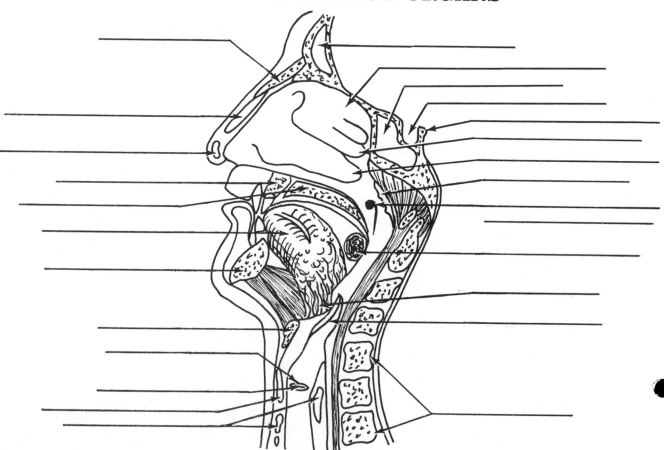

SEPTUM OF NOSE (Right Side)

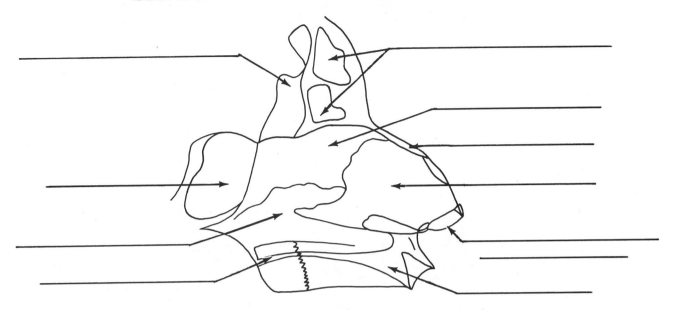

CARTILAGE OF NOSE (Inferior View)

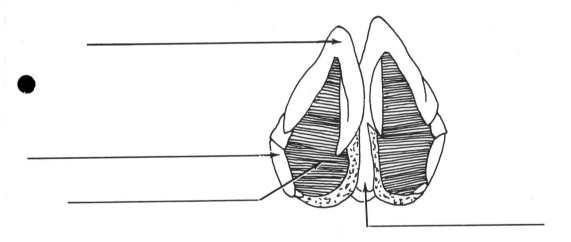

NASAL CAVITIES (Coronal Section)

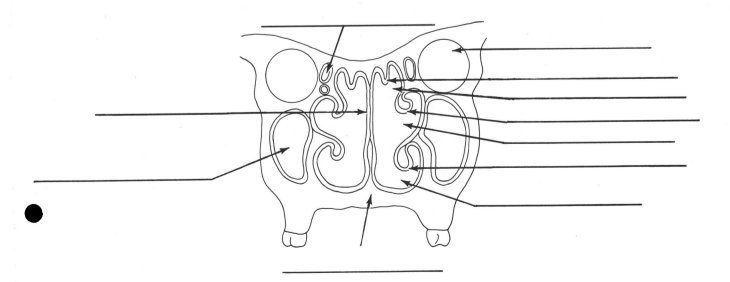

NERVE SUPPLY OF NOSE (Right Side)

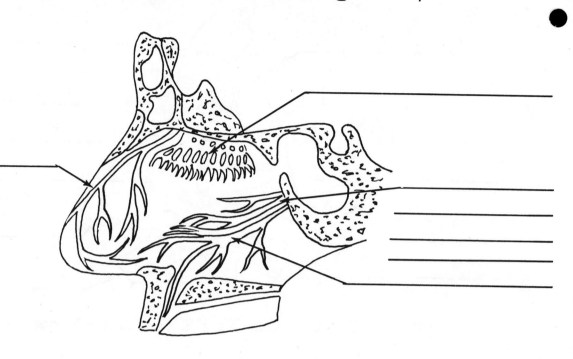

THE LUNGS (Anterior View)

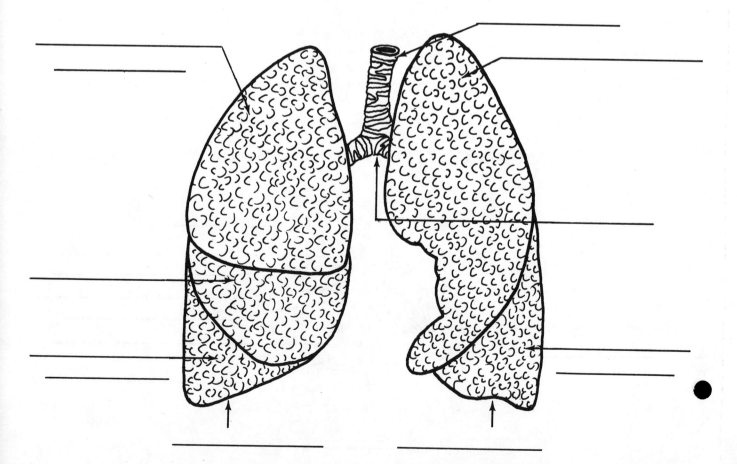

BRONCHIAL TREE AND TRACHEA
SHOWING CARTILAGINOUS STRUCTURE

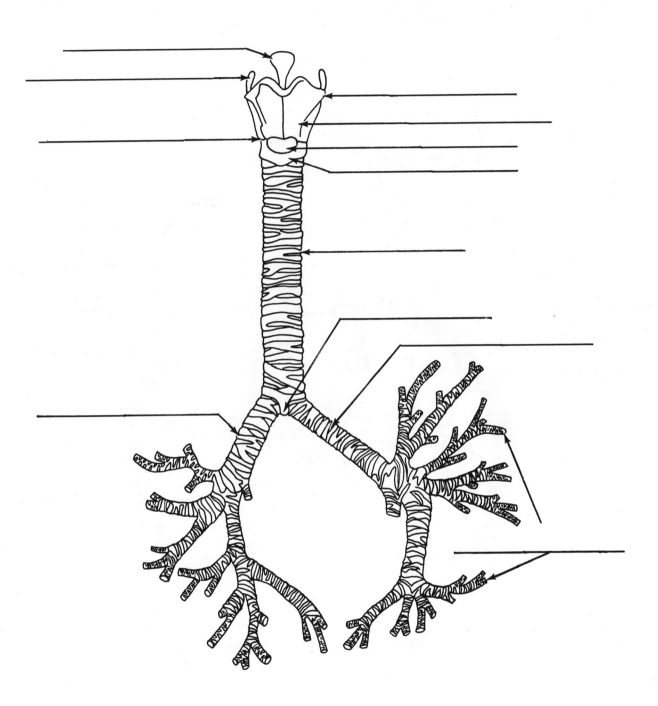

LOBULE OF THE LUNG

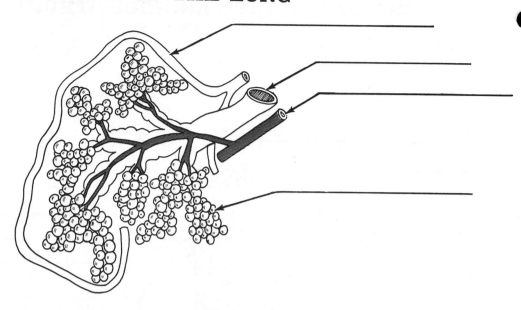

PLEURA AND MEDIASTINUM

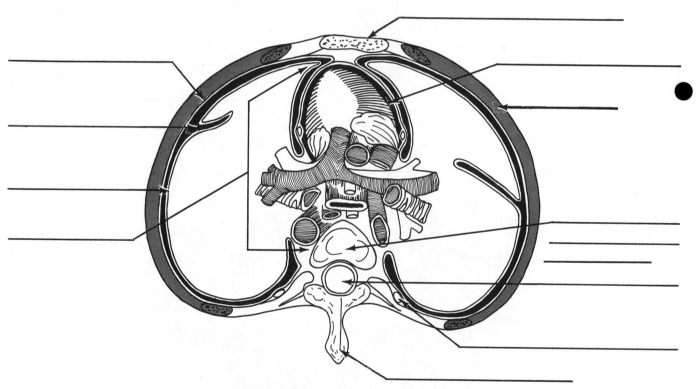

Contents of Mediastinum (shown here)

CHAPTER VIII

GASTRO-INTESTINAL SYSTEM

ALIMENTARY ORGANS

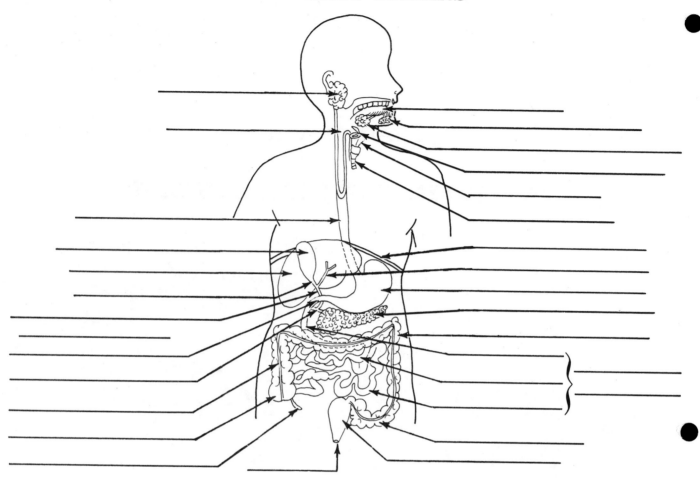

SALIVARY GLANDS AND RELATED STRUCTURES

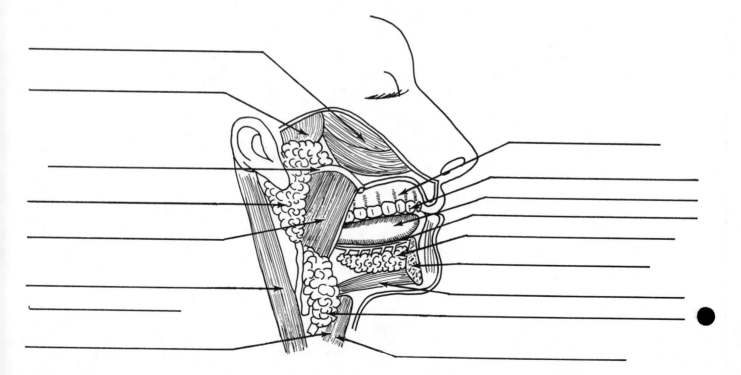

MOUTH CAVITY

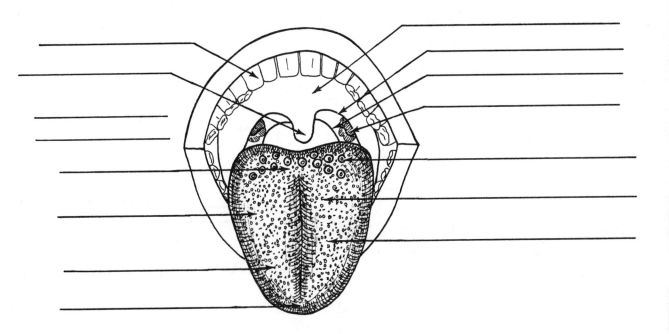

LONGITUDINAL SECTION OF TOOTH
AND SUPPORTING STRUCTURES

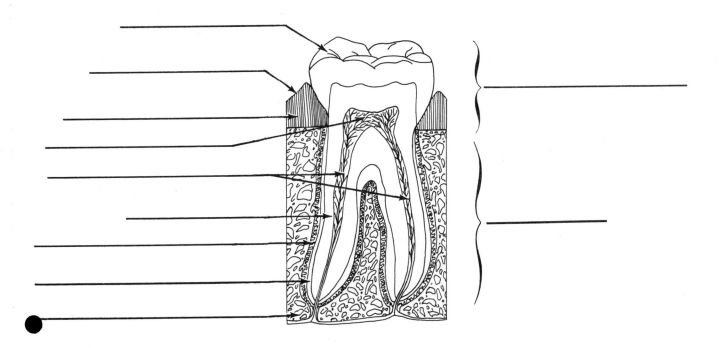

UPPER JAW OF ADULT (Right Side)

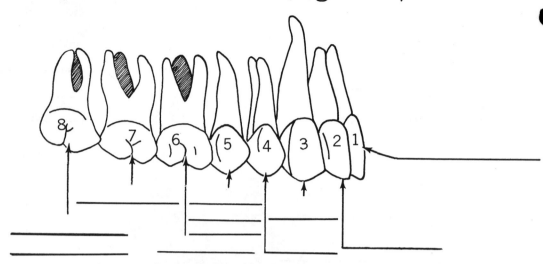

SIDE VIEW OF CHILD'S TEETH
(Deciduous Teeth)

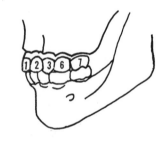

Write in the names of the teeth in the lines provided.

1. _____
2. _____
3. _____
6. _____
7. _____

SIDE VIEW OF ADULT TEETH

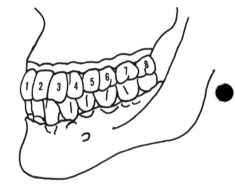

1. _____   5. _____
2. _____   _____
3. _____   6. _____
4. _____   7. _____
   _____   8. _____
   _____

INTERIOR OF STOMACH

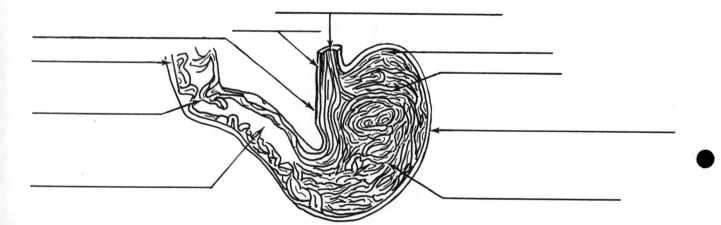

RELATIONSHIP OF STOMACH, DUODENUM AND PANCREAS

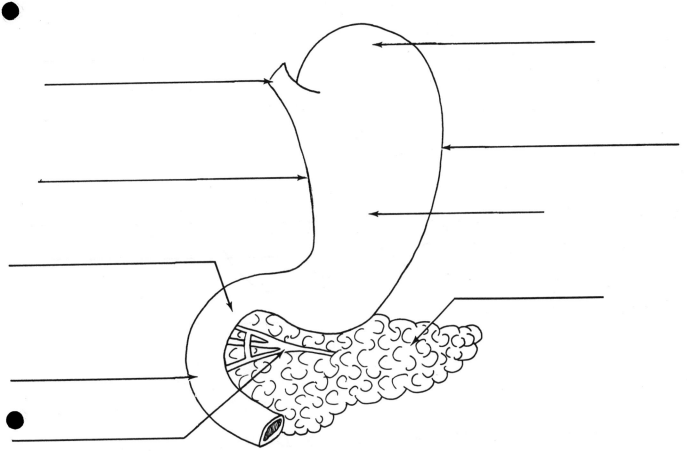

ANTERIOR SURFACE OF LIVER

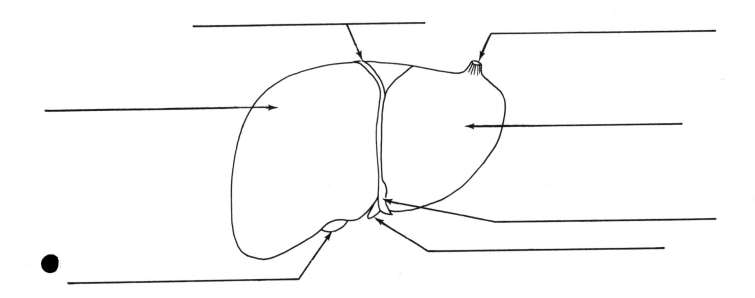

DORSAL (VISCERAL) SURFACE OF LIVER

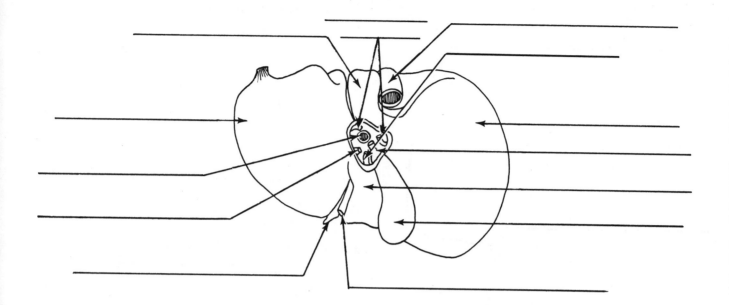

VISCERAL IMPRESSIONS ON DORSAL SURFACE
OF LIVER

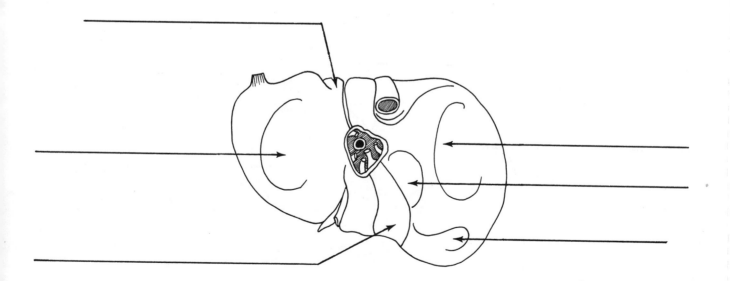

GALLBLADDER AND DUCTS

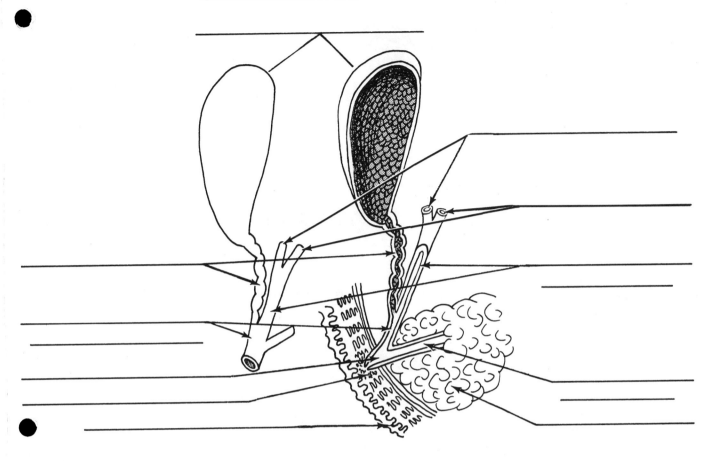

THE BILIARY SYSTEM

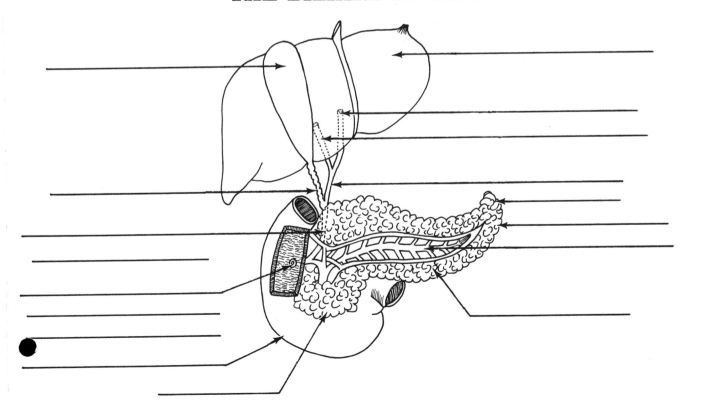

A VILLUS

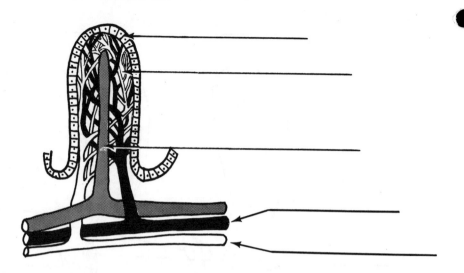

COATS OF THE INTESTINE (Cross Section)

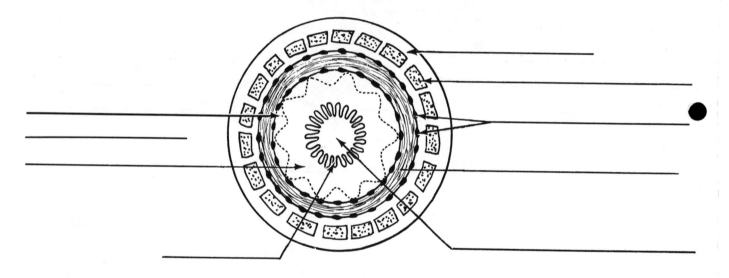

COATS OF THE INTESTINE SHOWN IN LAYERS

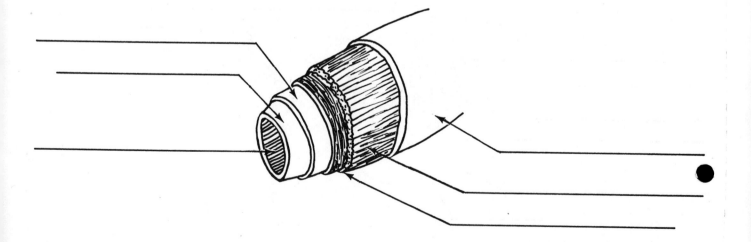

CECUM AND ACCESSORY ORGANS

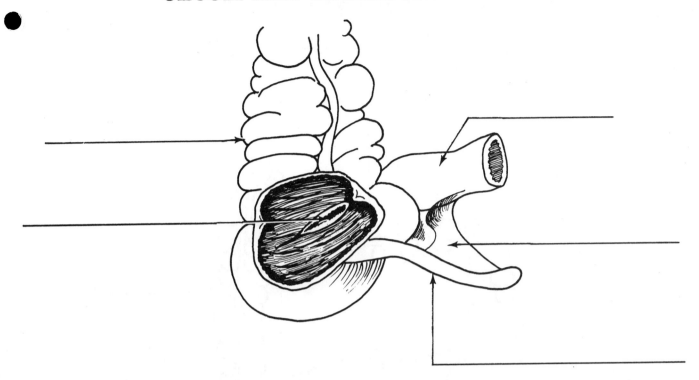

THE OMENTUM

SECTION OF SMALL INTESTINE SHOWING MESENTERY AND ARTERIAL SUPPLY

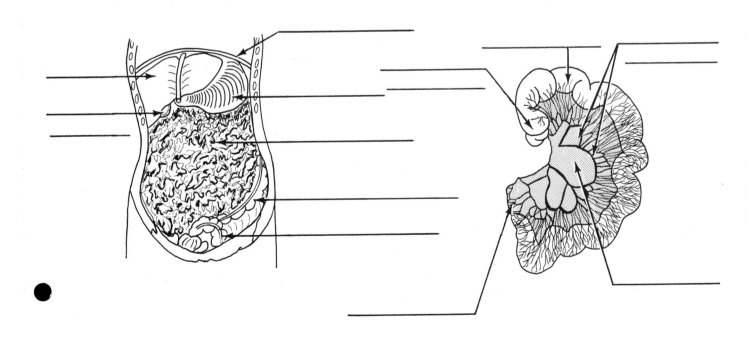

THE LARGE INTESTINE

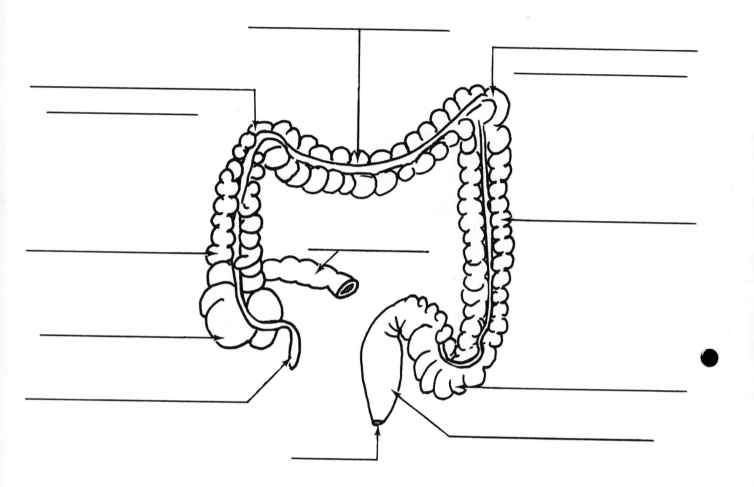

CHAPTER IX

GENITO-URINARY SYSTEM

RELATIONSHIP OF URINARY ORGANS, FEMALE
REPRODUCTIVE ORGANS AND RECTUM

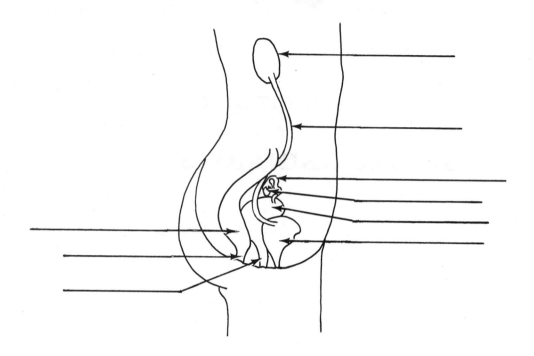

FEMALE REPRODUCTIVE SYSTEM
AND RELATED STRUCTURES (Sagittal View)

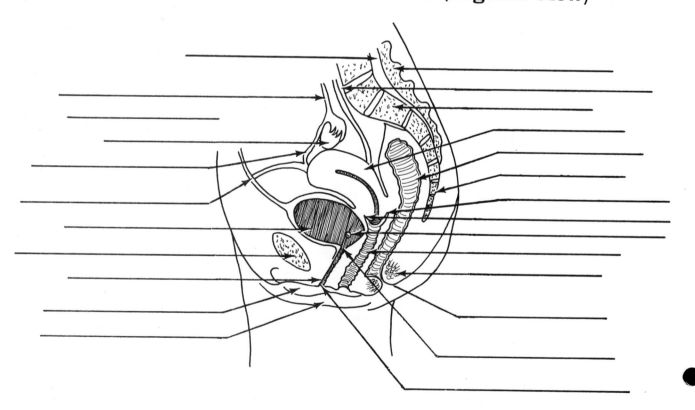

EXTERNAL GENITALIA OF FEMALE

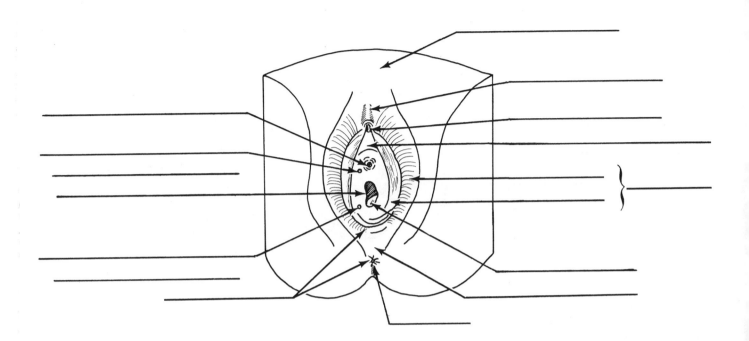

SECTION THROUGH UTERUS AND VAGINA

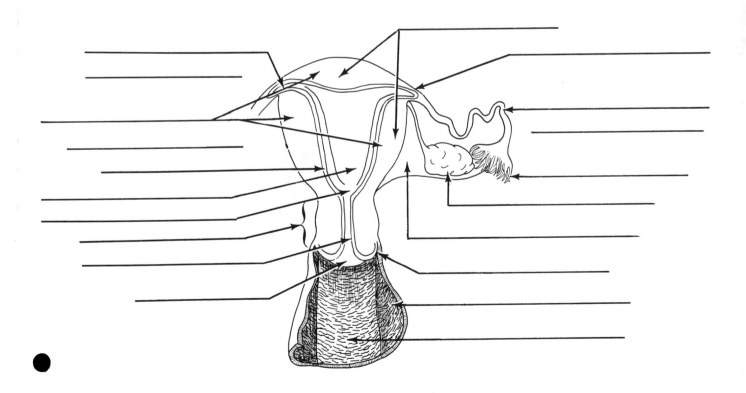

OVARY SECTIONED

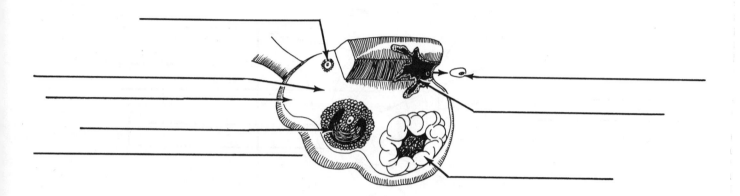

LOWER HALF OF BREAST DISSECTED
TO SHOW LACTIFEROUS DUCTS

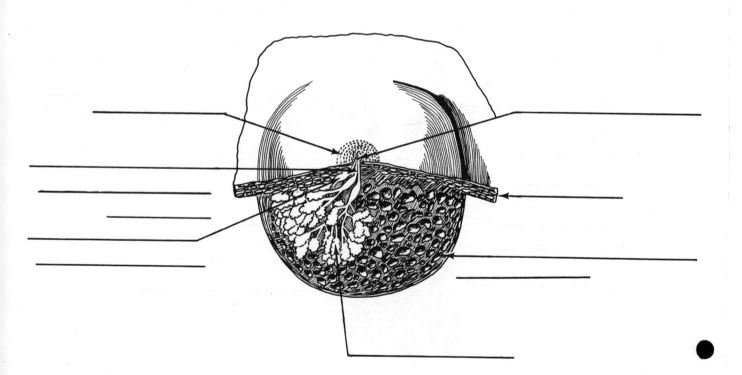

SAGITTAL SECTION OF MALE REPRODUCTIVE AND URINARY ORGANS

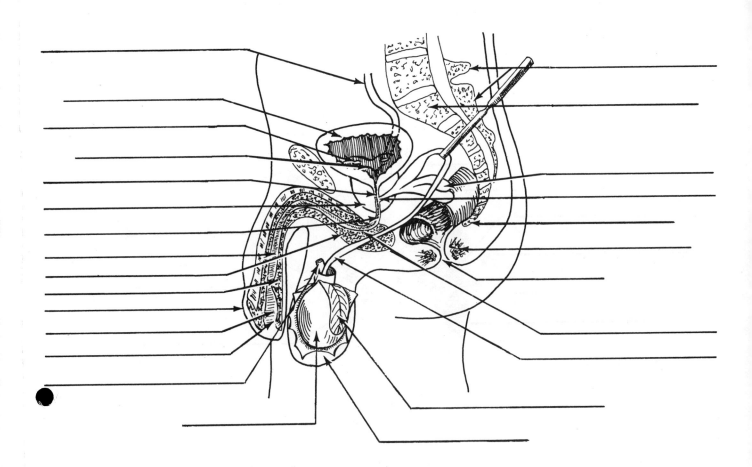

DORSAL VIEW OF BLADDER AND ACCESSORY MALE ORGANS

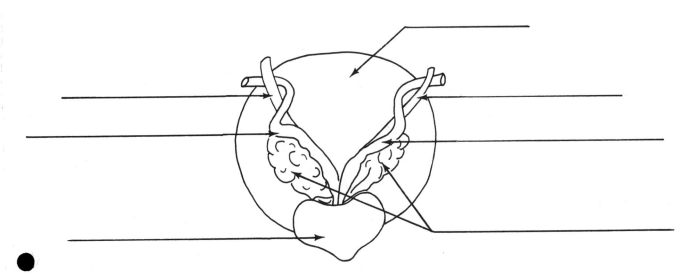

MALE ORGANS OF REPRODUCTION

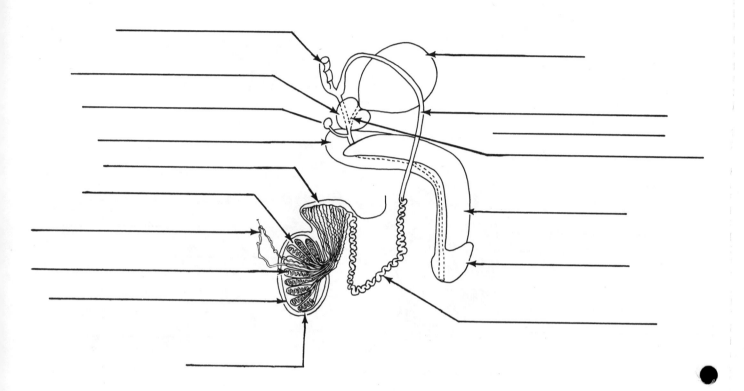

ANATOMICAL LOCATION OF URINARY SYSTEM

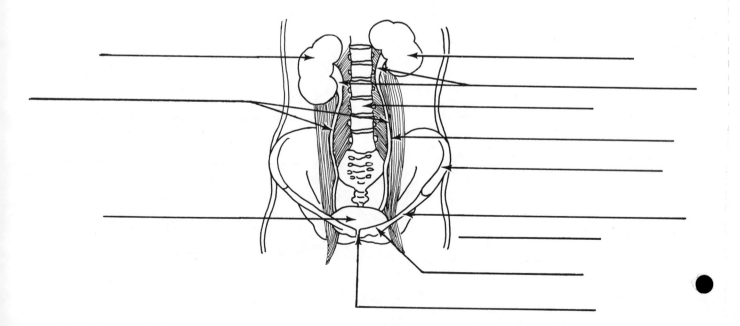

URINARY SYSTEM – MALE

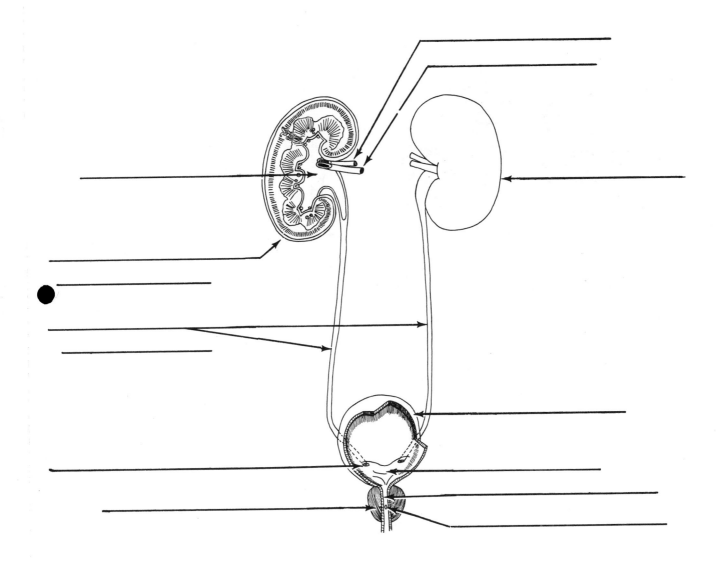

RIGHT KIDNEY SECTIONED

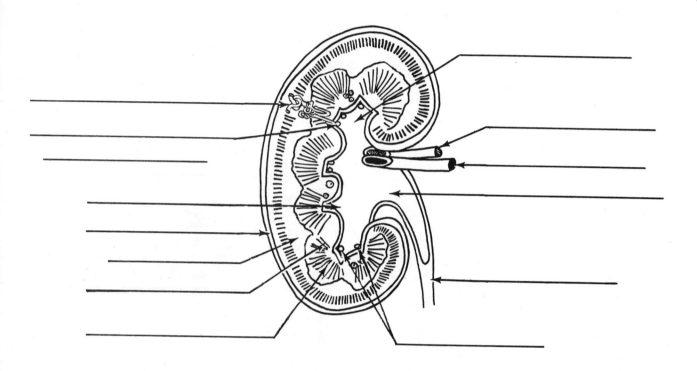

NEPHRON UNIT OF KIDNEY

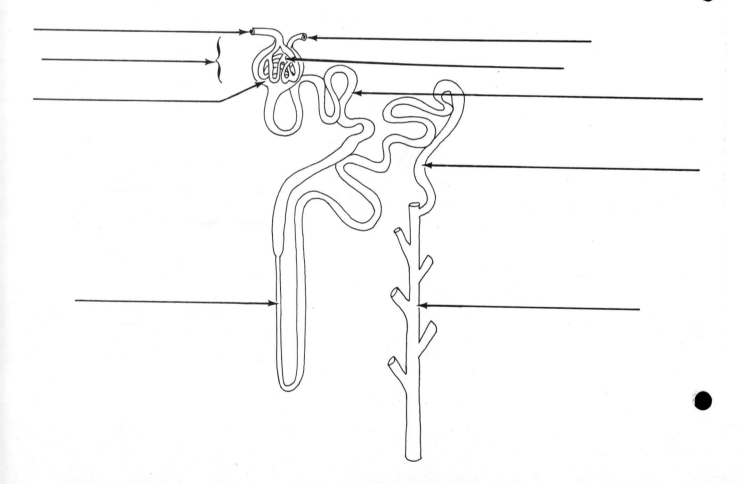

CHAPTER X

SENSES

TOUCH – THE SKIN (INTEGUMENTARY SYSTEM)

HAIR FOLLICLE

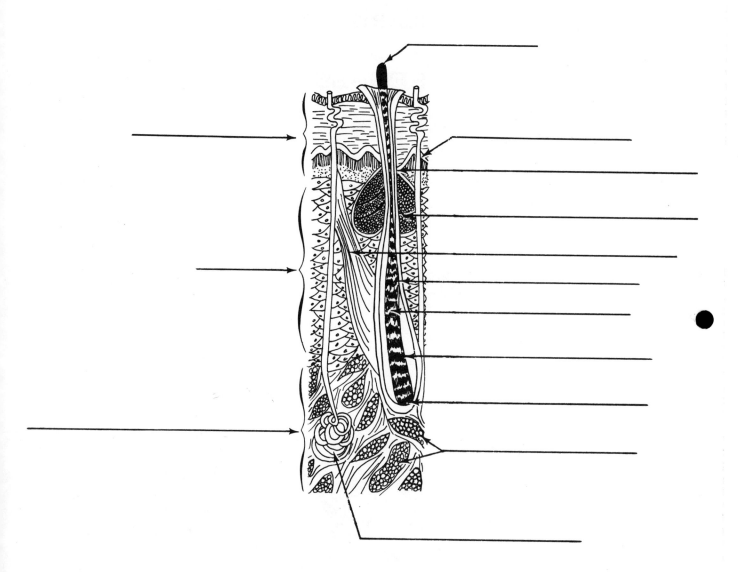

STRUCTURE OF SKIN

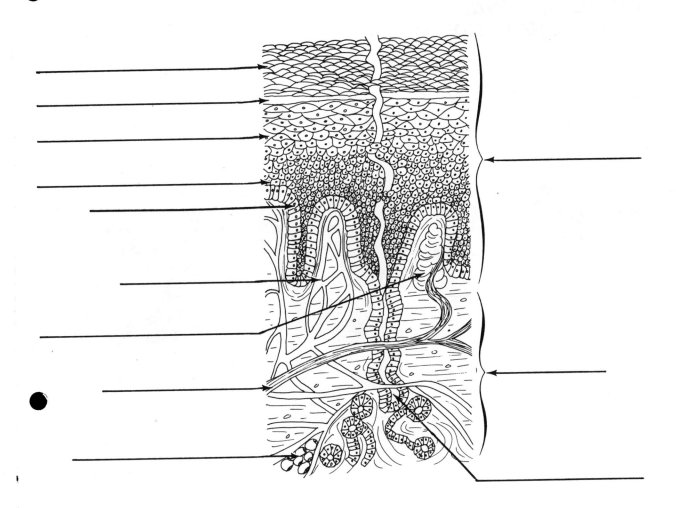

STRUCTURE OF FINGERNAIL

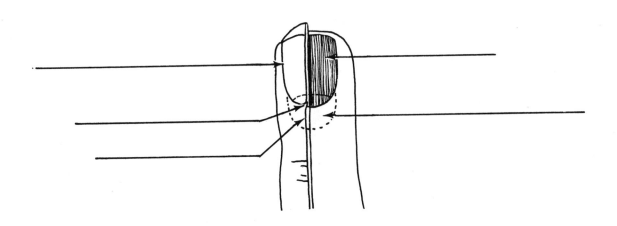

HEARING–(THE EAR)

AURICLE (PINNA)
OF EXTERNAL EAR

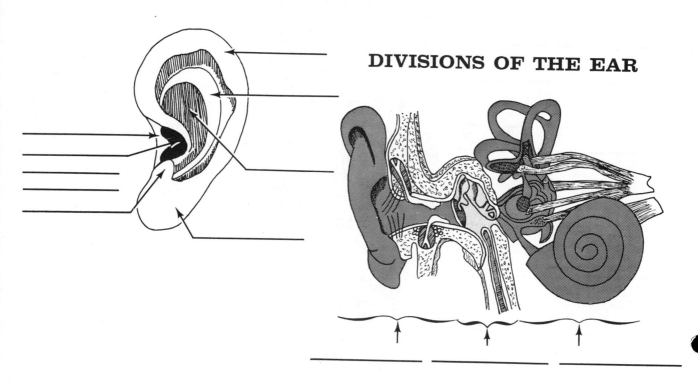

DIVISIONS OF THE EAR

STRUCTURE OF THE EAR

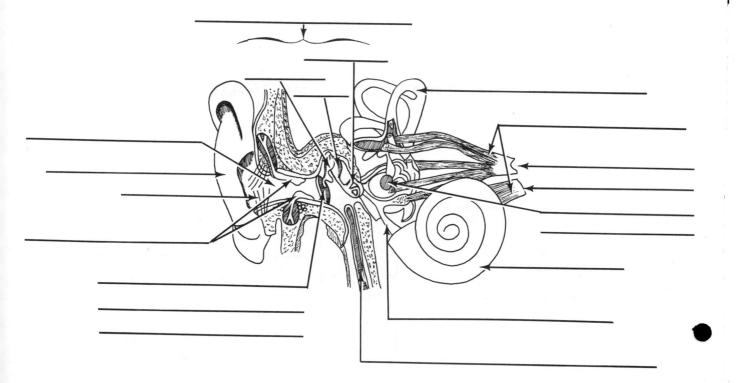

SIGHT – THE EYE

FRONT VIEW OF RIGHT EYE

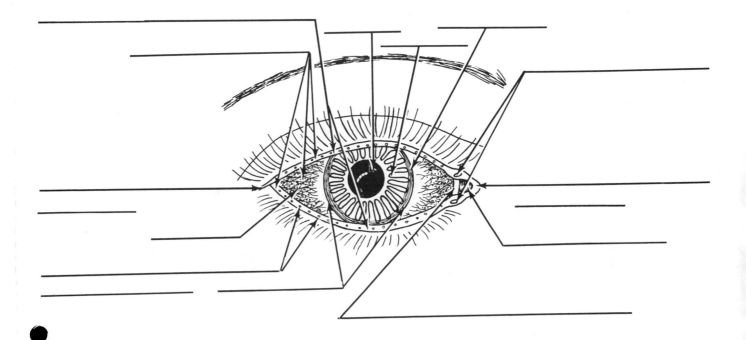

STRUCTURE OF LID AND TEAR APPARATUS

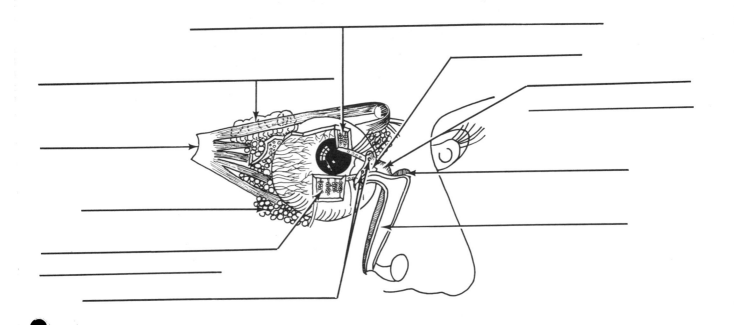

HORIZONTAL SECTION OF LEFT EYE
VIEWED FROM ABOVE

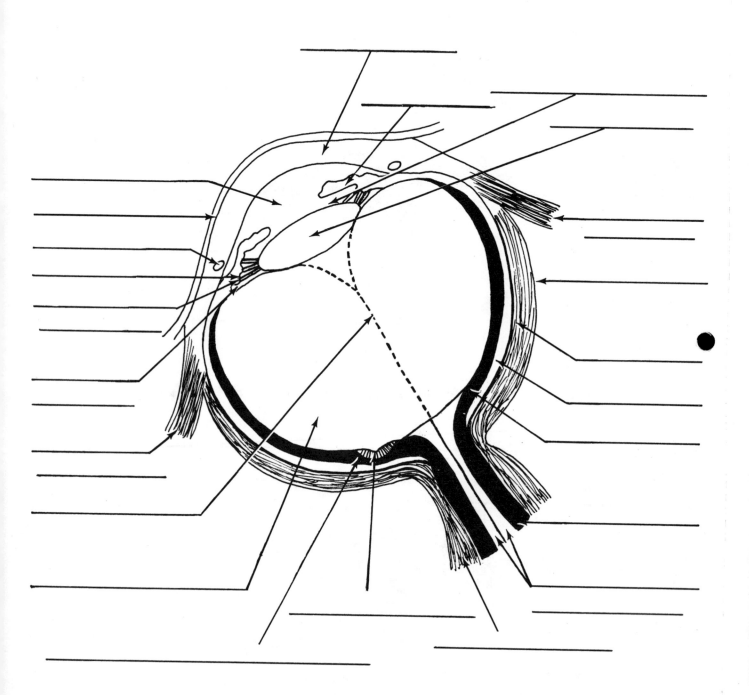